AF803523

AGRICULTURAL BIOTECHNOLOGY

Issues, Challenges, and Recent Developments

AGRICULTURAL BIOTECHNOLOGY

Issues, Challenges, and Recent Developments

Mohd. Sayeed Akhtar, PhD
Khalid Rehman Hakeem, PhD
Editors

AAP | APPLE ACADEMIC PRESS

First edition published 2025

Apple Academic Press Inc.
1265 Goldenrod Circle, NE,
Palm Bay, FL 32905 USA

760 Laurentian Drive, Unit 19,
Burlington, ON L7N 0A4, Canada

CRC Press
2385 NW Executive Center Drive,
Suite 320, Boca Raton FL 33431

4 Park Square, Milton Park,
Abingdon, Oxon, OX14 4RN UK

© 2025 by Apple Academic Press, Inc.

Apple Academic Press exclusively co-publishes with CRC Press, an imprint of Taylor & Francis Group, LLC

Reasonable efforts have been made to publish reliable data and information, but the authors, editors, and publisher cannot assume responsibility for the validity of all materials or the consequences of their use. The authors, editors, and publishers have attempted to trace the copyright holders of all material reproduced in this publication and apologize to copyright holders if permission to publish in this form has not been obtained. If any copyright material has not been acknowledged, please write and let us know so we may rectify in any future reprint.

Except as permitted under U.S. Copyright Law, no part of this book may be reprinted, reproduced, transmitted, or utilized in any form by any electronic, mechanical, or other means, now known or hereafter invented, including photocopying, microfilming, and recording, or in any information storage or retrieval system, without written permission from the publishers.

For permission to photocopy or use material electronically from this work, access www.copyright.com or contact the Copyright Clearance Center, Inc. (CCC), 222 Rosewood Drive, Danvers, MA 01923, 978-750-8400. For works that are not available on CCC please contact mpkbookspermissions@tandf.co.uk

Trademark notice: Product or corporate names may be trademarks or registered trademarks and are used only for identification and explanation without intent to infringe.

For Product Safety Concerns and Information please contact our EU representative GPSR@taylorandfrancis.com Taylor & Francis Verlag GmbH, Kaufingerstraße 24, 80331 München, Germany

Library and Archives Canada Cataloguing in Publication

Title: Agricultural biotechnology : issues, challenges, and recent developments / Mohd. Sayeed Akhtar, PhD, Khalid Rehman Hakeem, PhD, editors.
Other titles: Agricultural biotechnology (Palm Bay, Fla.)
Names: Akhtar, Mohd. Sayeed, editor | Hakeem, Khalid Rehman, editor
Description: First edition. | Includes bibliographical references and index.
Identifiers: Canadiana (print) 20250137364 | Canadiana (ebook) 20250137380 | ISBN 9781774919200 (hardcover) | ISBN 9781774919217 (softcover) | ISBN 9781003638087 (ebook)
Subjects: LCSH: Agricultural biotechnology.
Classification: LCC S494.5.B563 A37 2025 | DDC 338.1/6—dc23

Library of Congress Cataloging-in-Publication Data

CIP data on file with US Library of Congress

ISBN: 978-1-77491-920-0 (hbk)
ISBN: 978-1-77491-921-7 (pbk)
ISBN: 978-1-00363-808-7 (ebk)

Dedication

This book is dedicated to:

Syed Iqrar Ahmad (1926–1985)

About the Editors

Mohd. Sayeed Akhtar, PhD
Senior Assistant Professor, Gandhi Faiz-e-Aam College, Shahjahanpur, U.P., India

Mohd. Sayeed Akhtar, PhD, is a Senior Assistant Professor at Gandhi Faiz-e-Aam College, Shahjahanpur, U.P., India. He received his PhD degree from Aligarh Muslim University (AMU), Aligarh, India, in 2008, prior to conducting postdoctoral research at the Botanical Institute, University of Basel (BIB), Switzerland (2008–2010), and Chonbuk National University (CBNU), Republic of Korea, in 2011. He worked as an Assistant Professor at Jimma University in Ethiopia from 2011 to 2014 and as a fellow researcher at the Universiti Putra Malaysia's (UPM) Institute of Tropical Agriculture from 2014 to 2015. He has been listed among the top 2% of scientists in the world consecutively for the past three years, according to Stanford University in the United States. Dr. Akhtar is the author and co-author of more than 150 articles in peer-reviewed journals, conference proceedings, and book chapters, and has edited 21 books with international publishers. His numerous honors include the Best Researcher of the Year Award (2016) from the APSF, DST, India; the Outstanding Scientist in Plant Pathology Award (2017) from Venus International Research Awards, Chennai, India; the Global Teachers Award (2018) from the AKS Education Foundation, New Delhi, India; and the Albert Nelson Marquis Award (2018) from the Marquis Who's Who Society, USA; as well as the Teacher Innovation Award (2019 and 2022) from Zero Investment Innovation for Education Initiatives (ZIIEI), Sri Arvindo Society, Puducherry, India. In addition to this, he is a member of the International Society for Development and Sustainability, Japan; a Fellow of the Plantae Group, American Society of Plant Biologists; and a member of the International Society of Environmental Relationship and Sustainability, Chennai, India, and the International Service for the Acquisition of Agri-Biotech Application (ISAAA), Manila, Republic of the Philippines. He is serving the scientific community as an editorial board member and reviewer for several high-impact international journals. His current work focuses on molecular biotechnology, bioremediation, biomineralization, nano-fertilizers, and nanobiotechnology of rhizospheric plant-microbe interactions.

Khalid Rehman Hakeem, PhD
Professor, King Abdulaziz University, Jeddah, Saudi Arabia

Khalid Rehman Hakeem, PhD, is a Professor at King Abdulaziz University, Jeddah, Saudi Arabia. After completing his doctorate (botany, specialization in plant eco-physiology and molecular biology) from Jamia Hamdard, New Delhi, India, he worked as a lecturer at the University of Kashmir, Srinagar, India, and at Universiti Putra Malaysia, Selangor, Malaysia, where he was a Post-doctoral Fellow and Fellow Researcher (Associate Professor) for several years. Dr. Hakeem has more than 10 years of teaching and research experience in plant eco-physiology, biotechnology, molecular biology, medicinal plant research, plant-microbe-soil interactions, as well as in environmental studies. He is the recipient of several fellowships at both national and international levels. He has served as a visiting scientist at Jinan University, Guangzhou, China. Currently, he is involved with several international research projects with various government organizations. To date, Dr. Hakeem has authored and edited more than 60 books with international publishers. He also has to his credit more than 115 research publications in peer-reviewed international journals and 55 book chapters in edited volumes. At present, Dr. Hakeem serves as an editorial board member and reviewer for several high-impact international scientific journals. He is included in the advisory board of Cambridge Scholars Publishing, UK. He is also a fellow of the Plantae Group of the American Society of Plant Biologists, a member of the World Academy of Sciences, a member of the International Society for Development and Sustainability, Japan, and a member of the Asian Federation of Biotechnology, Korea.

Contents

Contributors

Elsayed Fathi Abd_Allah
Plant Production Department, College of Food and Agricultural Sciences, King Saud University, Riyadh, Saudi Arabia

Ghlomreza Abdi
Department of Biotechnology, Persian Gulf Research Institute, Persian Gulf University, Bushehr, Iran

Manshi Agrawal
Department of Microbiology, Dr. Harisingh Gour Vishwavidyalaya (A Central University), Sagar, Madhya Pradesh, India

Mohd. Sayeed Akhtar
Department of Botany, Gandhi Faiz-e-Aam College, Shahjahanpur, Uttar Pradesh, India

Fatima Akmal
Agriculture Extension Wing, Department of Agriculture Extension, Kamalia, Punjab, Pakistan

Khalid F. Almutairi
Plant Production Department, College of Food and Agricultural Sciences, King Saud University, Riyadh, Saudi Arabia

Mohamad Alnafissa
Department of Agricultural Economics, College of Food and Agricultural Sciences, King Saud University, Riyadh, Saudi Arabia

Volkan Altay
Hatay Mustafa Kemal University, Faculty of Arts and Sciences, Department of Biology, Hatay, Türkiye

Rabia Amir
Atta-Ur-Rahman School of Applied Biosciences, National University of Sciences and Technology, Islamabad, Pakistan

Mustansar Aslam
Department of Environmental Sciences, Faculty of Life Sciences, University of Okara, Pakistan

Marjan Assefi
University of North Carolina, Greensboro, USA

Chadha Ayed
Research Laboratory "Agrobiodiversity and Ecotoxicology" LR21AGR02, High Agronomic Institute of Chott Mariem, Sousse, Tunisia

Tuba Sharf Batool
Department of Plant Biotechnology, Atta-Ur-Rahman School of Applied Biosciences, National University of Sciences and Technology, Islamabad, Pakistan

Himanshu Bhardwaj
Department of Biochemistry, Patna University, Patna, Bihar, India

Bibekananda Bhoi
Department of Microbiology, Dr. Harisingh Gour Vishwavidyalaya (A Central University), Sagar, Madhya Pradesh, India

Pedro Garcia Caparros
Department of Superior School Engineering, University of Almería, Almería, Spain

Natasha Erum
Department of Biochemistry, Faculty of Sciences, University of Agriculture, Faisalabad, Pakistan

Nosheen Fatima
Atta-Ur-Rahman School of Applied Biosciences, National University of Sciences and Technology, Islamabad, Pakistan

Sarah Fatima
Atta-Ur-Rahman School of Applied Biosciences, National University of Sciences and Technology, Islamabad, Pakistan

Dibyangana Ganguly
Foundation for Innovative New Diagnostics India, New Delhi, India

Alvina Gul
Atta-Ur-Rahman School of Applied Biosciences, National University of Sciences and Technology, Islamabad, Pakistan

Abeer Hashem
Botany and Microbiology Department, College of Science, King Saud University, Riyadh, Saudi Arabia

Noor Ul Huda
Atta-Ur-Rahman School of Applied Biosciences, National University of Sciences and Technology, Islamabad, Pakistan

Fareeha Iqbal
Atta-Ur-Rahman School of Applied Biosciences, National University of Sciences and Technology, Islamabad, Pakistan

Warda Jabbar
Department of Plant Biotechnology, Atta-Ur-Rahman School of Applied Biosciences, National University of Sciences and Technology, Islamabad, Pakistan

Arooma Jannat
Atta-Ur-Rahman School of Applied Biosciences, National University of Sciences and Technology, Islamabad, Pakistan

Muhammad Raza Kaleem
Atta-Ur-Rahman School of Applied Biosciences, National University of Sciences and Technology, Islamabad, Pakistan

Fadime Karabulut
Department of Biology, Faculty of Science, Firat University, Elazig, Turkey

Maryam Khan
Atta-Ur-Rahman School of Applied Biosciences, National University of Sciences and Technology, Islamabad, Pakistan

Nadeem Kizilbash
Department of Chemistry, Faculty of Physical Science, Quaid-i-Azam University, Islamabad, Pakistan

Maneesh Kumar
Department of Biotechnology, Magadh University, Bodh Gaya, Bihar, India

Pratik Kumar
Jhpiego, India Country Office, New Delhi, India

Komal Kumari
Department of Biochemistry, Patna University, Patna, Bihar, India

Lovely Kumari
Department of Biotechnology, Jaipur National University, Jaipur, Rajasthan, India

Laila-Tul-Qadar
Atta-Ur-Rahman School of Applied Biosciences, National University of Sciences and Technology, Islamabad, Pakistan

Arti Manhas
Department of Botany, University of Jammu, Jammu Camp, Kashmir, India

Shafia Maryam
Atta-Ur-Rahman School of Applied Biosciences, National University of Sciences and Technology, Islamabad, Pakistan

Suryanarayan Bhaskar Meghana
Department of Biotechnology, East West First Grade College of Science, Bangalore, Karnataka, India

Sohila Nankali
North Central University, Minneapolis, USA

Syeda Gul Nisa
Atta-Ur-Rahman School of Applied Biosciences, National University of Sciences and Technology, Islamabad, Pakistan

Munir Ozturk
Ege University, Department of Botany and Center for Environmental Studies, Izmir, Türkiye

Javid A. Parray
Department of Higher Education, Government Degree College, Eidgah, Srinagar, Jammu and Kashmir, India

Boregowda Purushotham
Department of Biotechnology, East West First Grade College of Science, Bangalore, Karnataka, India

Zuhra Qayyum
Atta-Ur-Rahman School of Applied Biosciences, National University of Sciences and Technology, Islamabad, Pakistan

Khola Rafique
Pest Warning and Quality Control of Pesticides, Agriculture Department, Lahore, Punjab, Pakistan

Gulzar A. Rather
The Institute of Plant Sciences, Department of Ornamental Plants and Agricultural Biotechnology, The Volcani Center, ARO, Israel

Shafiq Ur Rehman
Department of Environmental Sciences, Faculty of Life Sciences, University of Okara, Pakistan

Umair Riaz
Soil and Water Testing Laboratory for Research, Bahawalpur, Agriculture Department, Government of Punjab, Pakistan

Javid Shaghaghi
North Central University, Minneapolis, USA

Noopur Sharma
Department of Botany, University of Jammu, Jammu Camp, Kashmir, India

Muhammad Shoaib
Institute of Molecular Biology and Biotechnology, Bahauddin Zakariya University, Multan, Pakistan; State Key Laboratory of Plant Cell and Chromosome Engineering, Institute of Genetics and Development Biology/Innovative Academy of Seed Design, Chinese Academy of Sciences, Beijing, China

Muhammad Irfan Sohail
Department of Environmental Sciences, Faculty of Life Sciences, University of Okara, Pakistan; High Tech Lab-II, Sub Campus Depalpur–Okara, University of Agriculture Faisalabad, Okara, Pakistan

Fazilet Vardar Sukan
Faculty of Engineering and Natural Sciences; Director of SUNUM Nanotechnology Research and Application Center, Sabanci University, Tuzla, Istanbul, Türkiye

Sambamurthy Kumaraswamy Sukrutha
Department of Microbiology, East West First Grade College of Science, Bangalore, Karnataka, India

Mallappa Kumara Swamy
Department of Biotechnology, East West First Grade College of Science, Bangalore, Karnataka, India

Maryam Abbasi Tarighat
Faculty of Nano- and Bio-Science and Technology, Persian Gulf University, Bushehr, Iran

Bengu Turkyilmaz Unal
Nigde Omer Halisdemir University, Faculty of Science and Arts, Department of Biotechnology, Niğde, Türkiye

Sidra Urooj
Atta-Ur-Rahman School of Applied Biosciences, National University of Sciences and Technology, Islamabad, Pakistan

Ab Waheedwani
Department of Horticulture, School of Agriculture, Lovely Professional University, Jalandhar, Punjab, India

Acknowledgments

The editors would like to express their sincere gratitude to each and every author for contributing their worthwhile thoughts and time to produce this edition of the book.

Preface

Biotechnological implements have played a crucial role in augmenting agricultural outputs and productivity. This has resulted in significant contributions to the economic development of multiple countries across the world, either directly or indirectly. It is noteworthy that agricultural biotechnology and industrial development are interdependent and mutually reinforcing, with both inputs and outputs being complementary.

In the present-day scenario, this field has made considerable strides towards enhancing the economic well-being of farmers and promoting the economic development of their respective countries. Agricultural biotechnology encompasses several fields, including genetic engineering, molecular markers and docking gears, proteomics, crop biotechnology, disease management, and the development of transgenic crops. These fields are employed to modify living organisms such as plants, animals, and microorganisms, as well as their metabolic and secondary metabolic products.

The present book expounds upon the latest strides made in the fields of agricultural biotechnology, molecular biology, genetic engineering, proteomics, molecular docking, and drug delivery. Overall, the book serves as a comprehensive repository of data pertaining to advanced aspects of agricultural biotechnology, crop biotechnology, proteomics, and molecular biology. Additionally, the book is poised to furnish the academic community, including students, teachers, and professionals engaged in research related to agricultural biotechnology, molecular biology, genetic engineering, crop physiology, and plant sciences, with invaluable insights.

—Editors

CHAPTER 1

Futuristic and Prospective Role of Agricultural Biotechnology in Sustainable Environment

MANEESH KUMAR,[1,*] MOHD. SAYEED AKHTAR,[2]
HIMANSHU BHARDWAJ,[3] BIBEKANANDA BHOI,[4] and PRATIK KUMAR[5]

[1]*Department of Biotechnology, Magadh University, Bodh Gaya, Bihar, India*

[2]*Department of Botany, Gandhi Faiz-e-Aam College, Shahjahanpur, Uttar Pradesh, India*

[3]*Department of Biochemistry, Patna University, Patna, Bihar, India*

[4]*Department of Microbiology, Dr. Harisingh Gour Vishwavidyalaya (A Central University), Sagar, Madhya Pradesh, India*

[5]*Jhpiego, India Country Office, New Delhi, India*

*Corresponding author.

ABSTRACT

Current innovations in biotechnology have enhanced agricultural productivity to meet the global food supply. Along with the basic necessity for human survival, agricultural biotechnology effectively infers the proper utilization of natural resources to maintain the cleanliness of the environment. Scientific communities are actively involved in meeting the requisite demand for the production of healthy crops. This would bring an end to the poverty and hunger issues, and make human life more secure, resilient, and sustainable. The sustainable development goals (SDGs) ensure the future development of new, improved, and competent cultivars to cope with the global food desire and energy crises. Innovative technologies in the world

Agricultural Biotechnology: Issues, Challenges, and Recent Developments. Mohd. Sayeed Akhtar and Khalid Rehman Hakeem (Eds.)
© 2025 Apple Academic Press, Inc. Co-published with CRC Press (Taylor & Francis)

of agronomy can provide genetic and molecular knowledge that enables agricultural productivity against biotic and abiotic stresses. With the help of scientific farming, many genetically modified (GM) crops can be sustainably grown well under given environmental conditions. The use of such crops can ease the workload and provide safety to farmers from carcinogens and other harmful xenobiotic chemicals. Therefore, the aim of this chapter is to focus on the revolutionary and forthcoming agricultural techniques that may fulfill the crop demand, and also manage the sustainability of the environment.

1.1 INTRODUCTION

New scientific techniques are always valued by human society. They apply scientific knowledge to satisfy their basic needs and improve their living culture. These justify economic growth and rationalize the proper utility of natural resources to achieve a sustainable environment (Rull, 2014). Scientists use cutting-edge methods to improve crop production and management to feed the world's growing population through the application of agricultural biotechnology (Struik and Kuyper, 2017). Agricultural biotechnology is a field that is growing rapidly and that could help solve some of the world's most important problems. The agricultural biotechnology sector is scientifically similar to the therapeutic biotechnology sector, with similar characteristics such as a long time to market for developing goods. However, the backdrop for innovation and entrepreneurship is substantially different when biotechnology is used in agriculture because of the unique difficulties, aims, and opportunities presented by this field (Kumar et al., 2016). It is a collection of numerous scientific approaches such as genetic engineering, molecular breeding, molecular diagnostics, tissue culture, and vaccinations that are used to improve crop and livestock productivity as well as the marketability of products obtained from these organisms. Instead, it refers to a set of scientific methods that have been developed to better the lives of organisms both large and small. Science has devised methods to boost crop yields based on deoxyribonucleic acid (DNA) knowledge. Breeder's ability to improve crops and livestock is bolstered by biotechnology, beginning with the ability to detect genes that may bestow advantages on certain crops and continuing with the ability to work with such features very accurately (Singh et al., 2018).

The progress made in the 21st century comes with many significant problems that can't be ignored. Climate change, the COVID-19 epidemic, the rise in population, the loss of natural resources, and very poor socioeconomic assumptions are some of the most important problems in agricultural production. All of these distressing circumstances ultimately result in degraded soil and a decline in plant and animal life. Without the help of computers and strategic thinking, none of them will be easy to do (Ben Ayed and Hanana, 2021). This decade has been marked by a move away from caring about other people. Global warming, ozone layer hazards, and the expansion of deserts over previously arable land are just a few of the critical, complex issues that scientists bring to our attention. In response, we seek more information and place the burden of solving the issues on organizations that lack the resources necessary to do so. Environmental deterioration, often viewed as a concern of rich nations and a side effect of economic development, has become a survival issue for poor countries. This is because it adds to the environmental and economic problems that many of the world's poorest countries are already facing. Significant approaches have been developed to improve the demand for sustainable feed items and simultaneously maintain the ecosystem (Reitz et al., 2020). Although the Green Revolution has accelerated agricultural production by several folds to meet the huge population demands, we ignored the environmental damages by using indiscriminate fertilizers. Climate change is one of the exclusive setbacks in the flourishing of agricultural systems. In the past decades, climate change has had a profound impact not just on natural ecosystems, but also on human ecosystems. These repercussions of climate change, regardless of their cause, highlight the susceptibility of natural and human ecosystems to alterations in climatic system function, component interaction, or external factors, whether natural or anthropogenic. These might lead to a high reduction in the yield of desirable crops. The indiscriminate proliferation of weeds and pests enforces the use of synthetic pesticides that would do no good for the soil. Climate change may affect weedy plants by making them move or spread to new areas, changing their life cycles, or changing how many of them there are and where they live. Weed migration will result in a difference in the structure and composition of weed communities in natural and managed ecosystems (Ramesh et al., 2017). From the perspective of global warming, Peters et al. (2014) identified three separate types of changes in weedy vegetation: range changes, niche changes, and trait changes. These changes occur across landscapes, communities, and population levels. As a result of climate change, weeds will undergo significant shifts in their biology, ecology, and interference potential, leading to complicated crop-weed interactions that call for novel adaptation mechanisms. The C_4 pathway is present in the major weeds of the planet, so it is assumed that they will become more competitive as the climate changes. However, due to adaptation processes in weedy species, this is not as straightforward as it sounds. As our climate continues to rise and the repercussions of this heat become more regular and severe, farmers and agricultural communities all over the world will face a growing number of obstacles. The heat waves also disturb the plant's life cycle and result in wilted (due to excessive transcription rates) plant growth (Lippmann et al., 2019; Poonam et al., 2017).

Flood and prolonged waterlogged conditions disconnect the breathing mechanism in the plants (Figure 1.1). This change is distributed unevenly around the world. For all such reasons, there is an urgent need to shift from the traditional paradigm of enhanced crop productivity to agricultural sustainability. In adopting this, crucial choice, especially the use of digital technologies might help in improving the current circumstances (Ben Ayed and Hanana, 2021). The combination of escalating climate change and an industrial structure that is already unstable constitutes a "perfect storm" that threatens farmer's livelihoods and our food supply. Fortunately, farmers can protect their businesses from climate change by adopting science-based farming practices that will stand the test of time. However, farmers encounter numerous challenges when attempting to alter their methods. Hence, politicians must move federal agriculture spending to support and speed up this transformation.

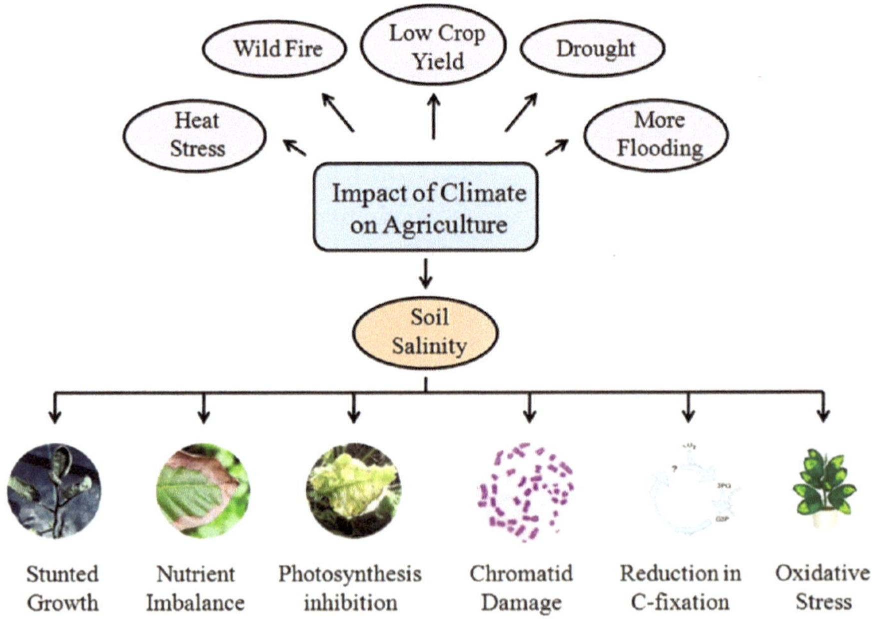

FIGURE 1.1 Current impacts on agriculture due to climatic changes.

In the future, the situation might become worse as the changes negatively affect the agricultural production in low latitudes countries whereas effects in northern latitudes may be negative or positive. The relationships between climate change, land degradation, and land use are intricate. Climate change has a big effect on land degradation because rising global temperatures speed up the hydrological cycle. This makes it rain harder, which is a major cause of soil erosion. As a result, with climate change, sustainable land management (SLM) becomes even more crucial. Land-use change in the form of forest clearance for rangeland and agriculture (e.g., to provide biofuels) and the cultivation of peat soils is a major source of greenhouse gas (GHG) emissions from biomass and soils. Many SLM activities (e.g., agroforestry, perennial crops, organic amendments, etc.) enhance soil and vegetation carbon content, giving local and immediate adaptation benefits and long-term global mitigation benefits, as well as social and economic co-benefits. Stopping, slowing, or reversing land degradation could help reduce climate change and make it easier for people to adapt to it. These changes later lead to food insecurity among helpless groups, such as the poor. These issues also jeopardized the biological integrity of agroecosystems by the severe use of agrochemicals, fossil fuels, natural resources, and machinery (Meena et al., 2021). Concerns have to be taken to maintain the growth of agriculture with a healthy ecosystem. On the whole, age-old traditional agricultural practices have lost their integrity among the farmers. To contend with the adverse effects governments are trying to develop traditional agriculture with new dimensions without any external input (Meena et al., 2021; Singh and Singh, 2017). Experts are now trying to develop efficient ways to do agriculture to maintain the agronomic, economic, environmental, social, trans-generational, and global conditions. Industrially, agriculture has now been strengthened economically to feed the world. Based on the priorities are set to lead to sustainable growth of agriculture that could be preserved for future use (Lencucha et al., 2020; Struik and Kuyper, 2017).

Bioscience research, translational education, and entrepreneurship are used to emphasize to development of sustainable agriculture. Precision agriculture (PA) under various key technologies started in the early 2000s including GPS, satellite imaging, and drone technology. This would be further boosted by applying modern techniques of biotechnology. It comprises lots of tools and techniques where we explore significant genotypic traits from vital crops for the betterment of humans. These traits enable transgenic crops to uphold extreme weather and protect against harmful pests. The modern perception of biotechnology began in the 19th century, when the Theory of Evolution was considered to explore more in molecular biology (Ashraf, 2016). Being a part of agriculture biotechnology, crop yield, and food safety has become a prime concern for farmers. It is a wide-ranging science that involves several scientists using their skills to recognize and control the genetic structure of different species in agricultural products. The inference of biotechnology here is to allow the breeding of plants and maintain stabilization against extreme abiotic stresses (Song et al., 2011). Many developing countries have been approaching to minimize the risk of malnourished children. They are trying to make improvements in the mode of agriculture they do. With the concern for hunger and malnutrition, these countries make efforts to upgrade the living standard but maintain the food and feed demands of their respective citizens. This might be a challenge for them. The economy and society demand ample food for survival with great hope and concern (Ranjha et al., 2022). The gain in productivity and technological advancement widely contribute to the use of natural resources and manage food safety. This highly impacts the food chain and other biotic factors in the environment. The consequences of genetically modified (GM) food crops are widely accepted accordingly. Although the techniques elaborate the concept of agriculture biotechnology, the possible risk factors are always with flora and fauna. Being a part of the Convention on Biological Diversity and Cartagena Protocol, India has also acquired the responsibility to strengthen the Biosafety move toward the environment (Singh et al., 2018). In past decades, biological sciences predominantly dominate agriculture to maintain the Biosafety of the environment. The upcoming escalation in breakthroughs, developments, and innovations, now become an essential part of agriculture biotechnology. Such changes behave as a toolbox that contains all the exciting solutions against the agricultural burden around the world. Genes were first introduced in corn in 1989 by using molecular techniques, and in the late 1990s, millions of acres were used in growing transgenic corn. This clarifies the potential role of biotechnology in agriculture in its formative years. This also encourages the commercialization of transgenic crops with the concern of a clean environment. Every particular phenotypic characteristic present in the transgenic plant interacts with the environment. This ensures the impact of recombinant DNA present in the transgenic (National Academies of Sciences, Engineering, and Medicine, 2002; Figure 1.2). Therefore, the aim of this chapter is to focus on the revolutionary and forthcoming agricultural techniques that may fulfill the crop demand, and also manage the sustainability of the environment.

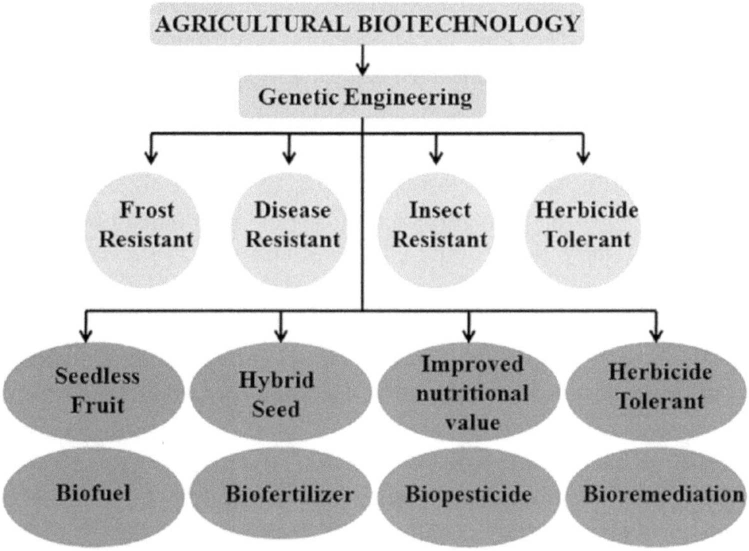

FIGURE 1.2 Significance of agricultural biotechnology in creating sustainable plant products against various stresses.

1.1.1 NEXT GENERATION FOOD CROPS

Based on the needs of the global market, genetic engineering plays a crucial role in agriculture that focuses on the concept of crop biotechnology. The genetic technologies include GM food (transgenics), genome editing (CRISPR, TALENs, ZFNs, and so forth), and breeding techniques that have been explored on a large scale in commercial farming in different global regions (Bao et al., 2019; Kumar et al., 2020). They are the key growth areas where skilled farmers try to utilize the new technologies. Commercial farming is exclusively popular in developed and in some of the developing countries.

The high global population and the demand for food along with alarming climatic changes create formidable confronts for the agriculture in coming future. This necessitates creating better environmental protection. Now, crop production is a hike at all times, but the quantity is insufficient for the whole population. Current global population growth and the associated increasing demands on farming, together with the threat of climate change and the need for better environmental protection, pose formidable challenges for the agriculture of the future. Crop productivity is already reaching high capacity but will have to increase perhaps by as much as 100% to sustain a world population of nearly 9–10 billion by the end of 2050 (Ben Ayed and Hanana, 2021;

De Fraiture and Wichelns, 2010; Martinis et al., 2020). The global challenges in agriculture are also not evenly distributed. The circumstances of sustainable agriculture are highly different between developed and developing countries. High urbanization, aged population, and stagnant cultivation are the conditions in developed countries whereas in developing countries population, infrastructure, and cultivating land are flourishing (Kruijssen et al., 2018; Martinis et al., 2020). Less developed countries, on the other hand, also face demographic growth but generally lack modern infrastructure and efficient farming skill practices that are required for the expansion of agricultural production. Climate change also afflicts different regions of the world with varying intensity and in very diverse ways, generating complex effects on agricultural systems, which remain hitherto unpredictable (Ganivet, 2020). In general, our growers face multiple challenges during crop production and its improvements. To optimize the diversity of the food production system, several new advancements have exclusively been integrated to maintain and resolve the multifaceted and interconnected physical and biological progressions that affect the yield across diverse scales. The modern technological proceeds encourage the farmers. The field of artificial intelligence (AI) emerged as one of the promising tools that includes machine learning (ML) algorithms, which are needed to explore the limitations of potential yield gap at the plot and field scales, and the proper harness (National Academies of Sciences Engineering Medicine, 2019). The ML approach enables the interpretation and prediction of variation in crop production concerning climatic changes.

1.1.1.1 ARTIFICIAL INTELLIGENCE (AI)

Artificial intelligence (AI) is a vast computational and robotic network. UN Food and Agriculture Organization has been planning to develop the network in agricultural production. As previously mentioned, the increase in the global population would put a huge burden on agriculture. There would be a shortage of cultivated land. Proper skill and robotic intelligence than direct all around to bring a paradigm shift from what we do in farming today (Zhai et al., 2020). AI comes up with lots of technologies including ML, vision, expert system, and robotic approach. Machine learning (ML) is one of the essential themes of AI. The skill enables the farmer to work more creatively and efficiently. It consists of statistical and mathematical methods that help in predicting the result and making a decision based on available data. The machine learner interacts with the biotic and abiotic conditions of the environment, collects information, and executes accordingly for better production (Jordan and Mitchell, 2015). The executions are performed with a computational vision that gives meaningful information including images, videos, and other illustration efforts (Kakani et al., 2020; Patrício and Rieder, 2018). During the execution process, expert systems are engaged in solving complex problems by applying reasoning via knowledge, and conventional procedural code (Bahrammirzaee, 2010). It also consists of natural language processing (NLP) to figure out human verbal communication, computational vision with an analog-to-digital translation such as video, and language detection and proficient systems to reproduce the judgment (Araújo et al., 2021; Jha et al., 2019). With proper planning and scheduling, the AI optimizes the progression of all sorts of agricultural issues for farmers. Through these activities, they follow the strategies and maintain the action sequences accordingly (Li et al., 2010; Figure 1.3).

In an approach to supporting sustainable agricultural practices, AI has played a pioneering role in the automation of agricultural practices (Eli-Chukwu, 2019; Jin et al., 2020). It minimizes the stress of the traditional mode of farming and modifies the variable outputs in food production. It is an essential need to make a predictive multidisciplinary approach to improve the food and agriculture sector. Artificial intelligence (AI) is an imaginative and powerful tool that includes human intelligence and processes different robotics and digital types of equipment. The cultivation of healthy crops is made possible by the intelligent application of AI within the agricultural sector. It evaluates the state of the soil and the crops and makes suggestions on fertilizers, monitoring the weather, and crop quality. The information would assist the farmers in improving their decision-making process and engaging in more productive farming techniques. It is based on the three cognitive skills of learning, thinking, and self-correction (Patel et al., 2021).

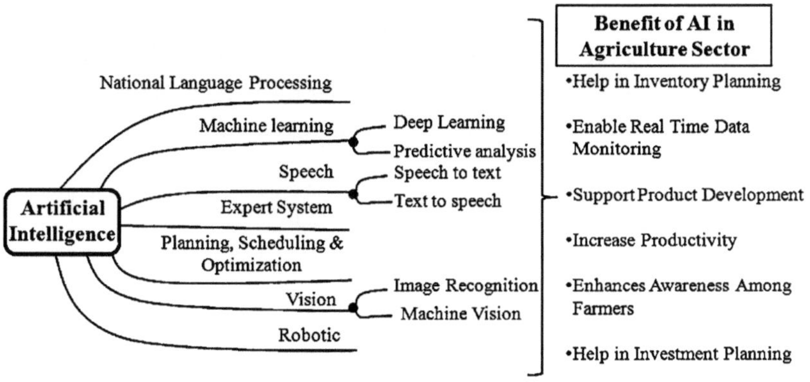

FIGURE 1.3 Benefits of artificial intelligence in the agricultural sector.

1.1.1.2 LEARNING

Learning is the fundamental part of AI solutions. From a conceptual perspective, learning improves the knowledge of an AI program by making observations in the agriculture sector. It is focused on the processing of collecting input-output pairs. It predicts the outputs from new input. Most of the information led to the basic kinds of literature that are studied, analyzed, and identified in both supervised and unsupervised groups (Uddin et al., 2019). However, that categorization is a generalization of real-world AI learning models and techniques. The AI learning models utilize two human learning processes that is, knowledge and feedback. The knowledge perspectives are constructed on the basis of the illustration of input and output data points, whereas feedback can be arranged in accordance with the interactions with the external environment, users, and various biotic and abiotic elements (Lei et al., 2020).

1.1.1.3 REASONING

Reasoning is a psychological process well associated with general thinking, intelligence, and cognition. It is one of the essential parameters to analyze AI and to predict a result from all the accessible information, details, and beliefs. It plays a great role in the logical process allowing the machine to draw conclusions, and predictions, and constructive with an approach based on rational thinking just like a human brain, and gives them the ability to act like humans (Muthu et al., 2020). In the agriculture sector, reasoning helps us to sort out the ecological problems with all facts and beliefs. It creates new knowledge from previously known knowledge. It would help us to find out the valid interference. There are many threats that farmers face while performing agriculture. During the performance, they have to manage the situation accordingly. Currently, reasoning is significantly being used in maintaining lots of agricultural activities and solving the problems regarding this.

1.1.1.4 SELF-CORRECTION

It is another expectation of AI programming. It is designed to continue to regulate algorithms on its own and provide the most precise and potential results. This model is fundamentally different from an active code that possibly builds the self-correction mechanism without the use of any external intelligent agent. It follows the properties of the intelligence code of the brain that actively interacts in words eliminating the errors during the processing (Gao et al., 2013; Ng and Mahkeswaran, 2021). Automation and AI maintain the aging agricultural workforce. By using self-driving agricultural machinery farmers can focus on sustainable harvests and profits. In the robotization of agriculture, farmers smartly manage all the data for future use and explore self-correction strategies to minimize the risk of any further mishappening during crop production.

1.2 IMPACT OF TRANSGENIC CROPS IN FUTURE

Many transgenic technologies exclusively impact the quantitative and qualitative production of crops. Incorporating genetic modification technology into agriculture could reduce the need for costly pesticides and herbicides. The elimination of indigenous pests will increase crop productivity and improve crop quality. As a result, pesticidal toxin genes from a bacterium can be inserted into the genome of the desired plant, protecting it from insect pests. This technique is used to create transgenic plants like *Bacillus thuringiensis* crops. Virus-resistant plants can also be produced by introducing viral coat proteins into plant cells (Kumar et al., 2012; Panter et al., 2012). The new era of crop improvement has become a crucial process where many concerns have been precluded for their extensive acceptance. During the concerns, many risk factors are involved in the cultivation of the transgenic crop. Scientists have been developing and analyzing the performance of genetically modified organisms (GMOs) and selecting useful genes for crop transformation. Other technologies like cisgenesis and genome editing, also facilitate the development of GMOs with multiple useful traits (Kamthan et al., 2016). Genetic engineering and plant transformation are played an essential role in crop improvement. In this new, beneficial foreign gene or silencing the expression of the endogenous gene(s) are introduced in crop plants. Many other important traits are also being used such as abiotic stress tolerance, disease resistance, herbicide tolerance, insect resistance (IR), and nutritional improvement. To date, more than 525 different transgenic activities have been performed in more than 32 crops. They all are approved and widely cultivated around the world. Transgenic technology plays a significant role in increasing crop yields, lessening pesticide, and insecticide use, cutting CO_2 emissions, and reducing the cost of crop production (Kumar et al., 2020).

1.2.1 GMOS SUPPORTS CLIMATIC CHANGES

Many researchers have conducted their experiments on the yield and potentiality of GM grains under different climatic changes. Genetic engineering and plant transformation have played a decisive role in gene introduction and crop improvement. The cultivation of GM crops has the potential to reduce the release of GHGs from farms. Increases in yield via GM have the potential to reduce emissions from both production and land-use change (Smyth, 2020). More widespread use of currently available GM crops in Europe has the potential to reduce agricultural GHG emissions by 7.5% of Europe's total (Sutherland et al., 2021). In countries of Africa and Asia, climate change has an impact lots and also, and they refuse to adopt the concept of GMOs in their regions. Due to this somehow fails to meet the food security in a warm climate (Van Aalst et al., 2008; Yadav and Lal, 2018). GM crops can help to reduce GHG emissions and maintain carbon sequestration in the soil by reducing tillage farming. Also, crop yield improvements can minimize the need to add new land to production, eliminating further CO_2 emissions from land-use change, while global demand for food production continues to rise. There is still a lot of controversy around genetically GM crops and new genomic breeding technologies, especially in Europe. Most of the criticism is based on unproven assumptions about potential harm rather than on any consideration of the existing or potential advantages. Higher crop yields, increased farm income, and, in certain circumstances, reduced chemical pesticide use have all been linked to the widespread adoption of GM crops, which have positive effects on the economy, the environment, and human health. Some studies also demonstrate that the use of GM crops can aid in the reduction of GHG emissions and the promotion of carbon sequestration in the soil (Blaustein-Rejto and Qaim, 2022; Kovak et al., 2021; Qaim, 2020).

1.2.2 GMOS ROLE IN INSECT AND PEST CONTROL

Carbon sequestration is critical for minimizing net GHG emissions. GHG emissions are reduced by moving CO_2 from the environment into soil storage pools via photosynthesis, where it becomes soil organic carbon (SOC). Previous studies have shown that these soil storage pools will be depleted roughly 15–20 years after the implementation of sustainable management practices, rendering the resulting carbon sequestration benefits null and void. However, new studies show that soil storage potential can be increased for decades after land management modifications are implemented if sustainable practices are consistently implemented (McConkey et al., 2020; Sutherland et al., 2021). The study was conducted on the GM herbicide-tolerant (HT) canola plant. The adoption of HT canola and conservation tillage has a synergistic connection, resulting in proportional decreases in GHG emissions. Indeed, smaller equipment, cheaper glyphosate, better crop input technologies, and advancements in crop genetics have all helped to reduce tillage practices, but there is growing evidence to suggest that sustained land-use changes would not be possible without GM HT crops. Because the technology is so good at controlling weeds, farmers can use less tillage and summer fallow on GMHT crops. This means that more carbon is being stored and less burden on the soil, while other GM characteristics have been created, IR and herbicide tolerance are the most frequently used globally (HT). These characteristics assist in reducing crop damage caused by insect pests and, accordingly, weeds, which increases effective yields. According to the findings of a global meta-analysis, the average yield benefits of GM crops are around 22%, with some variations between traits and geographical regions. The use of GM crops in industrialized countries located in temperate zones results in average yield increases of 10% for IR and 7% for HT, respectively. For the European Union (EU), these ideas seem to have some truth to them, as long as existing GM cultivars are used (Kovak et al., 2022; Qaim, 2020).

The combination of HT GM crops and glyphosate can boost soil carbon sequestration, preventing carbon dioxide (CO_2) from being released into the atmosphere and contributing to global warming. The study stresses that using tillage as a method of weed control contributes to soil disturbances, resulting in the release of carbon into the atmosphere. Substituting tillage for weed-control techniques that cause minimum soil disturbance and increase carbon absorption through continued crop production reduces carbon emissions. This research lends credence to the idea that the increased use of glyphosate on GM crops has a direct impact on the amount of carbon being stored in the soil. The results showed that farmers' ways of taking care of their land have changed a lot in the past 30 years. This is because of the spread of GM crops and HT types, which often involve using glyphosate (Sutherland et al., 2021). In countries of Africa and Asia, climate change has an impact lots and also, and they refuse to adopt the concept of GMOs in their regions. Due to this somehow fails to meet the food security in a warm climate (Van Aalst et al., 2008; Yadav and Lal, 2018). GM crops can help to reduce GHG emissions and maintain carbon sequestration in the soil by reducing tillage farming. Also, crop yield improvements can minimize the need to add new land to production, eliminating further CO_2 emissions from land-use change, while global demand for food production continues to rise.

1.3 PROSPECTIVE SIGNIFICANCE OF SUSTAINABLE AGRICULTURE

Sustainability faces severe problems in the 21st century, which humanity must overcome. The requirement to boost agricultural productivity to maintain food security for a worldwide population that is projected to reach 9–10 billion people in the next decades, while also contending with a changing climate that challenges sustainability, will place a strain on agricultural systems. There is a demand for crop production systems that produce more food with a higher nutritional value while also having less of an impact on the environment (Delgado and Berry, 2008; Delgado et al., 2019). During the 20th century, agricultural intensification was achieved by the extensive use of fertilizer, pesticides, and irrigation, at a huge environmental cost. These innovations contributed to the Green Revolution, which enabled billions of people to achieve food security. The problems of the 21st century are unique, and soil and water conservation will be vital to achieving food security. Sustainable precision agriculture and environmental (SPAE) will be needed so intensive agriculture and a changing climate does not cause extra impacts that could speed up climate change (Delgado et al., 2019; Fulton and Darr, 2017). For agriculture to be sustainable on a field level and across a watershed, the next generation of cropping systems must be created. These systems should combine biologically-based technologies (like plant-beneficial microorganisms and cover crops) with PA and precision conservation (PC). The employment of PA and PC in the development and deployment of technology with the ability to manage and communicate precise information and data at all levels of the agricultural ecosystem is essential to SPAE and the quick growth of these cropping systems. The relationship between PA, PC, and sustainable agriculture is inextricable. Sustainable agriculture and PC emphasize conservation effectiveness, environmental impact, and sustainability, whereas PA focuses on immediate cost savings in a particular area by maximizing input returns (Delgado et al., 2019). Sustainable agriculture is a relatively young field that requires more attention from scientists and researchers, and it is regarded as a necessity for human life. In recent decades, sustainable agriculture has attracted broad interest due to its ability to simultaneously achieve environmental and economic objectives. Green Chemistry is the design of chemical goods and processes to eliminate or limit the usage and production of hazardous compounds. Green chemistry is primarily founded on 12 principles and plays a crucial role in environmental conservation. Both sustainable agriculture and green chemistry are crucial for human progress.

1.4 IMPORTANCE OF GREEN CHEMISTRY IN FUTURISTIC AGRICULTURE

Green chemistry has been around for more than 30 years and has given scientists and chemical engineers a way to work in the big field of global sustainability. As a natural next step in efforts to stop pollution, the idea of "greening chemistry" has come up in the business and government worlds. It emphasizes the idea that contributors must take a systems approach to reduce environmental effects through the conscious design of chemical products. This includes not only how raw materials are obtained and how industrial and consumer goods are made and used, but also how these materials and goods can be reused, repurposed, or upcycled (Ganesh et al., 2021). The main reason why green chemistry strategies are needed in the agrochemical field is to make sure those pesticides and fertilizers are used smartly. Since pesticide production has such a direct bearing on human and environmental health, it is particularly well-suited to the concepts of green chemistry. However, farming practices today are still based on methods of intensive production that use technologies created during the "green revolution" that are not sustainable (Becer, 2021; Zimmerman et al., 2020).

1.5 CONCLUSION

Climate change, population growth, natural resource depletion, and inadequate socioeconomic assumptions are major concerns in agriculture. All of these upsetting circumstances erode soil and reduce plant and animal life. Numerous studies have been conducted to elaborate on the role of agricultural biotechnology under such abiotic stresses last decades. Most have been discovered; they increase crop productivity, primarily by lowering production costs by reducing the need for pesticide inputs; they provide better rotations to conserve natural resources, and they improve agricultural production by developing varieties resistant to biotic and abiotic stresses. The relevance of biotechnology in the agricultural sector has been demonstrated by the production of disease-free plants with enhanced nutritional value. However, the cost and accessibility of next-generation sequencing technology will alter the area of applied genomics, thereby speeding up the ability to identify the genetic elements underlying agriculturally relevant traits. We seek further information and ask groups without the resources to solve the problems. Environmental deterioration, a worry of rich nations, and a side effect of economic progress have become a survival issue for poor countries. It adds to the environmental and economic issues many poor countries face. Ways have been found to manage the ecosystem and increase the demand for feed that is good for the environment.

Many developing countries want to reduce child malnutrition by improving their agriculture. Concerned about hunger and malnutrition, these countries work to improve living standards while meeting food and feed needs. This could be difficult. Economically and socially, people want enough food to survive. Productivity gains and technological advances help manage natural resources and food safety. This affects the food chain and other biomes. GMO food crops' implications are well understood. Although the techniques expanded agro-biotechnology, flora, and fauna are constantly at risk. As a member of the Convention on Biological Diversity and Cartagena Protocol, India must promote environmental biosafety. Researchers have the potential to learn more, quickly, with the assistance of AI in sustainable agriculture. As a consequence of this, agricultural biotechnology companies are increasingly looking to AI for assistance in order to breed healthier plants, maximize growth, and gain a better understanding of the complex interaction that exists between genetics and the environment. These would be a great business opportunity to create effective digital operations to achieve efficient and cohesive support of better agricultural products, through AI. Green chemistry is also significantly involved in reducing the use of nonrenewable resources like synthetic fertilizers and pesticides, making farms more profitable and improving soil quality. Methods like organic, natural, biodynamic, and ecological farming do not use chemicals. Insects, weeds, and other pests can be killed by natural bio-pesticides, which also make plants healthier. Biopesticides deteriorate over time. Green chemistry, crop protection, and agrochemicals are needed for long-term progress. Sustainability and green chemistry are compatible. Sustainable farmers employ green inputs. Agrochemicals, pharmaceuticals, and food use biocatalysts. Biocatalysts enhance output and reduce waste.

KEYWORDS

- **agricultural biotechnology**
- **ecological farming**
- **genetically modified crops**
- **nonrenewable resources**
- **scientific farming**
- **sustainability**
- **sustainable development goals**

REFERENCES

Alam, M. S., Kong, J., Tao, R., Ahmed, T., Alamin, M., Alotaibi, S. S., Abdelsalam, N. R., & Xu, J. H. (2022). CRISPR/Cas9 mediated knockout of the OsbHLH024 transcription factor improves salt stress resistance in rice (*Oryza sativa* L.). *Plants*, *11*(1184). https://doi.org/10.3390/plants11091184.

Alukumbura, D. W. G. A. S. (2022). Wheat-associated bacterial and fungal communities during the interaction of *Fusarium graminearum* and *Trichoderma gamsii* T6085 and infection of a CRISPR/Cas9 derived green-fluorescent protein labeled *Leptosphaeria maculans* strain on *Brassica napus* (M.Sc. thesis). The University of Manitoba, Canada.

Araújo, S. O., Peres, R. S., Barata, J., Lidon, F., & Ramalho, J. C. (2021). Characterizing the agriculture 4.0 landscape—Emerging trends, challenges, and opportunities. *Agronomy*, *11*(667). https://doi.org/10.3390/agronomy11040667.

Ashraf, M. A., & Sarfraz, M. (2016). Biology and evolution of life science. *Saudi Journal of Biological Sciences*, *23*(S1). https://doi.org/10.1016/j.sjbs.2015.11.015.

Aslam, M., Fakher, B., Greaves, J. G., Jakada, B. H., Qin, R., & Qin, Y. (2022). A CBL-interacting protein kinase, AcCIPK18, from *Ananas comosus* regulates tolerance to salt, drought, heat stress, and *Sclerotinia sclerotiorum* infection in *Arabidopsis*. *Environmental and Experimental Botany*, *194*, 104728. https://doi.org/10.1016/j.envexpbot.2022.104728.

Bahrammirzaee, A. (2010). A comparative survey of artificial intelligence applications in finance: Artificial neural networks, expert system, and hybrid intelligent systems. *Neural Computing and Applications*, *19*(8), 1165–1195. https://doi.org/10.1007/s00521-010-0362-z.

Bao, A., Burritt, D. J., Chen, H., Zhou, X., Cao, D., & Tran, L. S. (2019). The CRISPR/Cas9 system and its applications in crop genome editing. *Critical Reviews in Biotechnology*, *39*(3), 321–336. https://doi.org/10.1080/07388551.2019.1602544.

Becer, J. (2021). The role of green chemistry in agriculture. *Trends in Green Chemistry*, *7*(034). https://doi.org/10.21767/2471-9889.100034.

Ben Ayed, R., & Hanana, M. (2021). Artificial intelligence to improve the food and agriculture sector. *Journal of Food Quality*, *2021*, 5584754. https://doi.org/10.1155/2021/5584754.

Chen, J. Y., Sun, B., Khan, Q., Yang, L. T., Zhang, B. Q., Xing, Y. X., & Li, Y. R. (2022). Overexpression of the α-tubulin gene of sugarcane (*Saccharum spp.* hybrids), SoTUA, enhances tobacco tolerance to cold stress. *Sugar Tech, 24*(1680–1688). https://doi.org/10.1007/s12355-022-01166-0.

De Fraiture, C., & Wichelns, D. (2010). Satisfying future water demands for agriculture. *Agricultural Water Management, 97*(4), 502–511. https://doi.org/10.1016/j.agwat.2009.08.008.

De Martinis, D., Rybicki, E. P., Colonna, N., Benvenuto, E., & Llorente, B. (2020). Next generation agriculture: Understanding plant life for food, health, and energy. *Frontiers in Plant Science, 11*, 1238. https://doi.org/10.3389/fpls.2020.01238.

Delgado, J. A., & Berry, J. K. (2008). Advances in precision conservation. *Advances in Agronomy, 98*, 1–44. https://doi.org/10.1016/S0065-2113(08)00201-6.

Delgado, J. A., Short, N. M., Roberts, D. P., & Vandenberg, B. (2019). Big data analysis for sustainable agriculture on a geospatial cloud framework. *Frontiers in Sustainable Food Systems, 3*, 54. https://doi.org/10.3389/fsufs.2019.00054.

Ding, T., Tomes, S., Gleave, A. P., Zhang, H., Dare, A. P., Plunkett, B., Espley, R. V., Luo, Z., Zhang, R., Allan, A. C., & Zhou, Z. (2022). MicroRNA172 targets APETALA2 to regulate flavonoid biosynthesis in apples (*Malus domestica*). *Horticulture Research, 9*, uhab007. https://doi.org/10.1093/hr/uhab007.

Ding, W., Wang, Y., Qi, C., Luo, Y., Wang, C., Xu, W., & Qu, S. (2021). Fine mapping identified the gibberellin 2-oxidase gene CpDw leading to a dwarf phenotype in squash (*Cucurbita pepo* L.). *Plant Science, 306*, 110857. https://doi.org/10.1016/j.plantsci.2021.110857.

Eli-Chukwu, N. C. (2019). Applications of artificial intelligence in agriculture: A review. *Engineering, Technology & Applied Science Research, 9*(6), 4377–4383. https://doi.org/10.48084/etasr.3099.

Fonseca-García, C., López-García, C. M., Pacheco, R., Armada, E., Nava, N., Pérez-Aguilar, R., Solis-Miranda, J., & Quinto, C. (2022). Metallothionein1A regulates rhizobial infection and nodulation in *Phaseolus vulgaris*. *International Journal of Molecular Sciences, 23*(3), 1491. https://doi.org/10.3390/ijms23031491.

Fulton, J., & Darr, M. (2017). GPS, GIS, guidance, and variable-rate technologies for conservation management. In J. A. Delgado, G. F. Sassenrath, & T. Mueller (Eds.), *Precision Conservation: Geospatial Techniques for Agricultural and Natural Resources Conservation* (pp. 59–81). Wiley Science.

Ganesh, K. N., Zhang, D., Miller, S. J., Rossen, K., Chirik, P. J., Kozlowski, M. C., Zimmerman, J. B., Brooks, B. W., Savage, P. E., Allen, D. T., & Voutchkova-Kostal, A. M. (2021). Green chemistry: A framework for a sustainable future. *Environmental Science & Technology, 55*(15), 8459–8463. https://doi.org/10.1021/acs.est.1c02105.

Ganivet, E. (2020). Growth in human population and consumption both need to be addressed to reach an ecologically sustainable future. *Environment, Development and Sustainability, 22*(5), 4979–4998. https://doi.org/10.1007/s10668-019-00475-7.

Gao, G. Q., Zhou, H. Y., Niu, X. M., & Fang, Z. M. (2013). Research on intelligent decision-making system of variable spraying for mobile robots in the greenhouse. *Applied Mechanics and Materials, 241*, 1835–1838. https://doi.org/10.4028/www.scientific.net/AMM.241-244.1835.

Hong, Y., Ahmad, N., Zhang, J., Lv, Y., Zhang, X., Ma, X., Xiuming, L., & Na, Y. (2022). Genome-wide analysis and transcriptional reprogramming of the MYB super-family revealed positive insights into abiotic stress responses and anthocyanin accumulation in

Carthamus tinctorius L. *Molecular Genetics and Genomics*, *297*(1), 125–145. https://doi.org/10.1007/s00438-021-01858-5.

Jha, K., Doshi, A., Patel, P., & Shah, M. (2019). A comprehensive review on automation in agriculture using artificial intelligence. *Artificial Intelligence in Agriculture*, *2*, 1–12. https://doi.org/10.1016/j.aiia.2019.05.004.

Jiao, P., Jiang, Z., Wei, X., Liu, S., Qu, J., Guan, S., & Ma, Y. (2022). Overexpression of the homeobox-leucine zipper protein ATHB-6 improves the drought tolerance of maize (*Zea mays* L.). *Plant Science*, *316*, 111159. https://doi.org/10.1016/j.plantsci.2022.111159.

Jin, Y., Chen, B., Lampinen, B. D., & Brown, P. H. (2020). Advancing agricultural production with machine learning analytics: Yield determinants for California's almond orchards. *Frontiers in Plant Science*, *11*, 290. https://doi.org/10.3389/fpls.2020.00290.

Jordan, M. I., & Mitchell, T. M. (2015). Machine learning: Trends, perspectives, and prospects. *Science*, *349*(6245), 255–260. https://doi.org/10.1126/science.aaa8415.

Kakani, V., Nguyen, V. H., Kumar, B. P., Kim, H., & Pasupuleti, V. R. (2020). A critical review on computer vision and artificial intelligence in food industry. *Journal of Agriculture and Food Research*, *2*, 100033. https://doi.org/10.1016/j.jafr.2020.100033.

Kamthan, A., Chaudhuri, A., Kamthan, M., & Datta, A. (2016). Genetically modified (GM) crops: Milestones and new advances in crop improvement. *Theoretical and Applied Genetics*, *129*(9), 1639–1655. https://doi.org/10.1007/s00122-016-2752-9.

Kovak, E., Blaustein-Rejto, D., & Qaim, M. (2022). Genetically modified crops support climate change mitigation. *Trends in Plant Science*, *27*(7), 627–629. https://doi.org/10.1016/j.tplants.2022.04.007.

Kruijssen, F., McDougall, C. L., & van Asseldonk, I. J. (2018). Gender and aquaculture value chains: A review of key issues and implications for research. *Aquaculture*, *493*, 328–337. https://doi.org/10.1016/j.aquaculture.2018.05.003.

Kumar, A., Jaiwal, R., Sreevathsa, R., Chaudhary, D., & Jaiwal, P. K. (2021). Transgenic cowpea plants expressing *Bacillus thuringiensis* Cry2Aa insecticidal protein impart resistance to *Maruca vitrata* legume pod borer. *Plant Cell Reports*, *40*(5), 583–594. https://doi.org/10.1007/s00299-020-02659-4.

Kumar, K., Gambhir, G., Dass, A., Tripathi, A. K., Singh, A., Jha, A. K., Yadava, P., Choudhary, M., & Rakshit, S. (2020). Genetically modified crops: Current status and prospects. *Planta*, *251*(3), 1–27. https://doi.org/10.1007/s00425-020-03372-8.

Kumar, S., Raj, S. K., Sharma, A. K., & Varma, H. N. (2012). Genetic transformation and development of cucumber mosaic virus-resistant transgenic plants of *Chrysanthemum morifolium* cv. Kundan. *Scientia Horticulturae*, *134*, 40–45. https://doi.org/10.1016/j.scienta.2011.12.022.

Kumar, V., Poulose, L., Ahalya, P. S., & Thushima, P. S. (2016). Agricultural biotechnology: Current status and prospects. *Biodiversity, Conservation and Sustainable Development*, *251*, 91. https://doi.org/10.1007/s00425-020-03372-8.

Lan, Y., Zhang, K., He, T., Wang, H., Jiang, C., Yan, H., & Xiang, Y. (2022a). Systematic analysis of the serine/arginine-rich protein splicing factors (SRs) and focus on salt tolerance of PtSC27 in *Populus trichocarpa*. *Plant Physiology and Biochemistry*, *173*, 97–109. https://doi.org/10.1016/j.plaphy.2022.01.015.

Lan, Y., Zhang, K., Wang, Y., Wu, J., Lin, M., Yan, H., & Xiang, Y. (2022b). Comparative analysis of the stellacyanins (SCs) family and focus on drought resistance of PtSC18 in *Populus trichocarpa*. *Gene*, *813*, 146106. https://doi.org/10.1016/j.gene.2021.146106.

Lee, S. Y., Shin, J. Y., Kwon, D. H., Xu, J., Kim, J. H., Ahn, C. H., Jang, S., Kwon, O. H., Lee, H. J., & Kim, W. H. (2022). Preventing scattering of *Tetranychus urticae* in *Rosa hybrida* through dsCOPB2 expression. *Scientia Horticulturae*, *301*, 111–113. https://doi.org/10.1016/j.scienta.2022.110682.

Lei, Y., Yang, B., Jiang, X., Jia, F., & Li, N. (2020). Applications of machine learning to machine fault diagnosis: A review and roadmap. *Mechanical Systems and Signal Processing*, *138*, 106587. https://doi.org/10.1016/j.ymssp.2019.106587.

Lencucha, R., Pal, N. E., Appau, A., Thow, A. M., & Drope, J. (2020). Government policy and agricultural production: A scoping review to inform research and policy on healthy agricultural commodities. *Global Health*, *16*(11). https://doi.org/10.1186/s12992-020-0542-2.

Li, H., Wang, X. H., Li, Q., Xu, P., Liu, Z. N., Xu, M., & Cui, X. Y. (2022a). GmCIPK21, a CBL-interacting protein kinase, confers salt tolerance in soybean (*Glycine max* L.). *Plant Physiology and Biochemistry*, *184*, 47–55. https://doi.org/10.1016/j.plaphy.2022.01.011.

Li, L., Li, S., Ge, H., Shi, S., Li, D., Liu, Y., & Chen, H. (2022c). A light-responsive transcription factor SmMYB35 enhances anthocyanin biosynthesis in eggplants (*Solanum melongena* L.). *Planta*, *255*(1), 12. https://doi.org/10.1007/s00425-021-03698-x.

Li, Q., Qin, Y., Hu, X., Jin, L., Li, G., Gong, Z., Xiong, X., & Wang, W. (2022b). Physiology and gene expression analysis of potato (*Solanum tuberosum* L.) in salt stress. *Plants*, *11*(12), 1565. https://doi.org/10.3390/plants11121565.

Li, X., Gao, L., Zhang, C., & Shao, X. (2010). A review of integrated process planning and scheduling. *International Journal of Manufacturing Research*, *5*(2), 161–180. https://doi.org/10.1504/IJMR.2010.031842.

Lippmann, R., Babben, S., Menger, A., Delker, C., & Quint, M. (2019). Development of wild and cultivated plants under global warming conditions. *Current Biology*, *29*(24), R1326–R1338. https://doi.org/10.1016/j.cub.2019.10.016.

Liu, W., Huo, C., He, L., Ji, X., Yu, T., Yuan, J., Zhou, Z., Song, L., Yu, Q., Chen, J., & Chen, N. (2022a). The NtNRAMP1 transporter is involved in cadmium and iron transport in tobacco (*Nicotiana tabacum*). *Plant Physiology and Biochemistry*, *173*, 59–67. https://doi.org/10.1016/j.plaphy.2022.01.013.

Liu, Z., Abou-Elwafa, S. F., Xie, J., Liu, Y., Li, S., Aljabri, M., Zhang, D., Gao, F., Zhang, L., Wang, Z., & Sun, C. (2022b). A nucleoporin NUP58 modulates responses to drought and salt stress in maize (*Zea mays* L.). *Plant Science*, *320*, 111296. https://doi.org/10.1016/j.plantsci.2022.111296.

Lloret, A., Quesada-Traver, C., Conejero, A., Arbona, V., Gómez-Mena, C., Petri, C., Sánchez-Navarro, J. A., Zuriaga, E., Leida, C., Badenes, M. L., & Ríos, G. (2021). Regulatory circuits involving bud dormancy factor PpeDAM6. *Horticulture Research*, *8*(1), 261. https://doi.org/10.1038/s41438-021-00706-9.

Ma, L., Zhang, Y., Wen, H., Liu, W., Zhou, Y., & Wang, X. (2022). Silencing of MsD14 resulted in enhanced forage biomass through increasing shoot branching in alfalfa (*Medicago sativa* L.). *Plants*, *11*(7), 939. https://doi.org/10.3390/plants11070939.

Meena, A., Karwal, M., & Raghavendra, K. J. (2021). Sustainable and climate-smart agriculture: Challenges and opportunities in Indian perspective. *Scientific Agriculture and Allied Sectors*, *3*, 47–57.

Muthu, B., Sivaparthipan, C. B., Manogaran, G., Sundarasekar, R., Kadry, S., Shanthini, A., & Dasel, A. (2020). IoT-based wearable sensor for disease prediction and symptom analysis

in the healthcare sector. *Peer-to-Peer Networking and Applications, 13*(6), 2123–2134. https://doi.org/10.1007/s12083-020-00936-0.
National Academies of Sciences, Engineering, and Medicine. (2002). *Environmental Effects of Transgenic Plants: The Scope and Adequacy of Regulation.* Washington, DC: National Academies Press.
National Academies of Sciences, Engineering, and Medicine. (2019). *Science Breakthroughs to Advance Food and Agricultural Research by 2030.* Washington, DC: National Academies Press.
Ng, A. K., & Mahkeswaran, R. (2021). Emerging and disruptive technologies for urban farming: A review and assessment. *Journal of Physics, 2003*, 012008.
Panter, S., Chu, P. G., Ludlow, E., Garrett, R., Kalla, R., Jahufer, M. Z. Z., de Lucas Arbiza, A., Rochfort, S., Mouradov, A., Smith, K. F., & Spangenberg, G. (2012). Molecular breeding of transgenic white clover (*Trifolium repens* L.) with field resistance to alfalfa mosaic virus through the expression of its coat protein gene. *Transgenic Research, 21*, 619–632.
Patel, G. S., Rai, A., Das, N. N., & Singh, R. P. (2021). *Smart Agriculture: Emerging Pedagogies of Deep Learning, Machine Learning and Internet of Things.* London, UK: CRC Press.
Patrício, D. I., & Rieder, R. (2018). Computer vision and artificial intelligence in precision agriculture for grain crops: A systematic review. *Computers and Electronics in Agriculture, 153*, 69–81.
Peters, K., Breitsameter, L., & Gerowitt, B. (2014). Impact of climate change on weeds in agriculture: A review. *Agronomy for Sustainable Development, 34*, 707–721.
Poonam, A., Kumar, N., Chakraborty, P., & Kothari, R. (2017). Plant growth under stress conditions: Boon or bane. In V. Shukla, S. Kumar, & N. Kumar (Eds.), *Plant Adaptation Strategies in Changing Environment* (pp. 291–313). Springer. https://doi.org/10.1007/978-981-10-6745-0_15.
Qin, L., Zhang, H., Li, J., Zhu, Y., Jiao, G., Wang, C., & Wu, S. (2022). Down-regulation of *GhADF1* in cotton (*Gossypium hirsutum*) improves plant drought tolerance and increases fiber yield. *The Crop Journal, 10*, 1037–1048.
Raffeiner, M., Üstün, S., Guerra, T., Spinti, D., Fitzner, M., Sonnewald, S., Baldermann, S., & Börnke, F. (2022). The Xanthomonas type-III effector XopS stabilizes Ca WRKY40a to regulate defense responses and stomatal immunity in pepper (*Capsicum annuum*). *The Plant Cell, 34*, 1684–1708.
Ramesh, K., Matloob, A., Aslam, F., Florentine, S. K., & Chauhan, B. S. (2017). Weeds in a changing climate: Vulnerabilities, consequences, and implications for future weed management. *Frontiers in Plant Science, 8*, 95. https://doi.org/10.3389/fpls.2017.00095.
Ranjha, M. M., Shafique, B., Khalid, W., Nadeem, H. R., Mueen-Ud-Din, G., & Khalid, M. Z. (2022). Applications of biotechnology in food and agriculture: A mini-review. *Proceedings of the National Academy of Sciences, India Section B: Biological Sciences, 92*, 11–15.
Raza, A., Su, W., Jia, Z., Luo, D., Zhang, Y., Gao, A., Hussain, M. A., Mehmood, S. S., Cheng, Y., Lv, Y., & Zou, X. (2022). Mechanistic insights into trehalose-mediated cold stress tolerance in rapeseed (*Brassica napus* L.) seedlings. *Frontiers in Plant Science, 13*, 857980. https://doi.org/10.3389/fpls.2022.857980.
Reitz, R. D., Ogawa, H., Payri, R., Fansler, T., Kokjohn, S., Moriyoshi, Y., Agarwal, A. K., Arcoumanis, D., Assanis, D., Bae, C., & Boulouchos, K. (2020). The future of the internal combustion engine. *International Journal of Engine Research, 21*, 3–10.

Rull, V. (2014). The most important application of science: As scientists have to justify research funding with potential social benefits, they may well add education to the list. *EMBO Reports*, *15*, 919–922.

Samtani, H., Sharma, A., & Khurana, P. (2022). Overexpression of *HVA1* enhances drought and heat stress tolerance in *Triticum aestivum* doubled haploid plants. *Cells*, *11*, 912. https://doi.org/10.3390/cells11050912.

Singh, K. N., Kumar, D., Shamim, M., & Kumar, M. (2018). Biosafety for sustainable agriculture. In R. L. Singh & S. Mondal (Eds.), *Biotechnology for Sustainable Agriculture* (pp. 305–333). Woodhead Publishing, Elsevier.

Singh, R., & Singh, G. S. (2017). Traditional agriculture: A climate-smart approach for sustainable food production. *Energy, Ecology and Environment*, *2*, 296–316.

Smyth, S. J. (2020). The human health benefits from GM crops. *Plant Biotechnology Journal*, *18*, 887–888.

Song, G. Q., Hancock, J. F., & Kole, C. (2011). Wild crop relatives: Genomic and breeding resources: Plantation and ornamental crops. In C. Kole (Ed.), *Springer*, 303.

Struik, P. C., & Kuyper, T. W. (2017). Sustainable intensification in agriculture: The richer shade of green. A review. *Agronomy for Sustainable Development*, *37*, 39. https://doi.org/10.1007/s13593-017-0445-7.

Sun, Y., Hu, D., Xue, P., & Wan, X. (2022). Identification of the *DcHsp20* gene family in carnation (*Dianthus caryophyllus*) and functional characterization of *DcHsp17.8* in heat tolerance. *Planta*, *256*, 2. https://doi.org/10.1007/s00425-022-03915-1.

Sutherland, C., Gleim, S., & Smyth, S. J. (2021). Correlating genetically modified crops, glyphosate use, and increased carbon sequestration. *Sustainability*, *13*, 11679.

Uddin, S., Khan, A., Hossain, M. E., & Moni, M. A. (2019). Comparing different supervised machine learning algorithms for disease prediction. *BMC Medical Informatics and Decision Making*, *19*, 1–6.

Ulloa, J. L., Aguayo, P., Conejera, D., Rubilar, R., Balocchi, C., & Valenzuela, S. (2022). Transcriptomic response in foliar and root tissues of a drought-tolerant *Eucalyptus globulus* genotype under drought stress. *Trees*, *36*, 697–709.

Van Aalst, M. K., Cannon, T., & Burton, I. (2008). Community level adaptation to climate change: The potential role of participatory community risk assessment. *Global Environmental Change*, *18*, 165–179.

Wan, X., Sun, Y., Feng, Y., Bao, M., & Zhang, J. (2022). Heat stress transcription factor *DcHsfA1d* isolated from *Dianthus caryophyllus* enhances thermo-tolerance and salt tolerance of transgenic *Arabidopsis*. *Biologia Plantarum*, *66*, 29–38.

Wei, H., Zhao, H., Su, T., Bausewein, A., Greiner, S., Harms, K., & Rausch, T. (2017). Chicory R2R3-MYB transcription factors *CiMYB5* and *CiMYB3* regulate fructan 1-exohydrolase expression in response to abiotic stress and hormonal cues. *Journal of Experimental Botany*, *68*, 4323–4338.

Wróbel-Kwiatkowska, M., Słupczyńska, M., & Rymowicz, W. (2022). Overexpression of medium-chain-length polyhydroxyalkanoates induces significant salt tolerance and fungal resistance in flax. *Plant Cell, Tissue and Organ Culture*, *151*, 123–132.

Wu, G. Q., Xie, L. L., Wang, J. L., Wang, B. C., & Li, Z. Q. (2023). Genome-wide identification of CIPK genes in sugar beet (*Beta vulgaris*) and their expression under NaCl stress. *Journal of Plant Growth Regulation*, *42*, 260–274.

Yadav, S. S., & Lal, R. (2018). Vulnerability of women to climate change in arid and semi-arid regions: The case of India and South Asia. *Journal of Arid Environments*, *149*, 4–17.

Yang, J., Wang, M., Zhou, S., Xu, B., Chen, P., Ma, F., & Mao, K. (2022). The ABA receptor gene *MdPYL9* confers tolerance to drought stress in transgenic apples (*Malus domestica*). *Environmental and Experimental Botany, 194*, 104695.

Yarra, R., & Kirti, P. B. (2019). Expressing the Class I wheat NHX (*TaNHX2*) gene in eggplant (*Solanum melongena* L.) improves plant performance under saline conditions. *Functional & Integrative Genomics, 19*, 541–554.

Zeng, W., Hassan, M. J., Kang, D., Peng, Y., & Li, Z. (2021). Photosynthetic maintenance and heat shock protein accumulation relating to γ-aminobutyric acid (GABA)-regulated heat tolerance in creeping bentgrass (*Agrostis stolonifera*). *South African Journal of Botany, 141*, 405–413.

Zhai, Z., Martínez, J. F., Beltran, V., & Martínez, N. L. (2020). Decision support systems for Agriculture 4.0: Survey and challenges. *Computers and Electronics in Agriculture, 170*, 105256.

Zhang, H., Sun, Z., Feng, S., Zhang, J., Zhang, F., Wang, W., Hu, H., & Zhang, W., & Bao, M. (2022a). The C2H2-type zinc finger protein PhZFP1 regulates cold stress tolerance by modulating galactinol synthesis in *Petunia hybrida*. *Journal of Experimental Botany, 73*, 6434–6448.

Zhang, Q., Geng, J., Du, Y., Zhao, Q., Zhang, W., Fang, Q., Yin, Z., Li, J., Yuan, X., Fan, Y., & Cheng, X. (2022b). Heat shock transcription factor (Hsf) gene family in common bean (*Phaseolus vulgaris*): Genome-wide identification, phylogeny, evolutionary expansion and expression analyses at the sprout stage under abiotic stress. *BMC Plant Biology, 22*, 33. https://doi.org/10.1186/s12870-021-03417-4.

Zhang, Y., Xie, Z., Wang, F., Zhong, C., Liu, Y., Li, Z., Wang-Pruski, G., & Zhang, Z. (2022c). Genome-wide identification and characteristics analysis of melon (*Cucumis melo* L.) MYB transcription factors and their responses to autotoxicity and saline-alkali stress. *Tropical Plant Biology, 15*, 93–109.

Zhao, W., Wang, K., Chang, Y., Zhang, B., Li, F., Meng, Y., Li, M., Zhao, Q., & An, S. (2022). *OsHyPRP06/R3L1* regulates root system development and salt tolerance via apoplastic ROS homeostasis in rice (*Oryza sativa* L.). *Plant, Cell & Environment, 45*, 900–914.

Zhou, Q., Mao, P., Luo, D., Chai, X., Deng, H., Fang, Q., Fang, L., Nan, Z., Wen, J., & Liu, Z. (2022). Comparative transcriptome analyses reveal that the *MsNST1* gene affects lignin synthesis in alfalfa (*Medicago sativa* L.). *The Crop Journal, 10*, 1059–1072.

Zimmerman, J. B., Anastas, P. T., Erythropel, H. C., & Leitner, W. (2020). Designing for a green chemistry future. *Science, 367*, 397–400.

CHAPTER 2

Improvement in Waterlogging Tolerance in Plants via Melatonin

ABEER HASHEM,[1] KHALID F. ALMUTAIRI,[2] MOHAMAD ALNAFISSA,[3] JAVID A. PARRAY,[4] FADIME KARABULUT,[5] CHADHA AYED,[6] and ELSAYED FATHI ABD_ALLAH[2,*]

[1]*Botany and Microbiology Department, College of Science, King Saud University, Riyadh, Saudi Arabia*

[2]*Plant Production Department, College of Food and Agricultural Sciences, King Saud University, Riyadh, Saudi Arabia*

[3]*Department of Agricultural Economics, College of Food and Agricultural Sciences, King Saud University, Riyadh, Saudi Arabia*

[4]*Department of Higher Education, Government Degree College, Eidgah, Srinagar, Jammu and Kashmir, India*

[5]*Department of Biology, Faculty of Science, Firat University, Elazig, Turkey*

[6]*Research Laboratory "Agrobiodiversity and Ecotoxicology" LR21AGR02, High Agronomic Institute of Chott Mariem, Sousse, Tunisia*

*Corresponding author.

ABSTRACT

Water stress (drought and water logging) has a significant impact on plant growth and development. There has been minimal research on the bioactive plant hormone melatonin's impact on water logging stress situations, despite the fact that it has been widely investigated in the context of water shortage stress in a number of plant species. Melatonin controls the levels of nitrogen and reactive oxygen species (ROS), which modifies the molecular

Agricultural Biotechnology: Issues, Challenges, and Recent Developments. Mohd. Sayeed Akhtar and Khalid Rehman Hakeem (Eds.)
© 2025 Apple Academic Press, Inc. Co-published with CRC Press (Taylor & Francis)

defense mechanism and improves plant resistance to water stress. Melatonin and other phytohormone interaction is also a crucial part of plants under drought stress. The aim of this chapter is to provide recent updates on the role of melatonin in the water logging and drought tolerance mechanism by improving physiological, molecular, and anatomical changes and also to discusses the impacts of nitric oxide and hydrogen sulfide (H_2S) during the tolerance mechanism.

2.1 INTRODUCTION

Abiotic stress, specifically the water stress, that is, waterlogged or drought condition, is being severed due to alteration in environmental conditions in the agriculture industry, primarily in the dry regions, that is, arid and semi-arid regions, due to an average increase in the global temperature, greenhouse effect, rainfall inequality, and poor water management under field conditions (Stuart et al., 2011; Xoconostle-Cazares et al., 2010). According to Jackson and Colmer (2005) water logging is a significant environmental stressor that has an impact on crop survival, development, and productivity in regions with high water table variations, excessive rainfall, and inadequate soil drainage. Water logging stress caused by many factors, that is, heavy rainfall, flooding, and inadequate drainage, could pose a danger for food security. On the other side, the human population is growing with an increasing trend and expected to be 10 billion in 2050 (Hasanuzzaman et al., 2018). Higher plants need access to molecular oxygen to support their development and metabolism. However, too much water in the soil frequently leads to an insufficient supply of oxygen to the plant cells, which causes several phenotypic, physiological, and metabolic disturbances, such as the inhibition of shoot and root growth, a decrease in water and nutrient uptake, reduced efficiency of photosynthesis and photochemical reactions on the leaves, respiratory problems in the roots, and leaf senescence (Boru et al., 2003). Additionally, water logging frequently causes plants to produce significantly more ethylene. According to (Najeeb et al., 2017), the high ethylene content under water logging stress frequently results in a considerable reduction in shoot and root growth, as well as leaf senescence, and abscission, and photosynthesis (Pierik et al., 2007). S-adenosyl-L-methionine (SAM), which is produced from methionine by SAM synthetase, is the first step in the biosynthesis of ethylene. The formation of 1-aminocyclopropane-1-carboxylic acid (ACC) from SAM is subsequently catalyzed by ACC synthase (ACS), and ACC oxidase (ACO) converts ACC to ethylene (Najeeb et al., 2017). Low molecular weight and aliphatic nitrogenous substances known as polyamines (PAs) include spermidine (Spd), spermine (Spm), and putrescine (Put) (Groppa and Benavides, 2008). They have been considered a class of plant growth regulators and implicated in the growth and development of plants (Tavladoraki et al., 2012). Additionally, many studies have suggested links between PA metabolism and plant responses to environmental stress conditions, such as drought, high temperature, salinity, nutrient deficiency, and others (Zhao et al., 2017), due to their roles in preserving nucleic acid and protein structures, promoting membrane stability, and scavenging free radicals (Alcázar et al., 2010).

The major effect of drought stress is the large production of reactive oxygen species (ROS) and malondialdehyde (MDA), which damage the cell membrane, nucleic acids, and proteins directly and/or indirectly (Arbona et al., 2008). This has a negative impact on gas exchange and photosynthesis, as well as plant development, quality, and quantity (Hossain et al., 2009; Laxa et al., 2019). In practice, a global-scale examination of published studies on maize and wheat over the last four decades found that water scarcity was responsible for 20–40% of yield losses (Daryanto et al., 2016). Meanwhile, water logging has been predicted to cause a 40–80% loss in crop productivity over 1.7 billion hectares (Shabala et al., 2014; Voesenek and Sasidharan, 2013). Plants have evolved numerous techniques to deal with water stress. Plants counteract the negative consequences of drought by closing their stomata, accumulating solutes, and producing wax biosynthesis (Hasanuzzaman et al., 2018). Furthermore, plants increase their resistance to oxidative damage by activating anti-oxidative capacities and inducing some molecular chaperones (Elkeilsh et al., 2019). As a result, plants alter their metabolism to shift towards anaerobic, glycolytic, and fermentative metabolism. The antioxidant enzyme system, heat shock proteins (HSPs), and osmolyte production are activated due to anoxia (Tewari and Mishra, 2018). The diverse plant responses are facilitated using different key regulators in the form of phytohormones (Verma et al., 2016). According to Ahmad et al. (2002); and Arnao and Hernández-Ruiz (2019a), melatonin is an important plant regulator and antioxidant that controls a variety of physiological plant processes in difficult environmental settings and shields plants from oxidative stress in water-stressed environments. However, there have been a lot of studies on melatonin compounds and their health benefits that have been published recently. To fully understand the functions of melatonin under biotic and abiotic stress, however, more study is required (Bose and Howlader, 2020; Moustafa-Farag et al., 2019). The aim of this chapter is to provide recent updates on the role of melatonin in the water logging and drought tolerance mechanism by improving physiological, molecular, and anatomical changes and also to discusses the impacts of nitric oxide and hydrogen sulfide (H_2S) during the tolerance mechanism.

2.1.1 MELATONIN-INDUCED DROUGHT STRESS TOLERANCE

Melatonin (N-acetyl-5 methoxy tryptamine), an antioxidant occurring naturally, is an outstanding and potent antioxidant that effectively manages the stress under severe drought impacts in plants (Arnao and Hernández-Ruiz, 2019b). Melatonin a low molecular weight essential amino acid L-tryptophan (Koyama et al., 2013) and considered as general signal molecule in plants and mammals (Tan et al., 2012). Melatonin is therefore highly recommended for reducing drought stress in a variety of plant species, including *Arabidopsis*, and rice (Lee and Back, 2019; Yang et al., 2019), fruit crops (Xia et al., 2020), field crops (Cao et al., 2019; Su et al., 2019), vegetable crops (Campos et al., 2019; Karaca and Cekic, 2019; Kaya and Doganlar, 2019), medicinal and ornamental plants (Campos et al., 2019). The study also suggested the role of melatonin to control the PAs levels in human skin and rat brains (Lee et al., 2003). There haven't been many studies done to clarify how melatonin affects PA metabolism in plants under abiotic stress, including water logging (Zhang et al., 2019). There were different concentrations were reported in different field (1 mM) in maize and fruits crop such as grapes 50 nM for the improvement of drought stress (Fleta-Soriano et al., 2017). Furthermore, seed priming (Khan et al., 2019), seed coating (Wei et al., 2015), direct soil treatment (Ding et al., 2019), foliar application (Li et al., 2019), in nutrient solutions and hydroponic systems (Zou et al., 2019), coupled with fertilization (Xia et al., 2020), and application in the root zone at germination stage (Li et al., 2019) can all be used to decrease drought stress. Similarly, another study suggested that melatonin treated showed more production of melatonin absorption from the environment at post-transcriptional control relative to control (Zheng et al., 2017).

2.1.2 MELATONIN USED AS DROUGHT STRESS TOLERANCE

Chlorosis is a typical sign when intense water logging condition caused excess generation of O_2 and H_2O_2, which degrades chlorophyll and chloroplasts break down (Smethurst and Shabala, 2003). Previous research has investigated the protective benefits of melatonin against chlorophyll degradation and photosynthetic capacity under various abiotic conditions, that is, high temperature and waterlogged conditions (Wang et al., 2019). It has been established that the chloroplast is the primary location for melatonin synthesis, but it was also one of the organelles that was most adversely affected by ROS. To maintain its structure and functionality, plenty of melatonin is needed to regulate different metabolic and biological pathways. Therefore, the absorbed and *in vivo* production of melatonin can work in lessen damage due to water stress and helps to plants for survival under such circumstances (Zhao et al., 2017). Because of the widespread usage of melatonin in the treatment of drought stress, scientists have been looking into the direct signal of melatonin's role under waterlogged condition. Different melatonin biosynthesis genes that is, tryptophan decarboxylase (TDC), serotonin N-acetyltransferase (SNAT), N-acetyl serotonin methyltransferase (ASMT) and caffeic acid O-methyltransferase (COMT) were involved for abiotic stress tolerance against drought. These genes were triggered and up regulated under drought stress condition and alter the plant tolerance mechanism under such conditions. However, endogenous melatonin levels fluctuate in response to changes in plant development cycle. Under abiotic stress response including drought and water stress enhanced the melatonin levels, as a result protective mechanism initiate to withstand against drought stress (Ahmad et al., 2021; Arnao and Hernández-Ruiz, 2015). These biosynthesis genes were expressed and increased the endogenous melatonin levels. Similar results were reported by (Arnao and Hernández-Ruiz, 2013) suggested that drought influence the global crops such as lupins, barley, and potato at different growth stages, however, melatonin applications improved the growth and development at different stages. Different studies were reported on the application of melatonin on crop plants, that is, *Arabidopsis* (Yang et al., 2019), barley (Li et al., 2016), Bermuda grass (Shi et al., 2015), apple (Li et al., 2015), grapevine (Jiao et al., 2016), and rice (Wei et al., 2016) were reported under drought stress. Similarly, another study reported on drought stress-induced melatonin biosynthesis enzymes content between 2 and 6-fold in one or more transcripts suggested the role of melatonin under drought stress (Arnao and Hernández-Ruiz, 2019; Wei et al., 2016). The role of melatonin in different crop plants under drought condition was tabulated in Table 2.1.

TABLE 2.1 Melatonin's Role in Different Crop Plants Under Drought Stress Conditions

Crop Name	Drought Stress	Effects	References
Rice	Water drainage	Expression of DEGs, growth characteristics, mitochondria structure enhance	Lee and Back (2016)
Arabidopsis	Water withholding	Soluble sugars, and drought responsive genes	Shi et al. (2015)
Maize	Water with holding	PSII efficiency	Fleta-Soriano et al. (2017)
Maize	Artificial induced PEG drought	Physiological and biochemical effects	Ye et al. (2016)
Wheat	40–60% field capacity	Chloroplast structure, photo-synthesis, antioxidant enzyme, membrane damage	Cui et al. (2017)
Wheat	30% pot holding capacity	Recovery after rehydration, growth attributes, antioxidants enzyme, physiology attributes	Ye et al. (2015)
Barely	Drought and cold stress	Endogenous melatonin, ABA water status, photo system II, photosynthesis	Li et al. (2016)
Soybean	20% field capacity	Electrolyte leakage, radicle and plumule, germination status	Wei et al. (2015)
Cassava	20% artificial induced PEG	Antioxidant enzyme activity	Ding et al. (2019)
Cotton	10% artificial induced drought by PEG	Chlorophyll, stomatal closure, antioxidants enzyme, cellular redox	Bai et al. (2020)
Alfalfa	Water withholding	Chlorophyll, stomatal closure, antioxidants enzyme, cellular redox	Antoniou et al. (2017)
Rapeseed	4% PEG 6000	Osmo-protectants, antioxidants, plant growth	Li et al. (2018)
Tobacco	Water withholding	Drought tolerance, MDA	Lee and Back (2019)

2.2 MECHANISM OF TOLERANCE IN PLANTS VIA MELATONIN

2.2.1 *PHYSIOLOGICAL AND MORPHOLOGICAL MECHANISMS*

Melatonin's reported multi-dimensional regulators roles in plants under drought conditions at different growth stages that is, at morphological and physiological characteristics in recent years. Similarly, Wu et al.

(2016) reported that anatomical changes were recorded in plant organ at cell membrane levels, in chloroplast grana lamella (Cui et al., 2017) reduced chloroplast ultra-structural damage and enhanced the tolerance level (Meng et al., 2014; Zhang et al., 2013), managed the cell extension, improved leaf width, improved porous tissue, and enhanced the stomata closure (Khan et al., 2019; Meng et al., 2014), cuticle formation (Ding et al., 2018), and wax accumulation (Li et al., 2019). Similarly, the other way the action of melatonin acts as biosynthesis moderator improves the mitochondria and chloroplast structure condition demonstrates more defense measures as stress level increases (Back et al., 2016; Tan et al., 2013). By improving the antioxidant enzyme activity, melatonin improved the physiological characteristics under sever circumstances (Sharma et al., 2020; Xia et al., 2020), by limiting oxidative injury, lowering the ROS and reactive nitrogen species (RNS) (Antoniou et al., 2017). Furthermore, the following processed were also reduced such as electrolyte leakage (Lee and Back, 2016), toxic substance (Liu et al., 2015), reduced cellular redox interruption (Antoniou et al., 2017), relative conductivity, lipid peroxidation (malondialdehyde reduction) (Ma et al., 2018) and improved nitro-oxidative homeostasis (GSH).

Different enzymatic and non-enzymatic activities were regulated under stress condition. The enzymatic activity involved peroxidase (POD), ascorbate peroxidase (APX), catalase (CAT), and superoxide dismutase (SOD), as well as nonenzymatic antioxidants (plant polyphenol, vitamin C, E, glutathione, and carotenoids) and Osmo-protectants (Ahmad et al., 2019; Ding et al., 2019), and secondary metabolites, that is, flavonoids, phenolics, and phenylalanine ammonia lyase were involved. Melatonin improves the photosystem of plants, as evidenced by higher chlorophyll content (Liang et al., 2019), faster food production process (Huang et al., 2019), and higher transpiration rates (Campos et al., 2019). Melatonin has also been shown to improve photo-protection by increasing photosystem II efficiency (Fleta-Soriano et al., 2017). Melatonin, as a multifunctional molecule, modulates the cell's osmotic potential (Ye et al., 2016) by accumulating soluble carbohydrates and proline (Li et al., 2018). Furthermore, under drought conditions, one of the significant difficulties for melatonin to control is water status. Melatonin improves plant resilience by increasing stomatal conductance (Ye et al., 2016), cell turgor and water holding capacity (Ma et al., 2018), and maintain the stomatal closure (opening and closing) process (Wei et al., 2018). Furthermore, the results also shown to enhanced the germination activity (Li et al., 2019), growth and development of root

(Zhang et al., 2013), reproductive process (Wei et al., 2015; Zhang et al., 2019), quality characteristics (Ma et al., 2018), seed yield (Zou et al., 2019), leaf senescence (Liang et al., 2018), and speedy recovery after rehydration (Sharma et al., 2020) are all improved tolerance mechanism were involved after the application of melatonin.

2.2.2 MOLECULAR MECHANISMS

Melatonin's defensive mechanisms have also been investigated, with a particular focus on its capacity in response to drought stress, to protect plant cells from redox homeostasis disruption. Melatonin affects ROS and RNS levels as well as enhanced the antioxidant-related genes such as glutathione S-transferase (GSTP), glutathione reductase (GR), dehydroascorbate reductase (DHAR), APX, POD, SOD monodehydroascorbate reductase (MDHAR) and CAT (Cui et al., 2018; Sharma et al., 2020), by managing the proline and osmo-protective components using proline synthesis genes (Antoniou et al., 2017). Melatonin not only protects plants from oxidative damage, but it also controls the genes expression (up and down regulation) involved in carbohydrate/fatty acid and amino acid metabolism (Ding et al., 2019; Wei et al., 2015), the carbon metabolic pathway (Shi et al., 2015), nitrogen metabolism and transport (Ding et al., 2019; Liang et al., 2018), plant secondary metabolism (Sharma et al., 2020), energy production (Cui et al., 2018; Ding et al., 2019), carotenoid metabolism and photosynthesis (Sharma et al., 2020; Xia et al., 2020), and cuticle wax biosynthesis (Ding et al., 2018). Similarly, another study was done in soybean where melatonin was used in solution form with seed coating for the improvement of drought stress tolerance, the results suggested that the metabolism of carbohydrates and fatty acids contents were stimulated due to melatonin application under treated relative to control (Wei et al., 2015). Application of melatonin results suggested that the N uptake and metabolic genes were up-regulated in tree leaves due to melatonin application under water stress condition and acts as significant N regulators helps to improve the metabolism and translocation of photosynthates under drought stress condition (Moustafa-Farag et al., 2020).

Melatonin also regulates the gene expression of glycolytic protein and electron transport gene in the respirational chain, promoting energy production in water-scarce situations (Cui et al., 2018). Furthermore, melatonin regulates the photosynthesis process by regulating different genes (i.e., *PGK, TKT, FBA, RPI, FBP, GAPA, TIM, RPK, Rubisco, SEBP,* and *RPE*) regulate the CO_2 fixation and enzymatic activities (Liang et al., 2019), carbon fixation protein expression (Su et al., 2019), photosynthesis light reaction (cytochrome P450) (Ding et al., 2019), and tetrapyrrole pigment biosynthesis (Ding et al., 2019; Jiao et al., 2016). Another study also recorded similar results regulates the photosynthesis activity process was enhanced when soybean seeds were primed with melatonin solution that regulates certain mechanism under abiotic stress condition especially drought and salinity (Wei et al., 2015). Photosystem I and II subunits (PsbO and PsbP) combined with oxygen improved the oxygen enhancer protein due to melatonin (Wei et al., 2015). Ferredoxin (*PetF*) gene was recorded to regulate the ascorbate reduction that protects the chlorophyll contents from filth. The relative expression of *PetF* gene recorded higher expression under the melatonin application treatments. Similarly, another gene VTC4 also encodes ascorbates biosynthesis enzyme regulates L-galactose 1-P-phosphatase recorded similar function (Wei et al., 2015). Melatonin suppressed the gene regulation and enzyme activity of different genes (*PPH, Chlase,* and *Chl-PRX*) regulates the chlorophyll degradation under water stress condition (Ma et al., 2018). Different biosynthesis genes that is, KCS1 (fatty acid elongation), TSS1 (triterpenoids synthesis), and LTP1 (triterpenoids synthesis) for lipid transport were significantly evaluated. It was noted that drought stress elevated the transcripts of the four genes, which were then further induced by melatonin treatment, confirming the significance of melatonin used in wax production (Ding et al., 2018).

Melatonin also considered as plant phytohormone that promotes stomatal closure via the signaling route of PMTR1 (Phyto melatonin receptors), control the hydrogen peroxide (H_2O_2) generation and the Ca^{2+} signaling transduction cascade (Wei et al., 2018). PMTR1 interacts with subunit of G-protein A (GPA1) in the plasma membrane that regulate the process (Wei et al., 2018). The dissociation of G and G (heterotrimeric G-proteins) is triggered by phyto-melatonin-receptor interaction, which causes NADPH oxidase-dependent H_2O_2 release and promotes Ca^{2+} and K^+ efflux, resulting in stomata management (Wei et al., 2018). Furthermore, as a respiratory burst oxidase, NAPDH oxidase creates superoxide radicals, which are then dismutase to H_2O_2 either enzymatically or non-enzymatically reactions. Melatonin also modulates the activity of critical transcription factors including *Myb4*, *AP37*, and zinc finger (Lee and Back, 2016; Shi et al., 2015), as well as transporter proteins like proton transporter (UCP1), potassium transporter (HKT1), and water channel protein (PIP2;1) significant for drought stress tolerance (Lee and Back, 2016). Furthermore, melatonin signaling orchestrates controlled several stress-signaling genes such as Ca^{2+} and protein kinases-related, suggesting kinase signaling role in drought tolerance (Shi et al., 2015). When plants experience water shortage due to drought, melatonin's play a protective and regulatory role, in combination with other anti-stress measures, begins to avoid, reduce, or eliminate the stresses (Ahmad et al., 2021; Arnao and Hernández-Ruiz, 2019b). Under water stress condition stress signals were transferred to the nucleus through the cell membrane to manage with the harsh impacts of drought (Wei et al., 2018; Zhou et al., 2018), by triggering the melatonin-biosynthesis genes (Li et al., 2015; Ma et al., 2018). As a results, the nucleus quickly starts to initiate the melatonin biosynthesis pathway and its associated genes, that is, tryptophan, in mitochondria and chloroplasts and overcome the stress situation (Back et al., 2016; Tan et al., 2013; Wang et al., 2017). Surprisingly, melatonin begins by delivering stress feedback to the nucleus, which then triggers the suitable stress reaction (Cui et al., 2017; Ma et al., 2018). Numerous protective genes regulated agronomic, physiological traits directly or indirectly. Therefore, these mechanisms were controlled by many genes responsible for chlorophyll degradation, photosynthesis, photo-protection, antioxidants enzymes systems, cell membrane stability, hydration, oxidative cell damage and leaf senescence improved under water stress condition. As a result, the whole condition of the plant is improved, including germination, juvenile stage, flowering, and reproductive stages by improving the tolerance, while harmful compounds are reduced, resulting in drought tolerance.

2.3 MELATONIN-WATER STRESS RESEARCH: INTERACTIONS OF MELATONIN WITH OTHER COMPOUNDS

Melatonin application with micronutrients in the form of nitric oxide (NO), and H_2S also played a significant effect under drought circumstances and plant defense mechanism (Bhuyan et al., 2020). Melatonin regulates a variety of plant responses under water shortage conditions as explained above. NO is a basic signal that acts as a pro-oxidant and antioxidant in response to harmful settings. Their endogenous absorption and local production are harmful for plants (Zhu et al., 2019). H_2S is a major metabolic regulator in plants that regulates enzymatic antioxidants to alleviate the damaging effects of drought and water logging (Cheng et al., 2013; Jin et al., 2011). There was a strong connection was present among the melatonin, nitrogen oxide and H_2S to overcome the biotic and abiotic stress response drought (Mukherjee et al., 2014), and salinity (Antoniou et al., 2017), as well as melatonin was used in fruit ripening process (Mukherjee, 2019). Melatonin protects alfalfa plants from drought damage by altering nitro-oxidative equilibrium through enzymatic and/or transcript-level control of RO and NS metabolic enzymes (Antoniou et al., 2017). However, it is yet unknown how endogenous melatonin is involved with NO in the presence of water scarcity (Zhu et al., 2019). There is dire need to determine the exact mechanisms were involved using the plant omics and advance next generations sequencing (NGS) techniques to determine the conditions specific transcripts and genes involved for tolerance mechanisms.

2.3.1 PHYTOHORMONES

Through the control of phytohormone levels, melatonin-a key hormone regulating molecule increases plants' tolerance to drought stress. There were different types of phytohormones present in nature such as abscisic acid (ABA), auxin (Aux), gibberellic acid (GA), salicylic acid (SA), jasmonic acid (JA), and brassinosteroids (BRs). In response to drought stress, the main physiological processes that are heavily regulated by phytohormones include antioxidant metabolism, carbohydrate synthesis (carbon metabolism), stomatal movement, and leaf senescence (Burgess and Huang, 2016). There were different Phyto-response up-regulate under drought stress such as JA, ABA, and BRs (Burgess and Huang, 2016; Sharma et al., 2020), while downregulate the GAs and CKs. However, the exogenous application enhanced the CKs, GAs, JA, and BRs (Sharma et al., 2020), and downregulate the ABA (Sharma et al., 2020). A schematic diagram presents the role and function of different phytohormones under melatonin and drought stress condition (Figure 2.1). ABA generation is stimulated by a lack of water (Behnam et al., 2013; Elkeilsh et al., 2019), which in turn downregulates the primary metabolic pathways, helps to leaf senescence and stomata regulation (Burgess and Huang, 2016). Under water stress, cytokinin manufacture and transport are suppressed, which results in cytokinin decrease and accelerated leaf senescence.

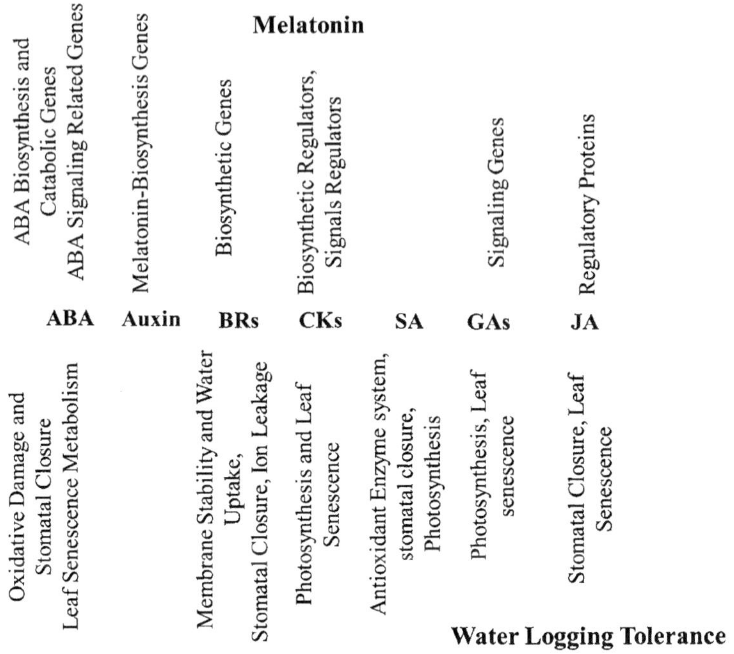

FIGURE 2.1 A schematic diagram showing the role and functions of each phytohormone along with melatonin under drought stress.

CKs are an essential phytohormones that prevent chlorophyll degradation and leaf senescence (Kudoyarova et al., 2007). Better photosynthetic efficiency and tolerance to drought-induced effects were recorded due to result of melatonin treatments and upregulation of CKs levels and several associated signaling pathways (Ma et al., 2018). Drought stress prevents the manufacture of gibberellin, which is greatly increased in response to melatonin therapy and results in drought tolerance (Sharma et al., 2020). Stomatal motility, photosynthesis, and the antioxidant defense system are all critically dependent on thc buildup of salicylic acid (SA) (Hasanuzzaman et al., 2018). Melatonin-treated plants have been shown to exhibit an increase (albeit not a substantial one) in the defense hormone salicylic acid in maize plants under dry circumstances (Fleta-Soriano et al., 2017). Through the over expression of melatonin-biosynthesis genes under drought conditions, which led to a decrease in IAA that may have been caused by competition for the same precursor, tryptophan, enhance drought tolerance was achieved using primarily transgenic plants (Yang et al., 2019; Zuo et al., 2014). Plant breeders should focus on enhancing the root's resilience to abiotic stressors, such as drought, since it is the first plant organ to interact with the environment. It's interesting to note that melatonin targets plant roots, acting similarly to auxin. In this context, Pelagio-Flores et al. (2012) offered strong relations in Arabidopsis thaliana results in the creation of lateral and adventitious roots, which confers a vast root system. Further research is required because there hasn't yet been a detailed study that reveals how melatonin interacts with ethylene or strigolactone in drought-like conditions. All the aforementioned information demonstrated that melatonin functions as an important regulator of numerous plant hormone components, or a "plant master regulator," making plants more tolerant of irrigation water shortages (Arnao and Hernández-Ruiz, 2020).

2.3.2 TOLERANCE TO MELATONIN-INDUCED WATER LOGGING STRESS

Though the application of melatonin is important for alleviating the abiotic stress especially salinity and drought condition, however, the research is at early stage via melatonin induced alleviating stress tolerance. Researchers have suggested that melatonin has a high capability to remove ROS, by lessen oxidative injury, resistance against water logging, and, as a result, overcome the yield, and quality losses (Chen et al., 2015).

Melatonin treatment increased endogenic melatonin levels, chlorophyll content, antioxidant enzyme activities, anaerobic respiration and photosynthesis, and aerophilic respiration while limit the malondialdehyde (MDA), ROS, and wilting (Zheng et al., 2017). Melatonin enhanced the biosynthesis enzyme genes (*MbT5H1*, *MbAANAT3*, and *MbASMT9*) expression under water stress (Zheng et al., 2017). They studied the effect of melatonin pretreatment on alfalfa on seed under water scarcity and suggested that melatonin might reduce stress damage while increasing plant growth, chlorophyll content, and PSII efficiency under drought due to up-regulation of stress related genes. Furthermore, melatonin also regulate the polyamine (putrescine, spermidine, and spermine) levels while decreasing ethylene, which is controlled by changes in gene expression (Zhang et al., 2019). Production of ethylene and polyamine (PAs) showing tolerance behavior under drought stress with melatonin application in crop plants (Figure 2.2). Melatonin application is a useful method for mitigating the severe effects of water logging. Zheng et al. (2017) reported that the melatonin mediated water logging mechanism was involved in apple seedlings studied. The studied suggested that melatonin regulates the aerobic respiration and photosynthesis by maintaining the ROS production and mitochondrial genes involved during the reaction. They also proposed another concept in alfalfa, their direct or indirect production of ethylene and polyamine (PAs) were recorded positive effect relative to control plants (Zhang et al., 2019). The results suggested that under stress condition the genes recorded (2–5) fold increase in melatonin (endogenous) level relative to control. However, the melatonin limit the ethylene production by downregulating the desired genes involved in ethylene production during the process and reduced the stress pressure by inhibiting chlorosis and premature senescence (Zhang et al., 2019). Similarly, melatonin enhanced the (PAs) levels by enhancing the polyamine metabolism enzymes genes expression. Melatonin enhanced water logging tolerance, at least in part, via modulating polyamines and ethylene production, resulting in more stable cell membranes, improved photosynthesis, and reduced ethylene response. Water stress increases ethylene production, melatonin (2–5 fold), polyamines (PAs), and ROS in plants. It was also suggested that melatonin production limit the ROS production. Similarly, another study also suggested that the pretreatment of melatonin enhanced the physiological, enzymatic, non-enzymatic and (PAs) in Alfalfa and enhanced the stress tolerance mechanism under drought. The study also suggested that melatonin enhanced the tolerance by reprograming PA and ethylene biosynthesis pathway. The PAs production was enhanced while ethylene production was decrease during the process. As a results, the membrane stability, photosynthesis, and leaf senescence (Zhang et al., 2019).

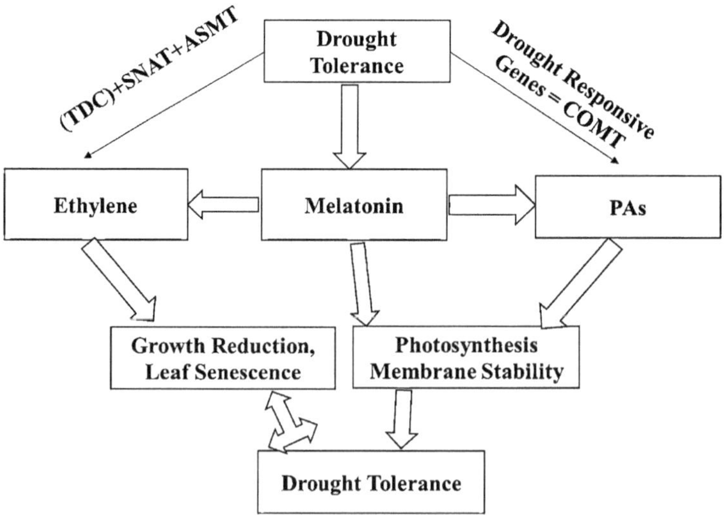

FIGURE 2.2 Role of melatonin on ethylene and PAs production and tolerance mechanism under drought stress.

Source: Adapted from Zhang et al., 2019

2.4 CONCLUSION

Tolerance against drought can be attributed towards functional and structural alteration at physiological and molecular levels. Improvements in the root area, number drought resistant genes involved in root zone area, enhanced growth, limit the oxidative injury, modulate osmotic potential, enhanced leaf water status, stomatal closure and osmolytes deposition resilient the stress affect and improved the yield under drought condition. Melatonin is a key component of the regulatory pathway network that controls all these systems, and it is a favorable substance for upcoming research and practical implications. Under water stress, the control of melatonin and its metabolic pathway is still unknown. There is dire need to figure the mechanism involved to determine the role of melatonin as nutrient and understand its exact mechanism would help us a lot and will provide us with a fantastic opportunity to exploit this beneficial chemical to improve plant tolerance and adaptability under drought condition. Furthermore, deep insight research employing microarray, transcriptome, and proteome analyses to clarify molecular pathways would aid in identifying the genes that regulate plant morphological, physiological, and biochemical characteristics due to exogenic melatonin usage under drought condition. Future research will investigate how novel receptor mediated Phyto-melatonin signaling played a significant role to enhance the physio-chemical properties and enhanced the tolerance mechanism. Furthermore, the most well-known facts on melatonin's interactions with other tiny signaling molecules, such as NO and H_2S, may be useful. The nitric oxide and H_2S signaling molecules with the association of melatonin signals molecules enhanced the tolerance mechanism and has made tremendous progress in the recent decades. As a result, more research needed for melatonin allocation in plant parts with respect to H_2S and NO application alone or in combination. The molecular processes exposing the interaction between the melatonin and other antioxidant enzymes with phytohormones like strigolactone and ethylene in improving drought tolerance warrant further research using mutagenesis or genome editing technique used to engineer the plant with novel characteristics and become a model plant under low water status. Because of the complexity of plants' interactions with a variety of environmental agents, future study on the association among melatonin and different stress combinations should be considered. Finally, using synthetic melatonin, a very inexpensive molecule, or Phyto-melatonin-rich extracts to improve plant tolerance may be an attractive strategy.

KEYWORDS

- **abiotic stress**
- **drought**
- **melatonin**
- **phytohormones**
- **phyto-melatonin**
- **tolerance mechanism**
- **water logging**

REFERENCES

Ahmad, S., Cui, W., Kamran, M., Ahmad, I., Meng, X., Wu, X., Su, W., Javed, T., El-Serehy, H. A., & Jia, Z. (2021). Exogenous application of melatonin induces tolerance to salt stress by improving the photosynthetic efficiency and antioxidant defense system of maize seedling. *Journal of Plant Growth Regulation, 40*, 1270–1283.

Ahmad, S., Kamran, M., Ding, R., Meng, X., Wang, H., Ahmad, I., Fahad, S., & Han, Q. (2019). Exogenous melatonin confers drought stress by promoting plant growth, photosynthetic capacity, and antioxidant defense system of maize seedlings. *PeerJ, 7*, e7793.

Alcázar, R., Altabella, T., Marco, F., Bortolotti, C., Reymond, M., Koncz, C., Carrasco, P., & Tiburcio, A. F. (2010). Polyamines: Molecules with regulatory functions in plant abiotic stress tolerance. *Planta, 231*, 1237–1249.

Antoniou, C., Chatzimichail, G., Xenofontos, R., Pavlou, J. J., Panagiotou, E., Christou, A., & Fotopoulos, V. (2017). Melatonin systemically ameliorates drought stress-induced damage in *Medicago sativa* plants by modulating nitro-oxidative homeostasis and proline metabolism. *Journal of Pineal Research, 62*, e12401.

Arbona, V., Hossain, Z., López-Climent, M. F., Pérez-Clemente, R. M., & Gómez-Cadenas, A. (2008). Antioxidant enzymatic activity is linked to waterlogging stress tolerance in citrus. *Physiologia Plantarum, 132*, 452–466.

Arnao, M. B., & Hernández-Ruiz, J. (2013). Growth conditions determine different melatonin levels in *Lupinus albus* L. *Journal of Pineal Research, 55*, 149–155.

Arnao, M. B., & Hernández-Ruiz, J. (2015). Functions of melatonin in plants: A review. *Journal of Pineal Research, 59*, 133–150.

Arnao, M. B., & Hernandez-Ruiz, J. (2019). Melatonin as a chemical substance or as Phyto melatonin-rich extracts for use as plant protector and/or bio stimulant in accordance with EC legislation. *Agronomy, 9*, 570.

Arnao, M. B., & Hernández-Ruiz, J. (2019a). Melatonin and reactive oxygen and nitrogen species: A model for the plant redox network. *Melatonin Research, 2*, 152–168.

Arnao, M. B., & Hernández-Ruiz, J. (2019b). Melatonin: A new plant hormone and/or a plant master regulator? *Trends in Plant Science, 24*, 38–48.

Arnao, M. B., & Hernández-Ruiz, J. (2019c). Role of melatonin to enhance phytoremediation capacity. *Applied Sciences*, *9*, 5293.

Arnao, M. B., & Hernández-Ruiz, J. (2020). Is Phyto melatonin a new plant hormone? *Agronomy*, *10*, 95.

Back, K., Tan, D. X., & Reiter, R. J. (2016). Melatonin biosynthesis in plants: Multiple pathways catalyze tryptophan to melatonin in the cytoplasm or chloroplasts. *Journal of Pineal Research*, *61*, 426–437.

Bai, Y., Xiao, S., Zhang, Z., Zhang, Y., Sun, H., Zhang, K., Wang, X., Bai, Z., Li, C., & Liu, L. (2020). Melatonin improves the germination rate of cotton seeds under drought stress by opening pores in the seed coat. *PeerJ*, *8*, e9450.

Behnam, B., Iuchi, S., Fujita, M., Fujita, Y., Takasaki, H., Osakabe, Y., Yamaguchi-Shinozaki, K., Kobayashi, M., & Shinozaki, K. (2013). Characterization of the promoter region of an Arabidopsis gene for 9-cis-epoxycarotenoid dioxygenase involved in dehydration-inducible transcription. *DNA Research*, *20*, 315–324.

Bhuyan, M., Hasanuzzaman, M., Parvin, K., Mohsin, S. M., Al Mahmud, J., Nahar, K., & Fujita, M. (2020). Nitric oxide and hydrogen sulfide: Two intimate collaborators regulating plant defense against abiotic stress. *Plant Growth Regulation*, *90*, 409–424.

Boru, G., Vantoai, T., Alves, J., Hua, D., & Knee, M. (2003). Responses of soybean to oxygen deficiency and elevated root-zone carbon dioxide concentration. *Annals of Botany*, *91*, 447–453.

Bose, S. K., & Howlader, P. (2020). Melatonin plays multifunctional role in horticultural crops against environmental stresses: A review. *Environmental and Experimental Botany*, *176*, 104063.

Burgess, P., & Huang, B. (2016). Mechanisms of hormone regulation for drought tolerance in plants. In *Drought Stress Tolerance in Plants, Vol 1* (pp. 45–75). Springer.

Campos, C. N., Ávila, R. G., de Souza, K. R. D., & Azevedo, L. M., & Alves, J. D. (2019). Melatonin reduces oxidative stress and promotes drought tolerance in young *Coffea arabica* L. plants. *Agricultural Water Management*, *211*, 37–47.

Cao, L., Jin, X., & Zhang, Y. (2019). Melatonin confers drought stress tolerance in soybean (*Glycine max* L.) by modulating photosynthesis, osmolytes, and reactive oxygen metabolism. *Photosynthetica*, *57*, 812–819.

Chen, H., Feng, C., Kong, J., Wang, L., Wang, N., Zheng, X., Zhou, Y., & Chan, D. (2015). Use of product containing melatonin as effective component for improving waterlogging stress resistance in plants. *CN105076136-A, CN105076136-B, CN105076136-A.*

Cheng, W., Zhang, L., Jiao, C., Su, M., Yang, T., Zhou, L., Peng, R., Wang, R., & Wang, C. (2013). Hydrogen sulfide alleviates hypoxia-induced root tip death in *Pisum sativum*. *Plant Physiology and Biochemistry*, *70*, 278–286.

Cui, G., Sun, F., Gao, X., Xie, K., Zhang, C., Liu, S., & Xi, Y. (2018). Proteomic analysis of melatonin-mediated osmotic tolerance by improving energy metabolism and autophagy in wheat (*Triticum aestivum* L.). *Planta, 248*(1), 69–87.

Cui, G., Zhao, X., Liu, S., Sun, F., Zhang, C., & Xi, Y. (2017). Beneficial effects of melatonin in overcoming drought stress in wheat seedlings. *Plant Physiology and Biochemistry, 118*, 138–149.

Daryanto, S., Wang, L., & Jacinthe, P. A. (2016). Global synthesis of drought effects on maize and wheat production. *PLOS ONE, 11*(6), e0156362.

Ding, F., Wang, G., Wang, M., & Zhang, S. (2018). Exogenous melatonin improves tolerance to water deficit by promoting cuticle formation in tomato plants. *Molecules, 23*(7), 1605.

Ding, Z., Tie, W., Fu, L., Yan, Y., Liu, G., Yan, W., et al. (2019). Strand-specific RNA-seq based identification and functional prediction of drought-responsive IncRNAs in cassava. *BMC Genomics, 20*(1), 214. https://doi.org/10.1186/s12864-019-5585-5.

Elkeilsh, A., Awad, Y. M., Soliman, M. H., Abu-Elsaoud, A., Abdelhamid, M. T., & El-Metwally, I. M. (2019). Exogenous application of β-sitosterol mediated growth and yield improvement in water-stressed wheat (*Triticum aestivum*) involves up-regulated antioxidant system. *Journal of Plant Research, 132*(6), 881–901.

Fleta-Soriano, E., Díaz, L., Bonet, E., & Munné-Bosch, S. (2017). Melatonin may exert a protective role against drought stress in maize. *Journal of Agronomy and Crop Science, 203*(4), 286–294.

Groppa, M., & Benavides, M. (2008). Polyamines and abiotic stress: Recent advances. *Amino Acids, 34*(1), 35–45.

Hasanuzzaman, M., Mahmud, J. A., Anee, T. I., Nahar, K., & Islam, M. T. (2018). Drought stress tolerance in wheat: Omics approaches in understanding and enhancing antioxidant defense. In *Abiotic Stress-Mediated Sensing and Signaling in Plants: An Omics Perspective* (pp. 267–307).

Hossain, Z., López-Climent, M. F., Arbona, V., Pérez-Clemente, R. M., & Gómez-Cadenas, A. (2009). Modulation of the antioxidant system in citrus under waterlogging and subsequent drainage. *Journal of Plant Physiology, 166*(13), 1391–1404.

Huang, B., Chen, Y. E., Zhao, Y. Q., Ding, C. B., Liao, J. Q., Hu, C., Zhou, L. J., Zhang, Z. W., Yuan, S., & Yuan, M. (2019). Exogenous melatonin alleviates oxidative damages and protects photosystem II in maize seedlings under drought stress. *Frontiers in Plant Science, 10*, 677.

Jackson, M., & Colmer, T. (2005). Response and adaptation by plants to flooding stress. *Annals of Botany, 96*(4), 501–505.

Jiao, J., Ma, Y., Chen, S., Liu, C., Song, Y., Qin, Y., Yuan, C., & Liu, Y. (2016). Melatonin-producing endophytic bacteria from grapevine roots promote the abiotic stress-induced production of endogenous melatonin in their hosts. *Frontiers in Plant Science, 7*, 1387.

Jin, Z., Shen, J., Qiao, Z., Yang, G., Wang, R., & Pei, Y. (2011). Hydrogen sulfide improves drought resistance in *Arabidopsis thaliana. Biochemical and Biophysical Research Communications, 414*(3), 481–486.

Karaca, P., & Cekic, F. Ö. (2019). Exogenous melatonin-stimulated defense responses in tomato plants treated with polyethylene glycol. *International Journal of Vegetable Science, 25*(6), 601–609.

Kaya, A., & Doganlar, Z. B. (2019). Melatonin improves the multiple stress tolerance in pepper (*Capsicum annuum*). *Scientia Horticulture, 256*, 108509.

Khan, M. N., Zhang, J., Luo, T., Liu, J., Rizwan, M., Fahad, S., Xu, Z., & Hu, L. (2019). Seed priming with melatonin coping drought stress in rapeseed by regulating reactive oxygen species detoxification: Antioxidant defense system, osmotic adjustment, stomatal traits and chloroplast ultrastructure preservation. *Industrial Crops and Products, 140*, 111597.

Kudoyarova, G. R., Vysotskaya, L. B., Cherkozyanova, A., & Dodd, I. C. (2007). Effect of partial root-zone drying on the concentration of zeatin-type cytokinins in tomato (*Solanum lycopersicum* L.) xylem sap and leaves. *Journal of Experimental Botany, 58*(1), 161–168.

Laxa, M., Liebthal, M., Telman, W., Chibani, K., & Dietz, K. J. (2019). The role of the plant antioxidant system in drought tolerance. *Antioxidants, 8*(4), 94.

Lee, H. J., & Back, K. (2016). 2-Hydroxymelatonin promotes the resistance of rice plant to multiple simultaneous abiotic stresses (combined cold and drought). *Journal of Pineal Research, 61*(3), 303–316.

Lee, H. J., & Back, K. (2019). 2-Hydroxymelatonin confers tolerance against combined cold and drought stress in tobacco, tomato, and cucumber as a potent anti-stress compound in the evolution of land plants. *Melatonin Research, 2*(1), 35–46.

Li, C., Tan, D. X., Liang, D., Chang, C., Jia, D., & Ma, F. (2015). Melatonin mediates the regulation of ABA metabolism, free-radical scavenging, and stomatal behavior in two *Malus* species under drought stress. *Journal of Experimental Botany, 66*(11), 669–680.

Li, D., Zhang, D., Wang, H., Li, H., Song, S., Li, H., & Li, R. (2019). Effects of melatonin on germination and amino acid content in different wheat varieties seeds under polyethylene glycol stress. *BioRxiv*. https://doi.org/10.1101/710954.

Li, H., Mo, Y., Cui, Q., Yang, X., Guo, Y., Wei, C., Yang, J., Zhang, Y., Ma, J., & Zhang, X. (2019). Transcriptomic and physiological analyses reveal drought adaptation strategies in drought-tolerant and susceptible watermelon genotypes. *Plant Science, 278*, 32–43.

Li, J., Yang, Y., Sun, K., Chen, Y., Chen, X., & Li, X. (2019). Exogenous melatonin enhances cold, salt, and drought stress tolerance by improving antioxidant defense in tea plant (*Camellia sinensis* (L.) O. Kuntze). *Molecules, 24*(9), 1826.

Li, J., Zeng, L., Cheng, Y., Lu, G., Fu, G., Ma, H., Liu, Q., Zhang, X., Zou, X., & Li, C. (2018). Exogenous melatonin alleviates damage from drought stress in *Brassica napus* L. (rapeseed) seedlings. *Acta Physiologiae Plantarum, 40*(3), 1–11.

Li, X., Tan, D. X., & Jiang, D. (2016). Melatonin enhances cold tolerance in drought-primed wild-type and abscisic acid-deficient mutant barley. *Journal of Pineal Research, 61*(3), 328–339.

Liang, B., Ma, C., Zhang, Z., Wei, Z., Gao, T., Zhao, Q., Ma, F., & Li, C. (2018). Long-term exogenous application of melatonin improves nutrient uptake fluxes in apple plants under moderate drought stress. *Environmental and Experimental Botany, 155*, 650–661.

Liang, D., Ni, Z., Xia, H., Xie, Y., Lv, X., Wang, J., Lin, L., Deng, Q., & Luo, X. (2019). Exogenous melatonin promotes biomass accumulation and photosynthesis of kiwifruit seedlings under drought stress. *Scientia Horticulture, 246*, 34–43.

Liu, J., Wang, W., Wang, L., & Sun, Y. (2015). Exogenous melatonin improves seedling health index and drought tolerance in tomato. *Plant Growth Regulation, 77*(3), 317–326.

Ma, X., Zhang, J., Burgess, P., Rossi, S., & Huang, B. (2018). Interactive effects of melatonin and cytokinin on alleviating drought-induced leaf senescence in creeping bentgrass (*Agrostis stolonifera*). *Environmental and Experimental Botany, 145*, 1–11.

Meng, J. F., Xu, T. F., Wang, Z. Z., Fang, Y. L., Xi, Z. M., & Zhang, Z. W. (2014). The ameliorative effects of exogenous melatonin on grape cuttings under water-deficient stress: Antioxidant metabolites, leaf anatomy, and chloroplast morphology. *Journal of Pineal Research, 57*(2), 200–212.

Moustafa-Farag, M., Almoneafy, A., Mahmoud, A., Elkelish, A., Arnao, M. B., Li, L., & Ai, S. (2019). Melatonin and its protective role against biotic stress impacts on plants. *Biomolecules, 10*(1), 54.

Moustafa-Farag, M., Mahmoud, A., Arnao, M. B., Sheteiwy, M. S., Dafea, M., Soltan, M., Elkelish, A., Hasanuzzaman, M., & Ai, S. (2020). Melatonin-induced water stress tolerance in plants: Recent advances. *Antioxidants, 9*(9), 809.

Mukherjee, S. (2019). Recent advancements in the mechanism of nitric oxide signaling associated with hydrogen sulfide and melatonin crosstalk during ethylene-induced fruit ripening in plants. *Nitric Oxide, 82*, 25–34.

Mukherjee, S., David, A., Yadav, S., Baluška, F., & Bhatla, S. C. (2014). Salt stress-induced seedling growth inhibition coincides with differential distribution of serotonin and melatonin in sunflower seedling roots and cotyledons. *Physiologia Plantarum, 152*(4), 714–728.

Najeeb, U., Tan, D. K., Bange, M. P., & Atwell, B. J. (2017). Protecting cotton crops under elevated CO2 from waterlogging by managing ethylene. *Functional Plant Biology, 45*(3), 340–349.

Pelagio-Flores, R., Muñoz-Parra, E., Ortiz-Castro, R., & López-Bucio, J. (2012). Melatonin regulates *Arabidopsis* root system architecture likely acting independently of auxin signaling. *Journal of Pineal Research, 53*(3), 279–288.

Pierik, R., Sasidharan, R., & Voesenek, L. A. (2007). Growth control by ethylene: Adjusting phenotypes to the environment. *Journal of Plant Growth Regulation, 26*(2), 188–200.

Shabala, S., Shabala, L., Barcelo, J., & Poschenrieder, C. (2014). Membrane transporters mediating root signaling and adaptive responses to oxygen deprivation and soil flooding. *Plant Cell & Environment, 37*(10), 2216–2233.

Sharma, A., Wang, J., Xu, D., Tao, S., Chong, S., Yan, D., Li, Z., Yuan, H., & Zheng, B. (2020). Melatonin regulates the functional components of photosynthesis, antioxidant system, gene expression, and metabolic pathways to induce drought resistance in grafted *Carya cathayensis* plants. *Science of the Total Environment, 713*, 136675.

Shi, H., Jiang, C., Ye, T., Tan, D. X., Reiter, R. J., Zhang, H., Liu, R., & Chan, Z. (2015). Comparative physiological, metabolomic, and transcriptomic analyses reveal mechanisms of improved abiotic stress resistance in Bermuda grass (*Cynodon dactylon* (L.) Pers.) by exogenous melatonin. *Journal of Experimental Botany, 66*(3), 681–694.

Shi, H., Qian, Y., Tan, D. X., Reiter, R. J., & He, C. (2015). Melatonin induces the transcripts of CBF/DREB1s and their involvement in both abiotic and biotic stresses in *Arabidopsis*. *Journal of Pineal Research, 59*(3), 334–342.

Smethurst, C. F., & Shabala, S. (2003). Screening methods for waterlogging tolerance in lucerne: Comparative analysis of waterlogging effects on chlorophyll fluorescence, photosynthesis, biomass and chlorophyll content. *Functional Plant Biology, 30*(3), 335–343.

Stuart, M., Gooddy, D., Bloomfield, J., & Williams, A. (2011). A review of the impact of climate change on future nitrate concentrations in groundwater of the UK. *Science of the Total Environment, 409*(13), 2859–2873.

Su, X., Fan, X., Shao, R., Guo, J., Wang, Y., Yang, J., Yang, Q., & Guo, L. (2019). Physiological and iTRAQ-based proteomic analyses reveal that melatonin alleviates oxidative damage in maize leaves exposed to drought stress. *Plant Physiology and Biochemistry, 142*, 263–274.

Tan, D. X., Manchester, L. C., Liu, X., Rosales-Corral, S. A., Acuña-Castroviejo, D., & Reiter, R. J. (2013). Mitochondria and chloroplasts as the original sites of melatonin synthesis: A hypothesis related to melatonin's primary function and evolution in eukaryotes. *Journal of Pineal Research, 54*(2), 127–138.

Tavladoraki, P., Cona, A., Federico, R., Tempera, G., Viceconte, N., Saccoccio, S., Battaglia, V., Toninello, A., & Agostinelli, E. (2012). Polyamine catabolism: Target for antiproliferative therapies in animals and stress tolerance strategies in plants. *Amino Acids, 42*(1), 411–426.

Tewari, S., & Mishra, A. (2018). Flooding stress in plants and approaches to overcome. In *Plant Metabolites and Regulation under Environmental Stress* (pp. 355–366). Elsevier.

Verma, V., Ravindran, P., & Kumar, P. P. (2016). Plant hormone-mediated regulation of stress responses. *BMC Plant Biology, 16*, 86. https://doi.org/10.1186/s12870-016-0771-y.
Voesenek, L., & Sasidharan, R. (2013). Ethylene and oxygen signaling-drive plant survival during flooding. *Plant Biology, 15*, 426–435.
Wang, C., He, J., Zhao, T. H., Cao, Y., Wang, G., Sun, B., Yan, X., Guo, W., & Li, M. H. (2019). The smaller the leaf is, the faster the leaf water loses in a temperate forest. *Frontiers in Plant Science, 10*, 58.
Wang, L., Feng, C., Zheng, X., Guo, Y., Zhou, F., Shan, D., Liu, X., & Kong, J. (2017). Plant mitochondria synthesize melatonin and enhance the tolerance of plants to drought stress. *Journal of Pineal Research, 63*, e12429.
Wei, J., Li, D. X., Zhang, J. R., Shan, C., Rengel, Z., Song, Z. B., & Chen, Q. (2018). Phyto melatonin receptor PMTR1-mediated signaling regulates stomatal closure in *Arabidopsis thaliana*. *Journal of Pineal Research, 65*, e12500.
Wei, W., Li, Q. T., Chu, Y. N., Reiter, R. J., Yu, X. M., Zhu, D. H., Zhang, W. K., Ma, B., Lin, Q., & Zhang, J. S. (2015). Melatonin enhances plant growth and abiotic stress tolerance in soybean plants. *Journal of Experimental Botany, 66*, 695–707.
Wei, Y., Zeng, H., Hu, W., Chen, L., He, C., & Shi, H. (2016). Comparative transcriptional profiling of melatonin synthesis and catabolic genes indicates the possible role of melatonin in developmental and stress responses in rice. *Frontiers in Plant Science, 7*, 676.
Wu, Y., Lian, H., Mu, X., Wang, X., & Zhang, Y. (2016). Effects of foliar spraying exogenous melatonin on physiological and biochemical characteristics of *Dendranthema morifolium* 'Chuju' seedlings under drought stress. *Acta Botanica Boreali-Occidentalia Sinica, 36*, 2241–2246.
Xia, H., Ni, Z., Hu, R., Lin, L., Deng, H., Wang, J., Tang, Y., Sun, G., Wang, X., & Li, H. (2020). Melatonin alleviates drought stress by a non-enzymatic and enzymatic antioxidative system in kiwifruit seedlings. *International Journal of Molecular Sciences, 21*, 852.
Xoconostle-Cazares, B., Ramirez-Ortega, F. A., Flores-Elenes, L., & Ruiz-Medrano, R. (2010). Drought tolerance in crop plants. *American Journal of Plant Physiology, 5*, 241–256.
Yang, W. J., Du, Y. T., Zhou, Y. B., Chen, J., Xu, Z. S., Ma, Y. Z., Chen, M., & Min, D. H. (2019). Overexpression of TaCOMT improves melatonin production and enhances drought tolerance in transgenic *Arabidopsis*. *International Journal of Molecular Sciences, 20*, 652.
Ye, J., Deng, X., Wang, S., Yin, L., Chen, D., Xiong, B., et al. (2015). Effects of melatonin on growth, photosynthetic characteristics and antioxidant system in seedling of wheat under drought stress. *Journal of Triticeae Crops, 35*, 1275–1283.
Ye, J., Wang, S., Deng, X., Yin, L., Xiong, B., & Wang, X. (2016). Melatonin increased maize (*Zea mays* L.) seedling drought tolerance by alleviating drought-induced photosynthetic inhibition and oxidative damage. *Acta Physiologiae Plantarum, 38*, 48.
Zhang, M., He, S., Zhan, Y., Qin, B., Jin, X., Wang, M., et al. (2019). Exogenous melatonin reduces the inhibitory effect of osmotic stress on photosynthesis in soybean. *PLoS ONE, 14*, e0226542.
Zhang, N., Zhao, B., Zhang, H. J., Weeda, S., Yang, C., Yang, Z. C., Ren, S., & Guo, Y. D. (2013). Melatonin promotes water-stress tolerance, lateral root formation, and seed germination in cucumber (*Cucumis sativus* L.). *Journal of Pineal Research, 54*, 15–23.
Zhang, Q., Liu, X., Zhang, Z., Liu, N., Li, D., & Hu, L. (2019). Melatonin improved waterlogging tolerance in alfalfa (*Medicago sativa*) by reprogramming polyamine and ethylene metabolism. *Frontiers in Plant Science, 10*, 44.

Zhao, H., Zhang, K., Zhou, X., Xi, L., Wang, Y., Xu, H., Pan, T., & Zou, Z. (2017). Melatonin alleviates chilling stress in cucumber seedlings by up-regulation of CsZat12 and modulation of polyamine and abscisic acid metabolism. *Scientific Reports, 7*, 1–12.

Zheng, X., Zhou, J., Tan, D. X., Wang, N., Wang, L., Shan, D., & Kong, J. (2017). Melatonin improves waterlogging tolerance of *Malus baccata* (Linn.) Borkh. seedlings by maintaining aerobic respiration, photosynthesis and ROS migration. *Frontiers in Plant Science, 8*, 483.

Zhou, L., Zhou, J., Xiong, Y., Liu, C., Wang, J., Wang, G., & Cai, Y. (2018). Overexpression of a maize plasma membrane intrinsic protein ZmPIP1;1 confers drought and salt tolerance in *Arabidopsis*. *PLoS ONE, 13*, e0198639.

Zhu, Y., Gao, H., Lu, M., Hao, C., Pu, Z., Guo, M., Hou, D., Chen, L. Y., & Huang, X. (2019). Melatonin-nitric oxide crosstalk and their roles in the redox network in plants. *International Journal of Molecular Sciences, 20*, 6200.

Zou, J., Jin, X., Zhang, Y., Ren, C., Zhang, M., & Wang, M. (2019). Effects of melatonin on photosynthesis and soybean seed growth during grain filling under drought stress. *Photosynthetic, 57*, 512–520.

Zuo, B., Zheng, X., He, P., Wang, L., Lei, Q., Feng, C., Zhou, J., Li, Q., Han, Z., & Kong, J. (2014). Overexpression of MzASMT improves melatonin production and enhances drought tolerance in transgenic *Arabidopsis thaliana* plants. *Journal of Pineal Research, 57*, 408–417.

CHAPTER 3

Abiotic Stress Tolerance: New Breeding Strategies to Improve Drought Tolerance in Canola

ABEER HASHEM,[1] KHALID F. ALMUTAIRI,[2] MOHAMAD ALNAFISSA,[3] JAVID A. PARRAY,[4] and ELSAYED FATHI ABD_ALLAH[2,*]

[1]*Botany and Microbiology Department, College of Science, King Saud University, Riyadh, Saudi Arabia*

[2]*Plant Production Department, College of Food and Agricultural Sciences, King Saud University, Riyadh, Saudi Arabia*

[3]*Department of Agricultural Economics, College of Food and Agricultural Sciences, King Saud University, Riyadh, Saudi Arabia*

[4]*Department of Higher Education, Government Degree College, Eidgah, Srinagar, Jammu and Kashmir, India*

*Corresponding author.

ABSTRACT

Crop output is negatively impacted by drought stress, which has also resulted in sizable losses for the world economy. Due to the restriction or cessation of photosynthetic activity as well as other metabolic activities in plant cells, has a significant impact on canola (*Brassica napus* L.) yield and essential oil quality. Phytohormones, such as ABA, react to water-scarce condition causing stomatal closure and thereby decrease sweating. The drought-resistant breeding lines of canola are usually used for physiological and biochemical changes. In canola production, drought stress also affects carbon uptake and translocation, lowering yield at the reproductive stage. Conventional breeding, marker-assisted breeding, and other genomic approaches are utilized to define the genetic and metabolic mechanisms conferring drought tolerance in canola.

Agricultural Biotechnology: Issues, Challenges, and Recent Developments. Mohd. Sayeed Akhtar and Khalid Rehman Hakeem (Eds.)
© 2025 Apple Academic Press, Inc. Co-published with CRC Press (Taylor & Francis)

For genetic improvement, a number of quantitative loci (QTLs) and alleles were found and exploited. Additionally, processing technologies can be used to cultivate high-yield canola varieties that are more petroleum-efficient and tolerant of volatile weather conditions. New genome editing technologies can be applied in plant gene introgression, gene targeting, updating signaling regulatory networks, and high throughput sgRNA libraries. Crops that have been genetically engineered with clustered regularly interspaced short palindromic repeat (CRISPR) technology have a greater potential to reduce drought stress, which has improved qualitative characteristics as well as production. The aim of this chapter is to focus on the potential use of genomic-assisted breeding strategies to increase canola resistance under drought conditions. It also highlights the challenges and opportunities associated with abiotic stress, its process, management, and future endorsements.

3.1 INTRODUCTION

Plants are subjected to a variety of abiotic challenges, such as water stress, temperature stress, salt stress, and heavy metals. According to Kumar et al. (2020), drought, harsh weather, and salinity each year result in significant losses of US $ 1.43 billion, 186 million, and 354 million. Shubha and Tyagi (2006) reported that environmental pressures cause more than 50% of the major damage to crop crops. Growth and development can be harmed permanently by the extreme effects of drought stress, which result in extreme protoplasmic water loss (Harb et al., 2010). Drying either completely stops or significantly reduces the metabolic and photosynthetic activity of plant cells (Dubey et al., 2018, 2019; Hashem et al., 2019; Kumar et al., 2010, 2013a, b, 2014, 2015).

Depending on the crop type, some abiotic stresses can be tolerated (Rozema and Flowers, 2008). Following exposure to stress, especially drought, adaptive mechanisms are activated and up-regulated, including the expression of stress-related genes that code for proteins and promote the synthesis of protective metabolites (Fukao and Xiong, 2013) during the canola growth cycle, drought sensitivity. Low-premium canola oil can be found in Erusic acid and glucosinolate. The drought stress reduced the anticipated seed production projection to 31% (Elferjani and Soolanayakanachally, 2018). Additionally, the seed yield was reduced by drought by 35% during the flowering phases and by 18% during the grain-filling stage (Godarzi et al., 2017). This unfavorable effect on canola growth and productivity forced the researchers to develop strategies to improve canola's drought efficiency.

Different physiological parameters help Canola to adjust to water scarcity at different levels of development. Some dryness and salinity indicators in seed germination, seedling emergence, 50% blooming, and seed maturation help to detect and pick germplasm. Different physiological, biochemical, and molecular responses have been established to accommodate abiotic stresses in the plants. The genetic and biological paths that were validated to enhance drought tolerance are used by modern and classical genetic approaches, including classical breeding methods (Zhang et al., 2014). The availability of water and genes for biochemical, cellular, and evolutional dryness responsibilities are determined by different genes (Tuberosa and Sialvi, 2006). The use of genome-based approaches and unique morphological characteristics that convey tolerance to drought and the production of maximum yields under drought conditions can be used for the specific plant genes or alleles linked in QTLs (Shamsudin et al., 2016). The collection of dry features is expected to be supported in genomic methods and the implementation of molecular marker approaches. The requisite information on pedigree, genotype, and hybridization of QTL targets is available in sequence-based data (Das et al., 2017). The aim of this chapter is to focus on the potential use of genomic-assisted breeding strategies to increase canola resistance under drought conditions. It also highlights the challenges and opportunities associated with abiotic stress, its process, management, and future endorsements.

3.2 DROUGHT EFFECT ON CANOLA

Different species respond differently to drought stress. Plant metabolism changes as a result of drought stress in a number of ways (Lisar et al., 2012). Depending on the severity, length, and crop stage of the drought, seed output has been reduced by 14 to 50% (Li et al., 2009; Zscheischler et al., 2014). Water scarcity results in oxidative stress, cellular damage, and the production of reactive oxygen species (ROS). Enzymatic and non-enzymatic defense responses reduced the negative effects of ROS (Munns and Test, 2008). The resilience to drought depends on maintaining the ROS scavenging equilibrium (Noctor et al., 2014). Drought reduces plant development (Sairam and Srivastava, 2001) and makes cell growth more vulnerable due to a drop in turgor pressure (Nonami, 1998).

3.2.1 PLANT GROWTH AND DEVELOPMENT

Growth is a process that depends on several levels. It involves cell division and expansion and is influenced by morphological, ecological, physiological, and molecular measures (Figure 3.1). Water stress and drought can have an impact on all processes associated with the growth of the plant (Fahad et al., 2017). In higher plants cell elongation is suppressed as a result of water scarcity and water transportation disruptions are repressed from the xylem to neighboring cells and tissues (Hussain et al., 2008). Some crop species increase their proline and other amino acids in the branch chain in response to drought (Din et al., 2011), while carbon dioxide (CO_2) assimilation and net photosynthesis decrease due to the closing of the stomach. Concurrently a phytohormone raises the ABA concentration when plants suffer from drought or water stress conditions (Wilkinson and Davies, 2010). Water-limiting conditions affect the growth, yield, and biochemical composition of plants (Birun Ara et al., 2011; Din et al., 2011) observed various forms of drought-tolerant canola and reported a reduced grain yield in the flowering stage, because of drought conditions. When droughts occur during the reproductive stage, the delayed flora, dehiscent fruit, and reduction in seed yield affect canola plants (Birun Ara et al., 2011). The ROS produces a metabolite flux that raises tissue load oxidative pressure by photorespiration. Under the conditions of drought, injuries caused by ROS are major constraints to the growth of biological macromolecules (Farooq et al., 2009).

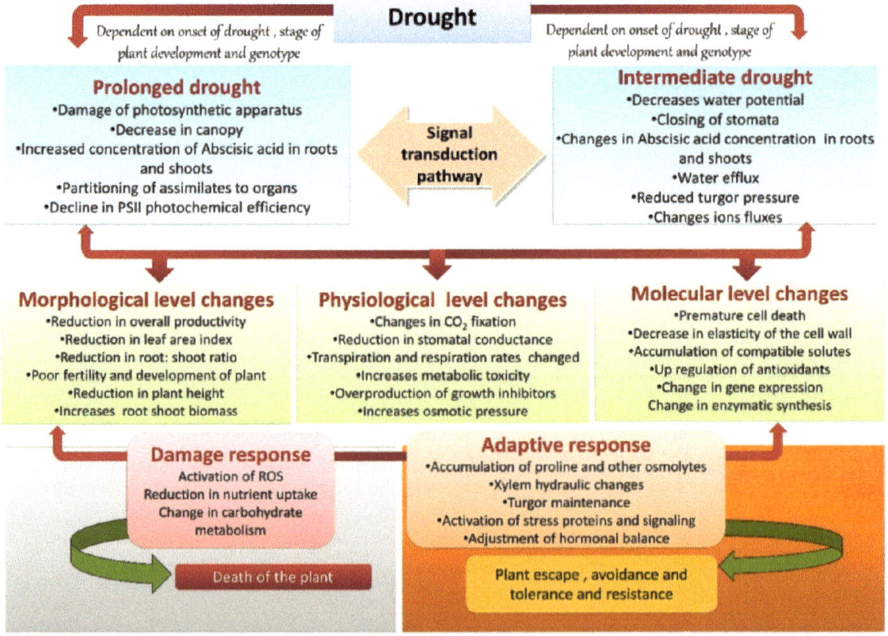

FIGURE 3.1 The diagram model describes different impacts of drought stress.

3.2.2 CARBON ASSIMILATION AND TRANSLOCATION

Drought has a very different impact on CO_2 diffusion and depends on organisms, stress intensity, and process. Carbon assimilation is related directly to the availability of carbohydrates and water. Water restriction (drought) reduced yield (Flexas and Medrano, 2002). In addition, stomata diffusion limitations reported reducing CO_2 carbon assimilation by about 55% due to drought (Yan et al., 2016). Carboxylation of Rubisco under abiotic stress is restricted by restricting diffusion of CO_2 by stomata or in mesophyll pathways. In addition, these stresses affect the catalytic reaction enzyme. Drought stress increases biomass distribution to roots and may increase water absorption potential (Leport et al., 2006). Furthermore, the export rate also decreased, with the reduction of both the photosynthate and the metabolism of carbohydrates and saccharose in the foods (Kim et al., 2000). Drought stress restricts photosynthesis stomatal conductivity and mesophyll and affects resource efficacy (water and nitrogen). This reduced the net photosynthesis and carbon assimilation rate. Furthermore, the rate of carbon assimilation and transport of electronics was reduced, and the resulting yield drought (Elferjani and Soolanayakanahally, 2018).

3.2.3 YIELD FORMATION (CARBOHYDRATE, PROTEIN, AND OIL)

Crop yield is significantly reduced when canola plants are exposed to drought at the blooming and grain-filling stage (Gan et al., 2004). In normal circumstances, in transgenic and non-transgenic canola, all the agronomic and yield characteristics are the same. However, in contrast to the non-transgenic genotype derived from the transgenic line, transgenic canola species perform better on yields under limited water source conditions (Wang et al., 2005). Table 3.1 identified the impact of drought on seed development, fatty acid content, and oil yield of species of *Brassica*. Experiments on the floral and pod-filling effects of drought showed that the leaf chlorophyll, protein, and proline content in all the tested canola varieties decreased. Some drought tolerance measures are included in *B. napus* species (Table 3.1). Drought stress induces sterility, reduced assimilation, and the flow of the developing ear during the flowering stage (Yadav et al., 2004). When canola plants experience drought stress during flowering, adverse events are reported. Drought drives the bloom in its reproductive process and consequently decreases the dehiscence of the fruit and eventually the final yield (Birun Ara et al., 2011).

TABLE 3.1 Indices for Drought Tolerance Influence of Drought Stress on Seed Yield, Oil Content, and Oil Yield in *B. napus*

Drought Indices	Stages	Agronomic or Physiological Traits	Drought Type	References
Biomass and water potential	Seedling	Biomass and osmosis/osmotic pressure	Greenhouse	Ashraf and Mahmood (1990)
DSI	Maturity	Yield and yield components	Field experiment	Chauhan et al. (2007)
LS, PSS, RVI	Germination	Biological traits	Petri dish along with induced PEG 6000 solution	Yang et al. (2007)
Growth stages (TDM, LAI, RGR, CGR)	Whole stages	Physiological traits	Field experiment	Moaveni et al. (2010)
DTI	Flowering and reproductive stages	Agronomy and physiology traits	Pot experiment	Din et al. (2011)
PCA, clustering	Flowering	Agronomy and morphology	Rain available water	Zhu et al. (2011)
Leaf wilting index	Seedling	Biological traits	Pots experiment	Li et al. (2012)

Abbreviations: PCA: Principal component analysis; DSI: drought susceptibility index; TDM: total dry matter; LAI: leaf area index; CGR: crop growth rate; and DTI: drought tolerance index.

A wide range of atmospheric conditions (heat stress) significantly decreased yield at the blooming and filling stage (Karaba et al., 2007). Studies have shown that many physiological responses to drought stress in *Brassica, that is,* protein synthesis, metabolite buildup, stomatal conductance (SC), and photosynthetic activity affected directly or indirectly and ultimately reduced the quality and yield attributes (Sangtarash et al., 2009). During all phases of development, *B. napus* is subtle to drought. Germination time was extended with a drastic reduction in germination rate in *B. napus* treated for osmotic stress induction with polyethylene glycol (PEG) (Willenborg et al., 2004). Yang et al. (2007) indicated that seed germination was affected in stress conditions with time-and-rate, vegetative growth is delayed and consequently yield losses were recorded on the *B. napus*. The heat and drought of the protein were raised, with oil quality decreasing when flowering and siliqua were subjected to dryness (Elferjani and Soolanayakanahally, 2018). Drought stress in the cultivation of canola can change breeding and planting. In addition, the supply of nutrients and storage of photosynthesis have been decreased because of the delayed growth of their reproductive structure (pollen tube extension) (Barnabes et al., 2008). Canola oil is more a seed processing income. Abiotic stresses affect oil thickness, contents, and composition (Si et al., 2003). Oil is derived from the assimilation of photosynthetic carbon of the sheet, later carbohydrate, converted by metabolic route into tricyclic glycerol (Baud and Lepiniec, 2010). The impact on oil quality of drought was varied due to the high interactions between G × E (Zhang et al., 2014). Although water stress during the blooming and silique production stage was reduced to 6–12% in canola (Champolivier and Merrien, 1996), oil content was decreased under drought conditions (Aslam et al., 2009).

3.2.4 WATER USE EFFICIENCY (WUE)

In these areas where crops depend on canal water or sub-surface water or where competition exists for sources of water between farm use and urban needs, water use efficiency (WUE) is important. The WUE characteristic helps plants to reduce slower soil negative effects due to drought and water depletion. One approach to WUE growth is to assimilate CO_2 alters into each unit of hydrochlorinated water (Chaerle et al., 2005). Efficiency can also increase the development of high grain filling and high seed yield beyond ground dry matter. It is therefore possible that WUE can be employed in genotypical selection as an indirect criterion for the yield of seed. In Brassica genotypes, there have been well documented genetic variations and water use quality. In addition, the resistance to dryness was limited to the canola varieties (Cowling, 2007).

In *Brassica napus* and *B. rapa* lines Fletcher et al. (2015) suggest a technique to maximize the use of water and the root system that supports the absorption of more root water. The true change between escape and drought avoidance helps to minimize the chance of drought tolerance between the genotypes, and the dry escape breeding strategy alone can do so. Kooyers (2015) notes that there is no invariant association between floral time and water quality, while in the wild some *Arabidopsis thaliana* L. genotypes exist that have short floral cycles and enhance WUE. These features also occur in *Brassica* species and increase tolerance among genotypes. These features are important for the enhancement of drought tolerance but rely on the climate. It also includes soil humidity content, local weather data, and irrigation methods. The moisture absorbance for transpiration is maximized during successful use of water by crops, non-stomatal transpiration is decreased, and soil evaporation is minimized. Drought tolerance wheat species enhances cell extension modification of osmotic and partially open pores to retain CO_2 attachment in extreme water scarcity (Ahmad et al., 2018). The harvest index (HI), a matrix for reproductive function and apportioning assimilation, is significant for the yield factor. The drought stress at the grain filling phase affects the reproductive feature, harvest index, and assimilated reproductive partitioning (Blum, 2009). Plant water status contributes to the enhancement of WUE through the continued assimilation of sources and plants leads to reproductive success, and thus increases the plant's final production (Passioura, 2007). Drought stress reduction in CO_2, photosynthetic rate, chlorophyll content, and sweat rate (Shafiq et al., 2014). Photosynthesis is taken up, and the production of biomass influences the early and high degree of stomata closure, while the soil's water quickly loses and thus loses yields after anthesis (Saradadevi et al., 2017).

3.2.5 PHYSIOLOGICAL CHANGES DUE TO DROUGHT

Phenotypical and molecular techniques provide unique possibilities for investigations and more precise use for yield under water stress conditions on the genetic and functional basis (Tardieu and Tuberosa, 2010). Parents with a range of complex characteristics cross through physiological reproductive technology to find the outcomes of compound genes that enhance their outcome. Also important for finding out genome selection are selected progeny using remote sensing techniques. Identification, as a fundamental genetic basis for drought and genetic transformation, is the identification of single genetic tools to improve yield components in the Calvin cycle expression. The physiological characteristics-based cross-test is designed to meet breeding targets. In addition, the genotype (G × E) interaction feed models help grow new genotypes in the new environment (Reynolds and Langridge, 2016). Direct selection or marker-assisted selection can determine the physiological characteristics of breeding programs; this depends heavily on genetic relationships with yield, genetic diversity, genotype, and environmental interactions. In direct selection, the breeding introgression for physiological properties is determined by using three different stages (pre-breeding, characterization, and early selection). A vast array of data sets for genetic resources are given for physiological breeding to achieve the best genetic characteristics that can also be used for separating yield generations. Table 3.2 provided several major measures of drought tolerance for agronomical and physiological characteristics. Under stress, plants accumulate metabolites including amino acids, are protein precursors, which play a role in growth and metabolism. Furthermore, proline is the key player of the antioxidant resistance molecule, metal chelator, and signaling molecule during stress (Hayat et al., 2012). Excess development of prolines occurs underwater deficit condition in plants, as a result, to the preservation of cell turgor, a softening cell membrane that helps to protect the leakage of electrolytes and rust that prevents oxidative eruption. Plant phenology can leverage genetic diversity under incremental moisture stress and confirm genes resistance to drought which can remain stable for dry yield (Trethowan and Reynolds, 2006). Drought condition decreased expression levels of *BnCOMT1* (lignin synthesis gene). *BNCOMT1* is responsible for the synthesis of a certain subunit of lignin due to the acquisition of complete lignin and the transcript *BNCOMT1* (Li et al., 2016).

TABLE 3.2 Some Gene Responses and Protein Functions Under Drought Conditions in *Brassica* and *Arabidopsis*

Plant	Genes	Encoding Protein	References
Arabidopsis	*DREB1A*	Dehydration response element binding protein	Kasuga et al. (1999)
Arabidopsis	*OsCDPK7*	Stress-induced Ca-dependent protein kinase	Saijo et al. (2000)
Arabidopsis	*ERECTA*	Transpiration efficiency	Masle et al. (2005)
Arabidopsis	*ERA1*	Negative regulator of ABA sensing	Wang et al. (2005)
Arabidopsis	*FTA*	Farnesyltransferase	Wang et al. (2009)
	AtFTB	β-subunit of Farnesyltransferase	
B. napus	*BnLAS*	Transcriptional regulator members in GRAS family	Yang et al. (2011)
B. napus	*AnnBn1*	Membrane-binding proteins Ca^{2+}	Xiao et al. (2012)

Multiple regulatory factors that favor dryness and other forms of abiotic strains, based on phylogenetic distances, have been included in physiological responses (Mizoi et al., 2012). Farnesyltransferase protein (subunits α and β) is negatively regulating ABA stomatal response units, and their efficacy has been established in the *Brassica napus* as an engineering objective of drought tolerance and yield safety (Wan et al., 2009). Drought-tolerance varieties in canola must grow, which can be used to increase the yield in drought-prone areas (Cattivelli et al., 2008).

Arabidopsis is a model plant compared for canola research helps to figure out the genes for ABA synthesis as a result many genetic mutants were developed. Scientists also figure out genes significant for phenotypic characters under water scarcity conditions. The three important (*abi*) genes were directly evaluated under 10 micronutrients ABA, and many phenotypical alterations were recorded under different *abi* mutants (Finkelsteins et al., 2002). A decreased functionality of the *abi* mutant, for example, induces a degradation of the signal of the guard cell and the reduction of Ca^{2+} caused by ABA, which eventually leads to stomach closure failure. A Petri plate experiment containing low concentrations of hormone and *ERA1* cloned gene was used as the first discovery of an ABA hypersensitive mutant (Cutler et al., 1996).

Pie et al. (1998) documented the detection of several hypersensitive ABA mutants by comparative experiments with wildlife that decreased dry stress wilting. As canola plants experience dryness, plant cells experience various physiological and chemical changes such as lipid peroxidation and enzyme activity. Artificial drought stress-related induction of PEG enhanced malondialdehyde (MDA) contents as the by-product of lipid peroxidation, which enhances roots and shoots enzyme activity (peroxidase, superoxide dismutase (SOD), and catalase (CAT)). Moreover, *B. napus* increased stress tolerance for amino-levulinic acid. In the seedlings *B. napus* L., the aminolevulinic acid, based on the spring biomass, the chlorophyll quantification contents, and the function of antioxidant enzymes, were improved in favor of stress growth under stress (Liu et al., 2011).

3.2.6 DROUGHT GENES ENHANCED DROUGHT TOLERANCE

When plants encounter drought stress, mechanisms for plant protection cause the accumulation of many genes that are immune to stress. Several genes have been correlated with the loss of functional analyses, such as *AtFTA*, and *ABI1* (Wang et al., 2005). Hypersensitivity during the germination of the guard cell was recorded in their mutant sense and increased stomach closure, a decrease in perspiration, and enhanced turgor pressure under drought conditions (Zhang et al., 2008) as similar to over-expression of *SNAC1* (Hu et al., 2006) and *AtMYB44* (Jung et al., 2008). In *Brassica* and *Arabidopsis,* some gene reactions and protein functions react under droughts in Table 3.3. The transcriptional regulatory and gene expression of *B. napus* L. abiotic stress networks are shown in Figure 3.2.

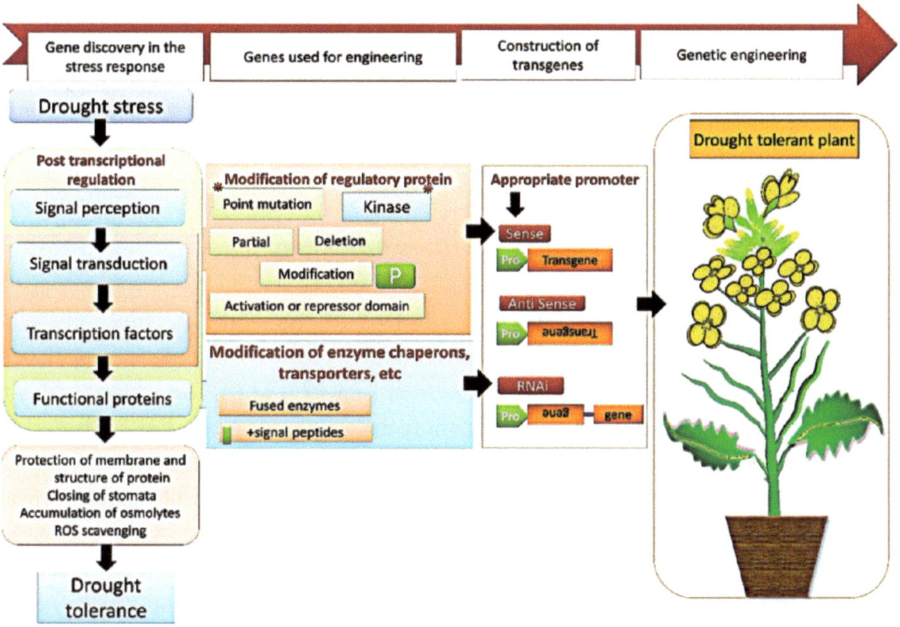

FIGURE 3.2 The diagram model imagines different transcriptional regulatory networks and gene expression under drought stress.

Cold, drought, and salt stress signal transduction pathways are subdivided into three major classes, namely ROS scavenging enzymes, and antioxidant compounds, or oxidative/osmotic stress, Ca^{2+} dependent signaling stimulates genes embryogenesis of form involved in stress-responsive protein manufacturing, and The excessively sensitive signaling (SOS) of Ca^{2+} based salt controls homeostasis affecting the anionic stress pathways (Xiong et al., 2002). Several stress-related genes (*ROB5*, *DNH4*, *SOD3*, and *CBF1*) have been identified and overexpressed in canola, by assessing the germination, seed intensity, and seed yield effects in drought, heat, and cold stress treatments under controlled conditions. The most productive genes assessed were *ROB5* and *SOD3* under high-stress conditions, as demonstrated by earlier germination, increased tiller number, flower initiation, ripening, improved seed sizes, and higher seed yield. *ROB5* performed better than control at various locations under moderate stress conditions (Xiao et al., 2008). The transcription factors as Factor Y and B subunits help protect against drought in both *Arabidopsis* and maize (Nelson et al., 2007). Another aid to the plant against drought tolerance is its expression of isopentenyl transference genes that require cytokinin biosynthesis (Rivero et al., 2007). This leads to faster stomach closure due to increased sensitivity of the signals of ABA under water deficiency (Wang et al., 2005; Table 3.2).

3.3 MANAGEMENT STRATEGIES

3.3.1 SELECTION AND BREEDING APPROACHES

Previous studies indicate that traditional plant reproductive methods with collaborative approaches have improved the yield of canola crops. Especially, the introduction of quantitative trait locus (QTLs), especially to resist both abiotic and biotic stress, was helpful to various advanced breeding methods. For desired characteristics, phenotyping included screening methods suitable according to respective environments and identifying more reactive genotype stress levels. To achieve desired breeding goals, the use of the combination of genomic and conventional breeding tools is very critical (Osterlund and Paterson, 2002). Because of its polyploidy stage, *B. napus* has a highly complicated genome; the QTL techniques alone cannot render actual markers for the transition between genotypes. However, the determination of the desired gene and allene interaction is critical through the new technology of molecular markers. Further sequencing generation and QTL techniques make it easier to identify the target gene and function as a gene link (Afzal et al., 2018). Functional genomics involves crop phenotyping and the precise discovery of the trait gene to be transferred to and subsequently used for gene functional check. The functional genomics strategies help to find several genes responsible for abiotic stress and provide a clear picture of the biological pathways under control and drought conditions. Furthermore, structural genomics is also significant in studying the chromosome (Morgante and Salamini, 2003). There are some breeding attempts to optimize crop adaptation with abiotic and there are also restrictions (Ishitani et al., 2004). Genes and alleles in a genome that act with resistance to drought stress as well as disease resistance are important for structure and functional analysis to identify using new breeding technologies (Kuraty et al., 2005). Positional cloning techniques permit a mutated gene or allele to be determined through an analysis of the association using markers that include genes or alleles within the genome. Sequencing of *Arabidopsis* and other crop genomes is possible and used by DNA polymorphisms to enhance crop plants (Buell and Last, 2010).

Many population studies on drought stress in crop plants have been published. Wu et al. (2012) said there were phenotypic trait shifts in more than 8% of the mutant population. In addition, *B. napus* has been predicted by Zhu et al. (2015) as being morphological and agricultural traits were categories based on a principal component's analysis during pre-flowering and post-flowering stages, under drought stress. The results of PCA predicted an important positive association between *B. napus* drought tolerance during the pre- and post-flowering period. Novel insights are provided by genetic studies including genome editing, genome-wide mapping, and study of genetic diversity. The introduction of new genes and gene combinations leads to drought tolerance and gradually increases yield. Expression profiling may be useful to classify the gene of mutants, which are stress-specific and favor the gene in drought (Gilchrist and Haughn, 2005).

3.3.2 BIOTECHNOLOGY AND GENOMICS

Tilling is another powerful, efficient technique for high-performance gene profiling that can be used to classify genes, alleles of mutants, or pairs of alleles in *B. napus* L. developed by Gilchrist et al. (2013). It would be helpful to identify new mutagens and genes of interest in important crop species from mutagenized genotypes and having high mutation density. The mutagenized population and DNA are used in the classical TILLING method, and only mutagenized components selected using mismatch endonuclease techniques were defined. Another successful tool that can be employed by researchers to decode the molecular mechanisms of drought stress in canola is sequencing for the next generation. In *B. rapa* a broad expression pilot genome study with a genome array was performed by Yu et al. (2012) and concluded that the dry response gene, its associated transcription factors, and the development of kinases that stabilize stress. Agricultural characteristics are correlated with coding transcripts of stress response. Using the next generation sequencing (NGS) and *de-novo* assembly technologies, the transcripts data collection in *B. juncea* during high temperatures and drought stress was also determined (Bhardwaj et al., 2015). The differentially expressed genes (DEGs) related to positive ion transport, ABA signaling, and wax production helped to sustain canola crops under extreme conditions by comparative genome study. In addition, certain DEGs were involved in the cell division and catalytic activity has been identified in the findings of gene ontology. The roots are the precursor and play a major part in drought and can detect when plants suffer from water deficit. To identify root hydrotropic-specific genes, a whole micro-array genome analysis has been conducted and 793 genes have been identified that display hydrotropic reactions out of the 22, 810 transcription-control genes (Moriwaki et al., 2010). Gene widespread association studies provide information to classify target genes that can help tolerate drought in crops (Liang et al., 2011). An overview of regulatory components in promoter regions used to classify genes involved in different osmotic stresses. Approximately 500 top-scoring genes with stress response were selected (Li et al., 2008).

3.3.3 CROP MANAGEMENT STRATEGIES

Blossoms and stress triggered the senescence of elderly leaves during vegetation cycles that occurred when the canola plant experienced droughts. The flowering process aborts flowers and the pot set in *B. juncea* is substantially reduced. Reports on the increased rates of nitrogen (N) application have been released in recent years. When N is delayed, the yield will not be substantially affected (Moody, 2008). Additional N applications can be used during canola growth in the reproductive phase effect under favorable seasonal conditions (Motley et al., 2001). Flowering time is susceptible to abiotic stresses, which can be handled efficiently by early maturity and rapid seeding. Besides the use of the various maturity classes, several varieties of canola may also be used, which have improved frost and drought resistance, and pests and pathogens (black leg disease). ABA-related plant phytohormones are used to control stomach sequence and to handle drought stress in canola under minimal water conditions (Hunt et al., 2003). Drought exposure affects the stomatal closure, lowered leaf water potential, relative water contents, and ABA. Increased ABA susceptibility reduces stomatal lead by stomach closure, resulting in a substantial drop in transpiration rates and a general impact on drought-tolerant plant photosynthesis (Nir et al., 2017).

Reduction of stomatal closures by targeting farnesyltransferase can be achieved in plants (Wang et al., 2009). The modification of the farnesylation protein of C15 carboxyl residues targets a protein linked to drought and heat stress after translation. These proteins are triggered in defense cells when the ABA response is negative. In Table 3.2, we report the use of protein farnesyl transferases to increase drought resistance in canola (Wang et al., 2009). An *Arabidopsis*-led, drought-inducible *rd29A* promoter, used to boost ABA sensitivity, shows that transgenic canola plants with an anti-sense construction of *ERA1* display a substantial reduction in stomatal conductance and water transpiration under severe drought stress conditions. Overexpression of the *BnFTA* gene, FTA-regulated genes, increases yield in the field under drought stress (Wang et al., 2009). In aquatic plants, elevated ABA levels cause potassium fluxes in stomatal-protecting cells, increasing turgor; consequently, stomatal closure occurs. Therefore, potassium is considered to contribute to minimizing water stress by specifically affecting stomatal closure. This role of stomatal closure may explain why dehydration is observed in plants in potassium-low soils (Benlloch et al., 2007). Several compounds, including brassinosteroids (BRs)—poly-hydroxy steroids involved in various growth processes—have been used to minimize stress-repressive crop conditions. The primary functions of BRs include cell division, seed germination, cell differentiation, vegetative growth, and apical dominance (Sasse, 2003). In addition, BRs increase resistance to abiotic stresses such as chilling, water, and high-temperature stresses, and pathogens (Kagale et al., 2007). Osmotic modification was a significant shift observed in the plant body, improving drought resilience due to osmolyte deposition as the plant was subjected to osmotic pressure changes (Dubey et al., 2018; Hashem et al., 2019; Kumar et al., 2013a, b, 2014, 2015). Osmolytes enable the plant to retain more external water, serving as a buffer to reduce internal water consumption (Jaleel et al., 2007). Similarly, Yancey (2001) found that osmolytes help improve protein structure and cell membrane stability under water-deficit conditions. Soluble sugars also protect against drought stress-induced damage to hydrophilic membranes, sugar, and protein hydroxyl groups (Garg et al., 2002). The overexpression of genes has been previously employed using transgenic approaches to enhance the stability of these sugars (Shinozaki et al., 2006). Various enzymes have been used to enhance drought resistance in *Arabidopsis* (Valliyodan and Nguyen, 2006), for example, trehalose phosphate synthase (TPS) or trehalose-6-phosphate phosphatase (TPP). Drought tolerance in canola has also been enhanced through molecular breeding. As a result, various researchers have identified important QTLs in different crops to enhance drought resistance (Table 3.3). For example, QTLs regulating heat tolerance in pollen have been established in maize (Yang et al., 2002), and genetic heterogeneity in thermal tolerance has been observed in wheat (Dhanda and Munjal, 2006). Kahraman et al. (2004) reported that lentils, rapeseed, and ryegrass possess QTLs for winter hardiness (Asghari et al., 2007). Significant data on drought, cold, and heat stress are available for quantitative trait loci (QTLs). Characterizing QTLs helps to identify drought-resistant genes and provides a foundation for enhancing stress tolerance in canola.

TABLE 3.3 Important Transgenic Canola Produced by Genetic Engineering for Enhancing Drought Tolerance

Target Gene	Phenotypes	Approach	References
COX (choline oxidase)	Moderate stress resistance	Gene expression	Haung et al. (2000)
DREB1A/C (repeat Binding factor 3)	Improving the drought tolerance	Transgenic plants	Avonce et al. (2004)
ERA1 improves the abscisic acid activity	Improved drought tolerance and yield protection	Antisense down-regulation	Wang et al. (2005)
AtGolS2 (drought inducible galactinol-synthetase gene)	Increased yield during drought stress	Overexpression	Umezawa et al. (2006)
DWARF4	Improve plant size and seed yield	Overexpression	Kagale et al. (2007)
BnDREB, BnCBF5	Enhanced drought tolerance	Transcription factors	
Phosphatidylinositol-phospholipase C_2	Protect at flowering and maturity	Gene expression	Georges et al. (2009)
BnFTA (farnesyltransferase)	Yield under drought stress	RNAi	Wang et al. (2009)

3.3.4 GENOME MODIFICATION TO OVERCOME DROUGHT

A potent genome excision method, inaccessible to bacterial immune system machinery, enabling scientists to modify genetic materials, CRISPR is a Cas-9/sg ribonucleic acid (RNA) system. This technique is a cost-effective, simple-to-use, and fast method for editing genomes, allowing for complex tissue genomic changes. The Cas-9 protein is an endonuclease oriented toward the RNA, applied to generate targeted double-stranded breaks (DSB) in short RNA sequences to assemble locations for recognition in specific plants and animals. Genome engineering with CRISPR allows for better precision in implementing desired modifications compared to conventional biotechnological approaches, which may introduce modifications to plants in different regions within the genome. Plant studies include gene function, regulatory reporting networks, and the screening of sgRNA libraries (Khatodia et al., 2016) for existing genome-editing efforts. Many benefits have been disclosed by targeted mutagenesis and transcriptional control models in crop plants using CRISPR (Khatodia and Khurana, 2014). In plants, there are three kinds of endonuclease-guided RNA applications. Cas-9 produces a double-strand break with a non-homologous end joint, used for physical and chemical mutagenesis under the target of DNA (Saika et al., 2014). Another model adjusts homologous recombination (HR) DSBs to produce target mutations or targeted gene introduction, gene substitution, and gene stacking at a specific position within the genome. The Cas-9 model is used to fix the

DSBs at an alternate location. However, multiplex genome editing can be carried out by targeting multiple sgRNAs together with Cas-9 nuclease at different sites in the third group. In the dissection of gene functions and the ultimate analysis of epistatic connections between them, multiplex genome editing can be used in plants (Xing et al., 2014). The CRISPRi (interference) system is used for stable and effective modulation by the fusion of effector domains to the transcription of targeted genes in plants using the inactivated *dCas-9* (Larson et al., 2013). It is widely used in genomics, gene expression, and regulation. However, gRNAs use specific DNA structures as a fusion protein with transcription factors, instigating or suppressing domains (Gilbert et al., 2013). By expressing sgRNA with a single site (Gilbert et al., 2015), the CRISPR activator (CRISPRa) regulates gene expression on a 1,000-fold scale. Several alternatives for genome editing, gene silencing, and transcriptional regulation, with improved and decreased targets in *Drosophila*, bacteria, yeast, humans, and plants, are now available. Zetsche et al. (2015) stated that light and agrochemical transcriptional activation can be used with the Cas-9 system in conjunction with a chemically inducible *dCas-9* system.

Random mutagenesis and genetic diversity approaches, combined with conventional plant breeding strategies, are capable of adding new features to cultivated genotypes. Targeted mutagenesis in plants through the CRISPR/Cas-9 method is now possible with the development of precise tools for genome editing (GE), leading to rapid gains in creative strategies to evolve desirable features in plants. Conventional approaches that use mutagenesis cannot access every gene for inactivation because gene integration is highly random in nature. CRISPR/Cas-9, however, allows for specific modifications in required areas of the genome, essential for generating mutants for previously inaccessible genes, mutant loci, and broad deletions. These new genetic instruments introduce specific modifications to a healthy and heritable plant genome. Furthermore, after targeted gene editing, transgenic areas can easily be removed to produce "transgene-free" plants, which can boost crop variant conditions (Kanchiswamy et al., 2015). Although genetically modified (GM) crops have been involved in environmental and health-related controversies, they have offered fast solutions for plant improvement (Hilbeck et al., 2011). With Cas-9/sgRNA, zinc finger nucleases (ZFNs), and transcription activator-like effector nucleases (TALENs), modern plant breeding techniques (NPBT) are much more specific than existing options for genome editing. New and improved breeding strategies can produce a split-line and bypass transgenic inserts more fully than traditional breeding techniques (Schaart et al., 2015). Importantly, modifications of plant content derived from NPBTs are classically indistinguishable and can be measured by the finished product instead of the production process (Woo et al., 2015).

3.4 CONCLUSION

In recent decades, tremendous advances have been made in canola improvement. Intensive and collaborative breeding activities have contributed to the production and commercialization of oil seed crops. An improvement in stress tolerance abilities in canola, as well as in their viability with oil and non-meal areas for oil seeds, was achieved via agronomic, physiological, and molecular efforts. There are, however, continuous changes, while we have more holes in the understanding of improving the quality and stability of oil seed canola crops. Nevertheless, ABA and brassino steroids are of great concern and have proven to significantly impact plant metabolism and help to regulate drought stress among plants. Stomatal closure is controlled by stress signals associated with brassino steroids and plant ABA levels. For stress signals and resistance, brassino steroids are essential, but their key function is not yet fully understood. Traditional crop reproduction and modern approaches towards plant omics are used in crop drought tolerance. A variety of drought-resistance genotypes, including MAS and genetic engineering, have been established that can resist harsh environments compared to current molecular breeding approaches. However, findings show that most species of *Brassica* are vulnerable to drought in the region. It is therefore important that phenotypical data related to drought information are diligently compiled, which helps scientists assist breeders in identifying genotypes resistant to drought. Gene mapping and QTL analyses offer qualitative and quantitative findings that are more indulgent to the casual genetic shows controlling such functions. This will use knowledge about these characteristics to effectively use genes to improve characteristics under conditions of stress in the field. The current state of the Brassica genome has opened the door to computer studies and reverse genetic engineering that can help increase tolerance to drought. The use, through micro-RNAs, of genomic tools to identify and use the motives behind drought-regulated genes would allow us to identify targeted genes and ultimately increase their yields. The main challenge is the information about QTL identification, using agricultural, physiological, MAS, and new molecular methods of breeding, to improve competition in technologies for producing drought resilience genotypes. The enhanced dryness stress tolerance in the QTL mapping and some limitations still exist to incorporate QTLs to identify dryness-reactive traits, along with various genes that contribute to the yield characteristics. Often, with desired alleles in elite germplasms, it is difficult and demanding. Therefore, for the identification of QTLs to verify signal transduction for stress response, a multidisciplinary approach must be adopted. Under regulated conditions, the transgenic lines using plant growth hormones, transcription factors, and plentiful proteins outperformed. Compared to natural areas, the output of the transgenic line under field stress conditions is not established since genotypes face varying levels of environmental stress under normal conditions. Genetic diversity in crop plants can contribute to improving stress mechanisms. However, much is expected with refining the editing multiplex tools to find more complex traits related to agronomic traits in crop plants. To understand quantitative characteristics, GOI, metabolic pathways, genetic functions and eventually adapting crops in a demanding climate, CRISPR/Cas-9 represents a better substitute for conventional breeding techniques. However, CRISPR technology is an immensely useful way of enhancing food safety and public understanding of GE varieties. Researchers can use more resilient GE and non-GE plants to abiotic and biotic stresses through the development of solid, advanced technologies, which will eventually contribute to more productive agricultural opportunities. Drought tolerance is a highly polygenic complex phenomenon and new approaches (e.g., omics, comprehensive genetic studies on natural variations, and genome-wide associations in the crop species can be performed directly in crop species) would help in the identification of genes of interest. Furthermore, these methods open new revenue, for drought-related studies aimed at improving the tolerance of drought in crop species including brassica. These reproductive strategies can be used for species production and adaptation to the dynamic environment to improve the capacity of over 9 billion people to provide food, fuel, and shelter by 2050.

KEYWORDS

- **abiotic stresses**
- **CRISPR/Cas-9**
- **drought**
- **genetic diversity**
- **genome editing technologies**
- **genotypes**
- **marker-assisted breeding**
- **quantitative trait loci**

REFERENCES

Afzal, M., Alghamdi, S. S., Rahman, M. H., Ahmad, A., Farooq, T., Alam, M., Khan, I. A., Ullah, H., Nasim, W., & Fahad, S. (2018). Current status and future possibilities of molecular genetics techniques in *Brassica napus*. *Biotechnology Letters, 40*(3), 479–492.

Ahmad, Z., Waraich, E. A., Akhtar, S., Anjum, S., Ahmad, T., Mahboob, W., Hafeez, O. B. A., Tapera, T., Labuschagne, M., & Rizwan, M. (2018). Physiological responses of wheat to drought stress and its mitigation approaches. *Acta Physiologiae Plantarum, 40*, 80. https://doi.org/10.1007/s11738-018-2651-6.

Ahmadi, M., & Bahrani, M. (2009). Yield and yield components of rapeseed as influenced by water stress at different growth stages and nitrogen levels. *American-Eurasian Journal of Agricultural and Environmental Sciences, 5*(6), 755–761.

Andersen, M. N., Asch, F., Wu, Y., Jensen, C. R., Naested, H., Mogensen, V. O., & Koch, K. E. (2002). Soluble invertase expression is an early target of drought stress during the critical, abortion–Sensitive phase of young ovary development in maize. *Plant Physiology, 130*, 591–604.

Asghari, A., Mohammadi, S. A., Moghaddam, M., & Mohammaddoost, H. R. (2007). QTL analysis for cold resistance-related traits in *Brassica napus* using RAPD markers. *Journal of Food, Agriculture & Environment, 5*(3–4), 188–192.

Ashraf, M. (2010). Inducing drought tolerance in plants: Recent advances. *Biotechnology Advances, 28*, 169–183.

Ashraf, M., & Akram, N. A. (2009). Improving salinity tolerance of plants through conventional breeding and genetic engineering: An analytical comparison. *Biotechnology Advances, 27*, 744–752.

Ashraf, M., & Mehmood, S. (1990). Response of four *Brassica* species to drought stress. *Environmental and Experimental Botany, 30*, 93–100.

Benlloch-González, M., Fournier, J. M., & Benlloch, M. (2009). K^+ deprivation induces xylem water and K^+ transport in sunflower: Evidence for a coordinated control. *Journal of Experimental Botany, 61*(1), 157–164.

Bhardwaj, A. R., Joshi, G., Kukreja, B., Malik, V., Arora, P., Pandey, R., Shukla, R. N., Bankar, K. G., Agrawal, S. K., Goel, S., Jagannath, A., Kumar, A., & Agrawal, M. (2015). Global insights into high temperature and drought stress-regulated genes by RNA-seq in economically important oilseed crop *Brassica juncea*. *BMC Plant Biology, 15*, 9. https://doi.org/10.1186/s12870-014-0405-1.

Birun Ara, A., Shekari, F., Hassanpouraghdam, M. B., Khorshidi, K. M., & Esfandyari, E. (2011). Effects of water deficit stress on yield components and phenology of canola (*Brassica napus* L.) at different growth stages. *Journal of Agricultural and Environmental Sciences, 9*(3–4), 506–509.

Blum, A. (2014). Genomics for drought resistance–Getting down to earth. *Functional Plant Biology, 41*, 1191–1198.

Buell, C. R., & Last, R. L. (2010). 21st-century plant biology: Impacts of the Arabidopsis genome on plant biology and agriculture. *Plant Physiology, 154*(2), 497–00.

Cattivelli, L., Rizza, F., Badeck, F. W., Mazzucotelli, E., Mastrangelo, A. M., Francia, E., Mare, C., Tondelli, A., & Stanca, A. M. (2008). Drought tolerance improvement in crop plants: An integrated view from breeding to genomics. *Field Crops Research, 105*(1–2), 1–14.

Christmann, A., Weiler, E. W., Steudle, E., & Grill, E. (2007). A hydraulic signal in root-to-shoot signaling of water shortage. *The Plant Journal, 52*(1), 167–174.

Collins, N. C., Tardieu, F., & Tuberosa, R. (2008). Quantitative trait loci and crop performance under abiotic stress: Where do we stand? *Plant Physiology, 147*, 469–486.

Cowling, W. A. (2007). Genetic diversity in Australian canola and implications for crop breeding for changing future environments. *Field Crops Research, 104*, 103–111.

Cutler, S., Ghassemian, M., Bonetta, D., Cooney, S., & McCourt, P. (1996). A protein farnesyl transferase involved in abscisic acid signal transduction in *Arabidopsis*. *Science, 273*(5279), 1239.

Das, G., Patra, J. K., & Baek, K. H. (2017). Insight into MAS: A molecular tool for the development of stress resistance and quality of rice through gene stacking. *Frontiers in Plant Science, 8*, 985.

Dhanda, S., & Munjal, R. (2006). Inheritance of cellular thermotolerance in bread wheat. *Plant Breeding, 125*(6), 557–564.

Din, J., Khan, S. U., Ali, I., & Gurmani, A. (2011). Physiological and agronomic response of canola varieties to drought stress. *Journal of Animal and Plant Sciences, 21*(1), 78–82.

Dubey, A., Kumar, A., Abd_Allah, E. F., Hashem, A., & Khan, M. L. (2018). Growing more with less: Breeding and developing drought resilient soybean to improve food security. *Ecological Indicators, 105*, 425–437. https://doi.org/10.1016/j.ecolind.2018.03.003.

Dubey, A., Malla, M. A., Khan, F., Chowdhary, K., Yadav, S., Kumar, A., & Khan, M. L. (2019). Soil microbiome: A key player for conservation of soil health under changing climate. *Biodiversity and Conservation, 28*, 2405–2429. https://doi.org/10.1007/s10531-019-01760-5.

Elferjani, R., & Soolanayakanahally, R. (2018). Canola responses to drought, heat, and combined stress: Shared and specific effects on carbon assimilation, seed yield, and oil composition. *Frontiers in Plant Science, 9*, 1224. https://doi.org/10.3389/fpls.2018.01224.

Fahad, S., Bajwa, A. A., Nazir, U., Anjum, S. A., Farooq, A., Zohaib, A., Sadia, S., Nasim, W., Adkins, S., & Saud, S. (2017). Crop production under drought and heat stress: Plant responses and management options. *Frontiers in Plant Science, 8*, 1147.

Fichtner, F., Castellanos, R. U., & Ulker, B. (2014). Precision genetic modifications: A new era in molecular biology and crop improvement. *Planta, 239*, 921–939.

Finkelstein, R. R., Gampala, S. S. L., & Rock, C. D. (2002). Abscisic acid signaling in seeds and seedlings. *The Plant Cell, 14*, S15–S45.

Fletcher, R. S., Mullen, J. L., Heiliger, A., & McKay, J. K. (2015). QTL analysis of root morphology, flowering time, and yield reveals trade-offs in response to drought in *Brassica napus*. *Journal of Experimental Botany, 66*, 245–256.

Flexas, J., & Medrano, H. (2002). Drought-inhibition of photosynthesis in C3 plants: Stomatal and non-stomatal limitations revisited. *Annals of Botany, 89*, 183–189. https://doi.org/10.1093/aob/mcf027.

Fukao, T., & Xiong, L. (2013). Genetic mechanisms conferring adaptation to submergence and drought in rice: Simple or complex? *Current Opinion in Plant Biology, 16*(2), 196–204.

Garg, A. K., Kim, J. K., Owens, T. G., Ranwala, A. P., Do Choi, Y., Kochian, L. V., & Wu, R. J. (2002). Trehalose accumulation in rice plants confers high tolerance levels to different abiotic stresses. *Proceedings of the National Academy of Sciences, 99*(25), 15898–15903.

Gilbert, L. A., Horlbeck, M. A., Adamson, B., Villalta, J. E., Chen, Y., Whitehead, E. H., & Weissman, J. S. (2015). Genome-scale CRISPR-mediated control of gene repression and activation. *Cell, 159*, 647–661. https://doi.org/10.1016/j.cell.2014.09.029.

Gilbert, L. A., Larson, M. H., Morsut, L., Liu, Z., Brar, G. A., Torres, S. E., & Qi, L. S. (2013). CRISPR-mediated modular RNA-guided regulation of transcription in eukaryotes. *Cell, 154*, 442–451. https://doi.org/10.1016/j.cell.2013.06.044.

Gilchrist, E. J., & Haughn, G. W. (2005). TILLING without a plow: A new method with applications for reverse genetics. *Current Opinion in Plant Biology, 8*, 211–215. https://doi.org/10.1016/j.pbi.2005.01.004.

Gilchrist, E. J., Sidebottom, C. H., Koh, C. S., MacInnes, T., Sharpe, A. G., & Haughn, G. W. (2013). A mutant *Brassica napus* (canola) population for the identification of new genetic diversity via TILLING and next-generation sequencing. *PLoS One, 8*(12), e84303.

Harb, A., Krishnan, P., Ambavaram, M. M. R., & Pereira, A. (2010). Molecular and physiological analysis of drought stress in *Arabidopsis* reveals early responses leading to acclimation in plant growth. *Plant Physiology, 154*(3), 1254–1271.

Hashem, A., Kumar, A., Al-Dbass, A. M., Alqarawi, A. A., Al-Arjani, A. B. F., Singh, G., & Abd_Allah, E. F. (2019). Arbuscular mycorrhizal fungi and biochar improve drought tolerance in chickpeas. *Saudi Journal of Biological Sciences, 26*(3), 614–624. https://doi.org/10.1016/j.sjbs.2018.11.005.

Hayat, S., Hayat, Q., Alyemeni, M. N., Wani, A. S., Pichtel, J., & Ahmad, A. (2012). Role of proline under changing environments: A review. *Plant Signaling & Behavior, 7*(11), 1456–1466.

Hilbeck, A., Meier, M., Römbke, J., Jänsch, S., Teichmann, H., & Tappeser, B. (2011). Environmental risk assessment of genetically modified plants–Concepts and controversies. *Environmental Sciences Europe, 23*, 1–12. https://doi.org/10.1186/2190-4715-23-13.

Hong, S. W., Lee, U., & Vierling, E. (2003). *Arabidopsis* hot mutants define multiple functions required for acclimation to high temperatures. *Plant Physiology, 132*(2), 757–767.

Hu, H., Dai, M., Yao, J., Xiao, B., Li, X., Zhang, Q., & Xiong, L. (2006). Overexpressing a NAM, ATAF, and CUC (NAC) transcription factor enhances drought resistance and salt tolerance in rice. *Proceedings of the National Academy of Sciences, 103*(35), 12987–12992.

Hunt, L., Mills, L. N., Pical, C., Leckie, C. P., Aitken, F. L., Kopka, J., & Gray, J. E. (2003). Phospholipase C is required for the control of stomatal aperture by ABA. *The Plant Journal, 34*(1), 47–55.

Ishitani, M., Rao, I., Wenzl, P., Beebe, S., & Tohme, J. (2004). Integration of genomics approach with traditional breeding towards improving abiotic stress adaptation: Drought and aluminum toxicity as case studies. *Field Crops Research, 90*(1), 35–45.

Jaleel, C. A., Gopi, R., Gomathinayagam, M., & Panneerselvam, R. (2008). Effects of calcium chloride on the metabolism of salt-stressed *Dioscorea rotundata*. *Acta Biologica Cracoviensia, Series Botanica, 50*(1), 63–67.

Jha, U. C., Chaturvedi, S. K., Bohra, A., Basu, P. S., Khan, M. S., & Barh, D. (2014). Abiotic stresses, constraints and improvement strategies in chickpea. *Plant Breeding, 133*(2), 163–178.

Jones, H. D. (2015). Regulatory uncertainty over genome editing. *Nature Plants, 1*, 14011. https://doi.org/10.1038/nplants.2014.11.

Jung, C., Seo, J. S., Han, S. W., Koo, Y. J., Kim, C. H., Song, S. I., & Cheong, J. J. (2008). Overexpression of AtMYB44 enhances stomatal closure to confer abiotic stress tolerance in transgenic *Arabidopsis*. *Plant Physiology, 146*(2), 623–635.

Kagale, S., Divi, U. K., Krochko, J. E., Keller, W. A., & Krishna, P. (2007). Brassinosteroids confer tolerance in *Arabidopsis thaliana* and *Brassica napus* to a range of abiotic stresses. *Planta, 225*, 353–364.

Kahraman, A., Kusmenoglu, I., Aydin, N., Aydogan, A., Erskine, W., & Muehlbauer, F. (2004). QTL mapping of winter hardiness genes in lentils. *Crop Science, 44*(1), 13–22.

Kanchiswamy, C. N., Sargent, D. J., Velasco, R., Maffei, M. E., & Malnoy, M. (2015). Looking forward to genetically edited fruit crops. *Trends in Biotechnology, 33*, 62–64. https://doi.org/10.1016/j.tibtech.2014.07.003.

Khatodia, S., & Khurana, S. M. P. (2014). Trending: The Cas nuclease mediated genome editing technique. *Biotechnology Today, 4*, 46–49. https://doi.org/10.5958/2322-0996.2014.00019.2.

Khatodia, S., Bhatotia, K., Passricha, N., Khurana, S., & Tuteja, N. (2016). The CRISPR/Cas genome-editing tool: Application in the improvement of crops. *Frontiers in Plant Science, 7*, 506.

Kim, J. Y., Mahé, A., Brangeon, J., & Prioul, J. L. (2000). A maize vacuolar invertase, IVR2, is induced by water stress: Organ/tissue specificity and diurnal modulation of expression. *Plant Physiology, 124*, 71–84.

Koornneef, M., Hanhart, C. J., Hilhorst, H. W., & Karssen, C. M. (1989). *In vivo* inhibition of seed development and reserve protein accumulation in recombinants of abscisic acid biosynthesis and responsiveness mutants in *Arabidopsis thaliana. Plant Physiology, 90*(2), 463–469.

Koornneef, M., Reuling, G., & Karssen, C. (1984). The isolation and characterization of abscisic acid–insensitive mutants of *Arabidopsis thaliana. Physiologia Plantarum, 61*(3), 377–383.

Kooyers, N. J. (2015). The evolution of drought escape and avoidance in natural herbaceous populations. *Plant Science, 234*, 155–162.

Kumar, A., Dames, J. F., Gupta, A., Sharma, S., Gilbert, J. A., & Ahmad, P. (2014). Current developments in arbuscular mycorrhizal fungi research and its role in salinity stress alleviation: A biotechnological perspective. *Critical Reviews in Biotechnology, 35*(4), 461–474. https://doi.org/10.3109/07388551.2014.899964.

Kumar, A., Gupta, A., Azooz, M. M., Sharma, S., Ahmad, P., & Dames, J. (2013b). Genetic approaches to improve salinity tolerance in plants. In *Salt Stress in Plants: Signaling, Omics, and Adaptations*. https://doi.org/10.1007/978-1-4614-6108-1_4.

Kumar, A., Sharma, S., & Mishra, S. (2010). Influence of arbuscular mycorrhizal (AM) fungi and salinity on seedling growth, solute accumulation, and mycorrhizal dependency of *Jatropha curcas* L. *Journal of Plant Growth Regulation, 29*(3), 297–306. https://doi.org/10.1007/s00344-009-9136-1.

Kumar, A., Sharma, S., & Mishra, S. (2015). Evaluating the effect of arbuscular mycorrhizal fungal consortia and *Azotobacter chroococcum* in improving biomass yield of *Jatropha curcas*. *Plant Biosystems, 105*, 1056–1064. https://doi.org/10.1080/11263504.2014.1001001.

Kumar, A., Sharma, S., Mishra, S., & Dames, J. F. (2013a). Arbuscular mycorrhizal inoculation improves the growth and antioxidative response of *Jatropha curcas* (L.) under Na2SO4 salt stress. *Plant Biosystems, 149*(2), 260–269. https://doi.org/10.1080/11263504.2013.845268.

Kumar, P., & Sharma, P. (2020). Soil salinity and food security in India. *Frontiers in Sustainable Food Systems, 4*, 533781. https://doi.org/10.3389/fsufs.2020.533781.

Larson, M. H., Gilbert, L. A., Wang, X., Lim, W. A., Weissman, J. S., & Qi, L. S. (2013). CRISPR interference (CRISPRi) for sequence-specific control of gene expression. *Nature Protocols, 8*, 2180–2196. https://doi.org/10.1038/nprot.2013.132.

Leport, L., Turner, N. C., French, R. J., Barr, M. D., Duda, R., & Davies, S. L. (2006). Physiological responses of chickpea genotypes to terminal drought in a Mediterranean-type environment. *European Journal of Agronomy, 11*, 279–291.

Li, W., Lu, J., Lu, K., Yuan, J., Huang, J., Du, H., & Li, J. (2016). Cloning and phylogenetic analysis of *Brassica napus* L. caffeic acid O-methyltransferase one gene family and its expression pattern under drought stress. *PLoS One, 11*(11), e0165975.
Li, Z., Wu, B., Lu, G., Cheng, Y., Zou, C., & Zhang, X. (2012). Differences in physiological responses of *Brassica napus* genotypes under water stress during the seedling stage. *Chinese Journal of Oil Crop Science, 34*(1), 33–39.
Lisar, S. Y., Motafakkerazad, R., Hossain, M. M., & Rahman, I. M. (2012). Water stress in plants: Causes, effects, and responses. In M. M. Ismail & H. Hasegawa (Eds.), *Water Stress* (pp. 1–14). InTech.
Liu, D., Pei, Z., Naeem, M. S., Ming, D., Liu, H., Khan, F., & Zhou, W. (2011). 5-Aminolevulinic acid activates the antioxidative defense system and seedling growth in *Brassica napus* L. under water-deficit stress. *Journal of Agronomy and Crop Science, 197*, 284–295.
Lu, G., Gao, C., Zheng, X., & Han, B. (2009). Identification of OsbZIP72 as a positive regulator of ABA response and drought tolerance in rice. *Planta, 229*, 605–615.
Malla, M. A., Dubey, A., Kumar, A., Yadav, S., Hashem, A., & Abd_Allah, E. F. (2019). Exploring the human microbiome: The potential future role of next-generation sequencing in disease diagnosis and treatment. *Frontiers in Immunology, 9*, 2868. https://doi.org/10.3389/fimmu.2018.02868.
Malla, M. A., Dubey, A., Yadav, S., Kumar, A., Hashem, A., & Abd_Allah, E. F. (2018). Understanding and designing the strategies for the microbe-mediated remediation of environmental contaminants using omics approaches. *Frontiers in Microbiology, 9*, 1132. https://doi.org/10.3389/fmicb.2018.01132.
Mirzaee, M., Moieni, A., & Ghanati, F. (2013). Effects of drought stress on the lipid peroxidation and antioxidant enzyme activities in two canola (*Brassica napus* L.) cultivars. *Journal of Agricultural Science and Technology, 15*(3), 593–602.
Mirzaei, A., Naseiri, R., Moghadam, A., & Esmailpour-Jahromi, M. (2013). The effects of drought stress on seed yield and some agronomic traits of canola cultivars at different growth stages. *Bulletin of Environmental Pharmacology and Life Sciences, 2*, 115–121.
Mizoi, J., Shinozaki, K., & Yamaguchi-Shinozaki, K. (2012). AP2/ERF family transcription factors in plant abiotic stress responses. *Biochimica et Biophysica Acta (BBA)–Gene Regulatory Mechanisms, 1819*, 86–96.
Moaveni, P., Ebrahimi, A., & Farahani, H. A. (2010). Physiological growth indices in winter rapeseed (*Brassica napus* L.) cultivars as affected by drought stress in Iran. *Journal of Cereals and Oilseeds, 1*, 11–16.
Moghadam, H. R. T., Zahedi, H., & Ghooshchi, F. (2011). Oil quality of canola cultivars in response to water stress and super absorbent polymer application. *Pesquisa Agropecuária Tropical, 41*(4), 579–586.
Moody, D. (2008). Timing of N application in canola. *BCG Trial Report*. Retrieved from: http://www.australianoilseeds.com/__data/assets/pdf_file/0019/4195/N_in_canola.pdf (accessed on 12 September 2024).
Morgante, M., & Salamini, F. (2003). From plant genomics to breeding practice. *Current Opinion in Biotechnology, 14*(2), 214–219.
Motley, K., Harbison, D., & Rice, A. (2001). Top dressing canola with urea: Amounts and timing. In *CWFS Compendium, 2000–2001*. Retrieved from: http://www.cwfs.org.au/fileadmin/CWFS_RC_2001 (accessed on 12 September 2024).
Nelson, D. E., Repetti, P. P., Adams, T. R., Creelman, R. A., Wu, J., Warner, D. C., & Donnarummo, M. G. (2007). Plant nuclear factor Y (NF-Y) B subunits confer drought

tolerance and lead to improved corn yields on water-limited acres. *Proceedings of the National Academy of Sciences, 104*(42), 16450–16455.

Osterlund, M. T., & Paterson, A. H. (2002). Applied plant genomics: The secret is integration. *Current Opinion in Plant Biology, 5*(2), 141–145.

Passioura, J. (2007). The drought environment: Physical, biological and agricultural perspectives. *Journal of Experimental Botany, 58*, 113–117.

Pei, Z. M., Ghassemian, M., Kwak, C. M., McCourt, P., & Schroeder, J. I. (1998). Role of farnesyltransferase in ABA regulation of guard cell anion channels and plant water loss. *Science, 282*(5387), 287–290.

Pérez-de-Castro, M., Vilanova, S., Cañizares, J., Pascual, L., Blanca, J. M., Diez, M. J., Prohens, J., & Picó, B. (2012). Application of genomic tools in plant breeding. *Current Genomics, 13*(3), 179–195.

Quetier, F. (2016). The CRISPR-Cas9 technology: Closer to the ultimate toolkit for targeted genome editing. *Plant Science, 242*, 65–76.

Reynolds, M., & Langridge, P. (2016). Physiological breeding. *Current Opinion in Plant Biology, 31*, 162–171.

Rivero, R. M., Kojima, M., Gepstein, A., Sakakibara, H., Mittler, R., Gepstein, S., & Blumwald, E. (2007). Delayed leaf senescence induces extreme drought tolerance in a flowering plant. *Proceedings of the National Academy of Sciences, 104*(49), 19631–19636.

Rozema, J., & Flowers, T. (2008). Crops for a salinized world. *Science, 322*(5907), 1478–1480.

Sah, S. K., Reddy, K. R., & Li, J. (2016). Abscisic acid and abiotic stress tolerance in crop plants. *Frontiers in Plant Science, 7*, 571.

Saika, H., Nishizawa-Yokoi, A., & Toki, S. (2014). The non-homologous end-joining pathway is involved in stable transformation in rice. *Frontiers in Plant Science, 5*, 560. https://doi.org/10.3389/fpls.2014.00560.

Sangtarash, M., Qaderi, M., Chinnappa, C., & Reid, D. (2009). Differential sensitivity of canola (*Brassica napus*) seedlings to ultraviolet-B radiation, water stress, and abscisic acid. *Environmental and Experimental Botany, 66*, 212–219.

Saradadevi, R., Palta, J. A., & Siddique, K. H. (2017). ABA-mediated stomatal response in regulating water use during the development of terminal drought in wheat. *Frontiers in Plant Science, 8*, 1251.

Sasse, J. M. (2003). Physiological actions of brassinosteroids: An update. *Journal of Plant Growth Regulation, 22*(4), 276–288.

Schaart, J. G., van de Wiel, C. C. M., Lotz, L. A. P., & Smulders, M. J. M. (2015). Opportunities for products of new plant breeding techniques. *Trends in Plant Science, 21*(5), 438–449. https://doi.org/10.1016/j.tplants.2015.11.006.

Shafiq, S., Akram, N. A., Ashraf, M., & Arshad, A. (2014). Synergistic effects of drought and ascorbic acid on growth, mineral nutrients and oxidative defense system in canola (*Brassica napus* L.) plants. *Acta Physiologiae Plantarum, 36*, 1539–1553.

Shahsavari, N., Jais, H., & Rad, A. S. (2014). Responses of canola oil quality characteristics and fatty acid composition to zeolite and zinc fertilization under drought stress. *International Journal of Agricultural Sciences, 4*, 49–59.

Tardieu, F., & Tuberosa, R. (2010). Dissection and modeling of abiotic stress tolerance in plants. *Current Opinion in Plant Biology, 13*, 206–212. https://doi.org/10.1016/j.pbi.2009.12.012.

Thameur, A., Ferchichi, A., & López-Carbonell, M. (2011). Quantification of free and conjugated abscisic acid in five genotypes of barley (*Hordeum vulgare* L.) under water stress conditions. *South African Journal of Botany, 77*, 222–228. https://doi.org/10.1016/j.sajb.2010.08.004.

Trenberth, K. E., Dai, A., van der Schrier, G., Jones, P. D., Barichivich, J., Briffa, K. R., & Sheffield, J. (2014). Global warming and changes in drought. *Nature Climate Change, 4*(1), 17–22.

Trethowan, R. M., & Reynolds, M. P. (2006). Drought resistance genetic approaches for improving productivity under stress. In *Proceedings of the 7th International Wheat Conference*, Mar del Plata, Argentina, November 27-December 2, 2005.

Tuberosa, R., & Salvi, S. (2006). Genomics-based approaches to improve drought tolerance of crops. *Trends in Plant Science, 11*(8), 405–412.

Valliyodan, B., & Nguyen, H. T. (2006). Understanding regulatory networks and engineering for enhanced drought tolerance in plants. *Current Opinion in Plant Biology, 9*(2), 189–195.

Verslues, P. E., Guo, Y., Dong, C. H., Ma, W., & Zhu, J. K. (2006). Mutation of SAD2, an importin β-domain protein in *Arabidopsis*, alters abscisic acid sensitivity. *The Plant Journal, 47*(5), 776–787.

Wan, J., Griffiths, R., Ying, J., McCourt, P., & Huang, Y. (2009). Development of drought-tolerant canola (*Brassica napus*) through genetic modulation of ABA-mediated stomatal responses. *Crop Science, 49*(5), 1539–1554.

Wang, Y., Beaith, M., Chalifoux, M., Ying, J., Uchacz, T., Sarvas, C., & Huang, Y. (2009). Shoot-specific down-regulation of protein farnesyltransferase (α-subunit) for yield protection against drought in canola. *Molecular Plant, 2*(1), 191–200.

Wang, Y., Ying, J., Kuzma, M., Chalifoux, M., Sample, A., McArthur, C., & Dennis, D. T. (2005). Molecular tailoring of farnesylation for plant drought tolerance and yield protection. *The Plant Journal, 43*(3), 413–424.

Wilkinson, S., & Davies, W. J. (2010). Drought, ozone, ABA, and ethylene: New insights from cell to plant to the community. *Plant Cell and Environment, 33*(4), 510–525.

Willenborg, C., Gulden, R., Johnson, E., & Shirtliffe, S. (2004). Canola: Germination characteristics of polymer-coated canola (*Brassica napus* L.) seeds subjected to moisture stress at different temperatures. *Agronomy Journal, 96*, 786–791.

Woo, J. W., Kim, J., Kwon, S. I., Corvalan, C., Cho, S. W., Kim, H., et al. (2015). DNA-free genome editing in plants with preassembled CRISPR-Cas9 ribonucleoproteins. *Nature Biotechnology, 33*, 1162–1164. https://doi.org/10.1038/nbt.3389.

Wu, H. J., Zhang, Z., Wang, J. Y., Oh, D. H., Dassanayake, M., Liu, B., & Wu, Y. (2012). Insights into salt tolerance from the genome of *Thellungiella salsuginea*. *Proceedings of the National Academy of Sciences, 109*(30), 12219–12224.

Xiao, Q., Zhang, X., Xu, B., Cheng, Y., Zheng, P., & Lu, G. (2012). Cloning and expression pattern of AnnBn1 gene in *Brassica napus*. *Chinese Journal of Oil Crop Science, 34*, 123–128.

Xing, H. L., Dong, L., Wang, Z. P., Zhang, H. Y., Han, C. Y., Liu, B., et al. (2014). A CRISPR/Cas9 toolkit for multiplex genome editing in plants. *BMC Plant Biology, 14*, 327. https://doi.org/10.1186/s12870-014-0327-y.

Xiong, L., Schumaker, K. S., & Zhu, J. K. (2002). Cell signaling during cold, drought, and salt stress. *The Plant Cell, 14*(S1), S165–S183.

Xiong, Y., Fei, S. Z., Arora, R., Brummer, E. C., Barker, R. E., Jung, G., & Warnke, S. E. (2007). Identification of quantitative trait loci controlling winter hardiness in an annual × perennial ryegrass interspecific hybrid population. *Molecular Breeding, 19*(2), 125–136.

Yadav, R., Hash, C., Bidinger, F., Devos, K., & Howarth, C. (2004). Genomic regions associated with grain yield and aspects of post-flowering drought tolerance in pearl millet across stress environments and tester background. *Euphytica, 136*(3), 265–277.

Yan, W., Zhong, Y., & Shangguan, Z. (2016). A meta-analysis of leaf gas exchange and water status responses to drought. *Scientific Reports, 6*, 20917. https://doi.org/10.1038/srep20917.

Yancey, P. H. (2001). Water stress, osmolytes, and proteins. *American Zoologist, 41*(4), 699–709.

Yang, C., Zhang, X., Zou, C., Cheng, Y., Zheng, P., & Li, G. (2007). Effects of drought simulated by PEG-6000 on germination and seedling growth of rapeseed (*Brassica napus* L.). *Chinese Journal of Oil Crop Science, 29*, 425–430.

Yang, J., Sears, R., Gill, B., & Paulsen, G. (2002). Quantitative and molecular characterization of heat tolerance in hexaploid wheat. *Euphytica, 126*(2), 275–282.

Zetsche, B., Volz, S. E., & Zhang, F. (2015). A split-Cas9 architecture for inducible genome editing and transcription modulation. *Nature Biotechnology, 33*, 139–142. https://doi.org/10.1038/nbt.3149.

Zhang, G. H., Su, Q., An, L. J., & Wu, S. (2008). Characterization and expression of a vacuolar Na+/H+ antiporter gene from the monocot halophyte *Aeluropus littoralis*. *Plant Physiology and Biochemistry, 46*(2), 117–126.

Zhang, X., Lu, G., Long, W., Zou, X., Li, F., & Nishio, T. (2014). Recent progress in drought and salt tolerance studies in Brassica crops. *Breeding Science, 64*(1), 60–73.

Zhu, Z., Zheng, W., & Zhang, X. (2011). Principal component analysis and comprehensive evaluation on morphological and agronomic traits of drought tolerance in rapeseed (*Brassica napus* L.). *Scientia Agricultura Sinica, 44*(9), 1775–1787.

CHAPTER 4

Mobility of Nutrients and Their Metabolism in Agriculture Crops

GULZAR A. RATHER,[1,*] ARTI MANHAS,[2] and NOOPUR SHARMA[2]

[1]*The Institute of Plant Sciences, Department of Ornamental Plants and Agricultural Biotechnology, The Volcani Center, ARO, Israel*

[2]*Department of Botany, University of Jammu, Jammu Camp, Kashmir, India*

*Corresponding author.

ABSTRACT

Crop productivity and sustainable agriculture both depend critically on the mobility of nutrients. Various complex processes are involved in nutrient intake, mobility, transport, and metabolism in a variety of crop plants. Some important macronutrients like nitrogen, phosphorus, potassium, and so forth, as well as micronutrients like iron, zinc, and manganese and so forth, might play a crucial role in the growth and development of plants. Therefore, the aim of the chapter is to focus on nutrient management techniques for maximizing crop output, quality, and resilience. It deals with the issues of nutritional imbalance, inadequacies, and environmental factors that affect nutrient availability and also the techniques for improving the effectiveness of nutrient utilization.

4.1 INTRODUCTION

Due to diminishing food resources, the globe is dealing with a number of issues that make it difficult to meet the demands of a growing global population with enough food (Rehman et al., 2020a). Nutrient deficiencies are an issue in many nations (Saeed et al., 2020). Numerous elements, such as soil

Agricultural Biotechnology: Issues, Challenges, and Recent Developments. Mohd. Sayeed Akhtar and Khalid Rehman Hakeem (Eds.)
© 2025 Apple Academic Press, Inc. Co-published with CRC Press (Taylor & Francis)

light, temperature, nutrient availability, and water availability, affect plant growth. The rhizobacteria also promote plant development and fix atmospheric nitrogen, solubilize insoluble phosphate, and secrete hormones like indole-3-acetic acid (IAA), kinetics, and gallic acid (Gas) are other microbes that are crucial to the plant's growth (Rehman et al., 2020b). One of the first scientists to demonstrate the importance of nutrients for plant growth was the German scientist Justus von Liebig. There are more than 100 chemical elements, according to studies, but researchers have found that only about 17 of them are important (Jones and Jacobsen, 2005). These minerals are essential for plant growth and development (Kalsoom et al., 2020). Some of the 17 essential elements, including carbon (C), hydrogen (H), and oxygen (O), are non-mineral nutrients since they are derived from water and air. The remaining 14 nutrients are made up of eight micronutrients, including boron (B), chlorine (Cl), copper (Cu), iron (Fe), manganese (Mn), nickel (Ni), and zinc (Zn), as well as six macronutrients, including nitrogen (N), phosphorus (P), potassium (K), calcium (Ca), magnesium (Mg), and sulfur (S) (Adnan et al., 2020; Brady and Weil, 2008).

The availability and uptake of soil-bound macronutrients and micronutrients affect the nutrition of plants (Acosta-Durán et al., 2007). Most plant nutrition experiments have focused on analyzing the impact of a single nutrient on plant growth; however, there have been few studies that have looked at the impact of many nutrients in a single experiment. Collaboration between the nutrients in this situation can be seen in the crop's response to adding extra nutrients and the consequences of increasing the nutrient concentration (Fageria, 2001). Due to the population's rapid growth and the need to satisfy rising food demands, it is essential for cultivated land to fulfill the maximum yield per unit area standards in order to achieve the highest level of productivity. Plant nutrition affects agricultural productivity and quality. The only way to provide these nutrients is through the use of fertilizers. On the other hand, excessive fertilizer application may result in the buildup of heavy metal, which can also cause crop distraction (Hayyat et al., 2020). Additionally, one of the biggest issues in the development of sustainable crops is imbalanced plant nutrition (Wasaya et al., 2019). They may cause nitrate buildup and eutrophication of the water (Savci, 2012). Similar to how excessive phosphorus application adversely affects water quality. Micronutrient deficits are also common in industrialized nations with high levels of development. Over 3 billion people worldwide experience Fe and Zn deficiency (Graham et al., 2001). Additionally, it is anticipated that between 6 and 10 billion people will live on Earth in 2050 (Byrnes and Bumb, 1998). A significant increase in food production is required to ensure human well-being and to feed a planet with rapidly growing populations. The projected increases in global population will put a significant strain on the currently cultivated area through intensification and urbanization for crop production. Given the circumstances, we must concentrate on sustainability because it has recently been a phrase that is used broadly in many aspects of our lives, especially in agriculture due to the impact that some crop production techniques have on the environment (Hanson et al., 2007). Agricultural sustainability is the management and consumption of an agricultural ecosystem in a way that preserves its productivity, biotic diversity, capacity for regeneration, functional vitality, and ability so that it can fulfill ecologically significant, social and economic functions now and in the future at the local, national, and international levels while not harming other ecosystems (Lewandowski et al., 1999). The present chapter deals with the types of nutrients, their roles, method of uptake, and deficiency symptoms which are discussed in detail.

4.2 TYPES OF MACRO- AND MICRO-NUTRIENTS AND THEIR ROLE

All plants require a variety of inorganic mineral elements from their surroundings in order to successfully develop their vegetative and reproductive organs. Based on the proportional amounts needed for plant growth, minerals can be separated into two categories. Nitrogen (N), potassium (K), calcium (Ca), magnesium (Mg), phosphorus (P), and sulfur (S) are among the macronutrients. Plants often contain these elements in amounts greater than 0.1% of dry tissue weight. Iron (Fe), zinc (Zn), manganese (Mn), copper (Cu), boron (B), chlorine (Cl), molybdenum (Mo), and nickel (Ni) are currently recognized micronutrients; they are often present at concentrations less than 0.01% of dry tissue weight. It is widely acknowledged that these 14 minerals, along with the elements carbon (C), hydrogen (H), and oxygen (O), are necessary for all plants to flourish. Despite the fact that more than half of the elements in the periodic table have been detected in plant tissue, many other elements are thought to enter plants non-selectively. The majority of these non-essential elements offer the plant no known benefits, and many like cadmium (Cd) or chromium (Cr) are even harmful to plant growth. A significant factor in the processes of eutrophication in the waters of both industrialized and developing countries is the use of N- and P-fertilizers in agriculture (Conley et al., 2009; White and Hammond, 2009).

4.2.1 CRITERIA FOR ESSENTIALITY

The need for plants for minerals has long piqued scientist's curiosity. Mineral enrichment of soils was known to encourage growth and increase agricultural yields as early as the early 1800s. Arnon and Stout's essentiality criteria were published in 1939. These can be rephrased as follows: (i) a particular plant must be unable to complete either the vegetative or reproductive stages of its life cycle in the absence of the mineral nutrient; (ii) the function of the element is unique to that element and cannot be substituted by another mineral element; and (iii) the element must directly affect a plant's metabolism or structural function rather than indirectly influencing plant growth by improving certain unfavorable soil microbiological or chemical conditions Solution culture is a technique used by plant nutritionists. In the hydroponics approach, plants are grown in an aerated liquid medium with a known chemical composition. Specific nutrients can be excluded from this medium to investigate how the absence of those nutrients affects plant metabolism and growth. Hydroponic culture has been used to determine all micronutrient minerals that have been determined to be essential thus far. Experiments sometimes required meticulous impurity removal from the chemicals used to create the hydroponic solutions, complete cleaning of the containers used to grow the plants, and/or filtration. Hydroponic culture has been used to determine the essentiality of every micronutrient mineral up to this point. However, some experiments called for meticulous impurity removal from the chemicals used to create the hydroponic solutions, meticulous cleaning of the plant-growing containers, and/or air filtration to avoid dust contamination. These initiatives have helped to build a greater knowledge of the common essential mineral nutrients.

4.2.1.1 NITROGEN

Because it is a component of all proteins and amino acids, as well as all enzymes, nitrogen plays a crucial part in cellular metabolism. Nitrogen has a crucial role in the transcription, translation, and replication of genetic information as a component of nucleotides and nucleic acids, including deoxyribonucleic acid (DNA) and ribonucleic acid (RNA). In the soil, nitrate and ammonium ions are the two forms in which nitrogen may be found. Nitrate is chemically converted to ammonium within plants before being incorporated into organic molecules. For some species, atmospheric nitrogen (N_2), which is obtained through the nitrogen fixation process, serves as a substitute source of nitrogen.

4.2.1.2 PHOSPHORUS

Through the pyrophosphate bond in ATP, phosphorus plays a crucial role in the transfer of energy, and the attachment of phosphate groups to a variety of sugars produces metabolic energy for respiration and photosynthesis. Two tiny purple acid phosphatases (PvPAP3, PvPAP4) from common beans were cloned, and Tian et al. (2016) described how P-starvation increases their expression in both leaves and roots. Plants primarily take up phosphorus in the form of $H_2PO_4^{2-}$ and HPO_4^- orthophosphate anions. Phosphorus is taken up by plant roots. Plant roots must forage for phosphate because it is found in the soil solution at incredibly low concentrations (White and Hammond, 2008).

4.2.1.3 POTASSIUM

Potassium is soluble in soil solutions. Since K^+ and its accompanying anions are the most prevalent cation in the cytoplasm and are not metabolized, they considerably increase the osmotic potential of cells. According to Ilan (1971), cytokinin treatments altered the K^+ and Na^+ selectivity in sunflower leaves and cotyledon cells. The regulation of turgor, which affects cell extension and growth, the regulation of stomatal opening/closing, which affects leaf gas exchange, and pressure-driven phloem translocation, which affects long-distance nutrient flow are all examples of how potassium functions in plant water relations processes. The usage of expensive K fertilizers in agriculture has been suggested to be reduced by breeding crops that uptake and/or utilize K more effectively (Fageria, 2009; Rengel and Damon, 2008). The potassium ion also aids in creating the electrochemical gradient across membranes, which aids in the movement of several chemical species across membranes.

4.2.1.4 CALCIUM

Particularly in the cell wall and on the surface of membranes, calcium forms intra and intermolecular connections. In the absence of calcium, cultured plants may continue to grow for a while by utilizing endogenous calcium (Eklund and Eliasson, 1990; Williams, 1995). Calcium is easily absorbed from soils as the abundant cation, but its free ion concentration in the cytoplasm is kept low by sequestration in the vacuoles, by complexing with calcium-binding proteins (e.g., calmodulin), and by interacting with calcium ions. As its release from intracellular pools stimulates numerous protein kinases, phosphatases, or phospholipases, whose target molecules subsequently regulate many physiological activities, calcium plays a crucial part in stimulus-response coupling involving signal transduction pathways (Bush, 1995).

4.2.1.5 MAGNESIUM

It is important for the formation of chlorophyll and enhances phosphorus mobility and use. Component and activator of several plant enzymes. It is also connected to grass tetany directly increases plants' use of iron and also influences the uniformity and earliness of maturity.

4.2.1.6 SULFUR

It is an essential component of amino acids. It aids in vitamin and enzyme development and promotes the growth of nodules on legumes. It also aids in the development of seeds. Chlorophyll synthesis requires it even if it is not one of the ingredients Tripeptide glutathione, engaged in the detoxification of oxygen radicals; phytochelatins and metallothioneins, involved in the detoxification of heavy metals; and the proteins thioredoxin and ferredoxin, involved in redox chemistry, are all significant sulfur compounds.

4.2.1.7 ZINC

It helps the plant's enzyme system and growth hormones. It is required for the formation of chlorophyll, synthesis of carbohydrates, and production of starch. It also helps the seed to develop.

4.2.1.8 MANGANESE

Mn^{2+} is the main source of manganese in soils, however, manganese can also be found in biological systems in oxidation states II, III, and IV. Manganese is crucial in redox processes because Mn^{2+} is easily converted into Mn^{4+} in the photosystem II-related protein complex that splits water. Several enzymes involved in the catalysis of oxidation-reduction, decarboxylation, and hydrolytic processes are activated by manganese, which also functions as a cofactor. Mn is needed by plants for a number of enzymes that catalyze redox, decarboxylation, and hydrolytic reactions as well as the manganese protein in photosystem II and the manganese-containing superoxide dismutase (SOD) (Marschner, 1995).

4.2.1.9 COPPER

In soils, copper can be found as either a divalent (Cu2) or a monovalent (Cu1) cation. It easily forms complexes with several organic molecules, including proteins, and because of its high monovalent electron affinity, it is a good choice for a variety of redox processes. According to Sandmann and Boger (1983), there are three different categories of copper-containing proteins: blue proteins, which function in one-electron transfer, like plastocyanin, a component of the photosystem's electron transport chain; non-blue proteins, which catalyze peroxidation reactions, like CuZn-SOD; and multi-copper proteins, which have at least four copper atoms per molecule and function as oxidases, like cytochrome oxidase.

4.2.1.10 CHLORINE

The monovalent chloride ion which represents chlorine in aqueous solutions is easily assimilated by plants. Despite the fact that plants contain a large number of chlorinated organic compounds, nothing is known about their functional use. Chloride can have a role in osmoregulation, particularly in stomatal guard cells, and it is necessary for the water-splitting protein complex of photosystem II. It also boosts the activity of the vacuolar proton-pumping ATPase.

4.2.1.11 MOLYBDENUM AND NICKEL

Molybdenum and nickel are the two elements that plants need the least for their growth and development. These minerals work in redox processes because they are transitional elements that can exist in various oxidation states. Only a small number of enzymes, including nitrate reductase, nitrogenase, and xanthine oxidase/dehydrogenase for molybdenum, and urease for nickel, have been verified to require these minerals. As a result, both of these minerals are crucial for nitrogen metabolism.

4.3 EFFECTS OF MINERAL NUTRIENTS ON PLANT GROWTH

Abiotic factors like pH, redox state, and temperature, as well as biotic factors like microbial release of organic acids and phenolic compounds produced metabolically or through the breakdown of soil organic matter, can both affect the speciation and solubility of minerals. By releasing protons, chelators, and/or chemical reductants into the rhizosphere, plant roots can also alter the rhizosphere to change the availability of nutrients. Plants have the ability to activate high-affinity transporters and other processes in their roots to help them satisfy their mineral nutrient needs when faced with a specific nutrient shortfall, such as limited availability. By restricting the biosynthesis or expression of essential elements of energy capture and/or metabolism, mineral shortages have an impact on plant growth.

4.4 NUTRIENT MOBILIZATION AND UPTAKE

The ability of the roots to absorb nutrients and the concentration of nutrients at the surface of the root are the factors affecting nutrient uptake by roots. The availability and movement of mineral elements in the soil can vary substantially over time and space due to various factors like the presence of rhizosphere bacteria, siderophores, organic acids, seasonal and climatic conditions, and soil qualities (Wenzel et al., 2003). In order to boost metal absorption, plants have developed a variety of techniques, including acidification, the secretion of organic chelators, and the development of high-affinity metal transporters (Rauser, 1999). To enhance the solubilization of cations like Fe, Cu, and Zn, protons are released into the rhizosphere in the first of these mechanisms (Palmer and Guerinot, 2009). In the plasma membrane of root cells, ATP-dependent proton pumps (H^+-ATPases) mediate this. Some members of this family, like *AtAHA1*, are expressed constitutively in *Arabidopsis thaliana*, while others, like *AtAHA2*, are activated in response to Fe deficiency (Santi and Schdmidt, 2009). Moreover, it has been observed that large populations of root-associated microorganisms are supported by the carbon-rich root exudates, and they enhance the weathering of minerals by excreting organic acids, phenolic compounds, protons, and siderophores (Draver and Vance, 1994). These activities ultimately mobilize the mineral nutrient thus increasing their availability to plants. For instance, through the excretion of organic acids, the ectomycorrhizal fungus can mobilize P, potassium (K), calcium (Ca), and magnesium (Mg) from solid mineral substrates

(Landeweert et al., 2001). A recent review by Tsai and Schmidt (2017) highlighted the previously underappreciated function of root-secreted coumarins in iron mobilization through reduction and chelation as part of a coordinated strategy developed to enhance the uptake of iron from recalcitrant pools.

For the uptake of any nutrient element, the root must come into contact with it in the soil. Therefore, the root's periphery serves as an essential transit barrier (Kochian and Lucas, 1983). Either apoplastic or symplastic routes (Figure 4.1) nutrients move from the surface of the roots to the long-distance water transport system, the xylem. In any case, for the nutrients to enter the xylem, they must pass through at least one plasma membrane. The plasma membrane of epidermal cells must be passed through for nutrients to be transported via the symplastic route. In order to pass the Casparian strip when nutrients are transported via the apoplastic pathway, they must first enter the plasma membrane of cortex or endodermis cells. Ferric oxidase-reductase (FRO) family membrane proteins chemically reduce redox-active metals like Fe, Cu, and Mn (Robinson et al., 1999). Due to the presence of a Casparian strip in the root stele, active nutrient uptake into the symplast is necessary (Chen et al., 2011). This involves the transport of Fe^{2+} by the primary high-affinity iron transporter IRT1 (Vert et al., 2002), the transport of Cu^{+} by COPT1 (Sancenon et al., 2004), and the transport of Zn, which is mostly carried out by ZIP family members (Guerinot, 2000). The ZIP family member with the best characterization, infrared thermal imaging (IRT1), may transfer additional cations in yeast and plants (Zn, Mn, Co, Ni, and Cd), demonstrating its broad substrate affinity (Nishida et al., 2011).

For nutrients to be carried by the transpiration stream across long distances to the shoot, they must first be absorbed into the root symplast and then released into the dead xylem components. Numerous transporters have been identified in the root pericycle, and it has been demonstrated in several instances that they play a role in nutrient loading into the xylem. The SKOR, BOR1, and SOS1 transporters are some examples that serve as illustrations. Root pericycle and stellar parenchyma cells display the outward-rectifying K^{+} channel encoded by the *SKOR* gene in the plasma membrane (Gaymard et al., 1998). Moreover, a boron transporter that is encoded by the *Arabidopsis BOR1* gene can be seen in the pericycle and on the inner surface of the endodermis when it is fused to the fluorescent reporter green fluorescent protein (Takano et al., 2002). After absorption in the root apoplast cations can either be translocated radially into the root steles or loaded into the xylem. Thereafter, the xylem sap is transported to the shoot by the transpiration stream so that metals can be distributed to the aerial tissues (Clemens et al., 2002;

Figure 4.2). Chelation appears to be necessary to prevent oxidative stress and to aid in ion translocation in the vasculature because transition metals are so reactive (Pich et al., 2001). Association with amino acids, organic acids, mugineic acids, and metallothioneins results in metal chelation (MTs). Metal-specific transporters control the flow of free or chelated ions into the vascular tissues. AtHMA5 is believed to control Cu xylem loading, whereas AtHMA2/AtHMA4 controls Zn/Cd xylem loading from root cells. In contrast to wild-type plants, the hma5 and hma2/hma4 mutants exhibit significantly higher levels of metal accumulation in the roots, which is consistent with their long-distance transport activity (Hussain et al., 2004).

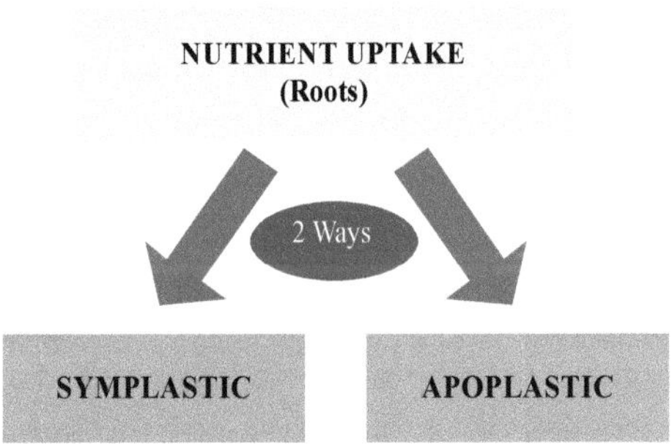

FIGURE 4.1 Two ways of nutrient uptake by plant roots.

The process of metals unloading from the xylem is not well understood. Although the shoot expresses several FRO and ZIP proteins, suggesting a potential function in cation mobilization towards aerial tissues, additional mechanisms for the long-distance delivery of metals are probably necessary for metal homeostasis (Grotz and Guerinot, 2006). The general mechanisms of chelation and ligand exchange during xylem unloading however are poorly understood (Kramer et al., 2007). Another vascular tissue called the phloem, found in plants, runs parallel to the xylem and transports, and redistributes nutrients, hormones, and other signal molecules throughout the entire plant body. Some nutrient ions go through a process known as "phloem recirculation," in which they quickly transfer from the xylem to the phloem in leaves and stems and then are re-translocated back to the roots. For the re-distribution of nutrients, this cycling is very crucial, especially for elements like nitrogen, potassium, phosphorus, and magnesium (Hall and Baker, 1972). As far as phloem transport is concerned apoplast loading into companion cells and sieve components, as well as unloading at the intended sink tissues, are the steps involved in it. This mechanism is carried out by members of the oligopeptide transporter family (OPT), which may transport amino acids that are bonded to metal (Curie et al., 2001). In *A. thaliana* the phloem-specific transporter OPT3 promotes the transfer of transition metals instead of the tiny peptides. This protein aids in the accumulation of Fe in growing tissues like seeds and the loading of Fe into the phloem of leaves. As a result, AtOPT3 contributes to Cd partitioning and Fe xylem-to-phloem redistribution in this tissue (Zhai et al., 2014).

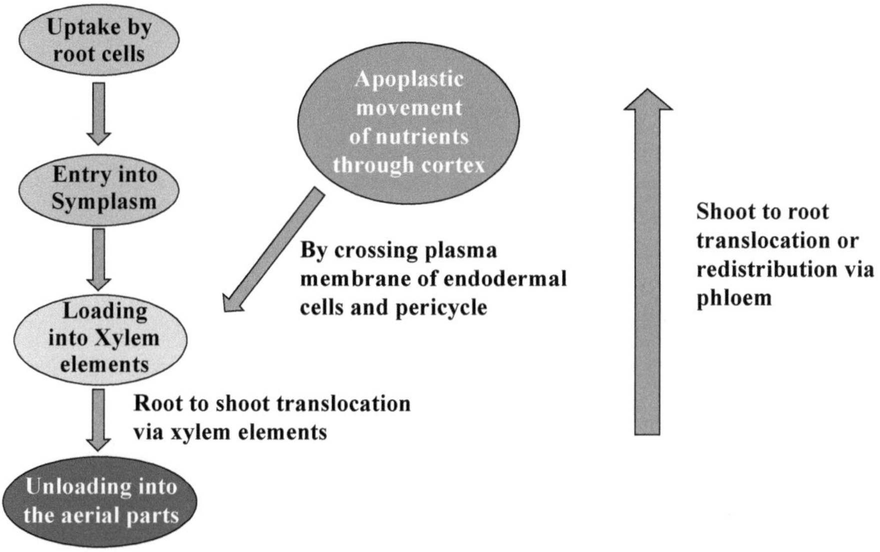

FIGURE 4.2 Schematic representation of metal uptake by plants.

4.5 DEFICIENCY SYMPTOMS OF NUTRIENT ELEMENTS

For crops to yield efficiently and bear grains and fruits of acceptable quality, it is quite important that both their macronutrient and micronutrient requirements are fulfilled. The dosage response curves for each micronutrient demonstrate that toxicity brought on by high doses of the same components can affect yields just as deficits of those nutrients can. Therefore, it is crucial to monitor soils and/or crops to make sure that the concentrations of micronutrients in soils are in the ideal range and are neither too low nor too high. Plants have visual signs on their leaves, stems, fruit, and roots that are typical for deficiency or toxicity when they receive too much or too little of the mineral components that serve as their nutrition. Understanding these visual signs can help us identify the underlying causes of the diseases so that they can be treated well on time. Since it is well known that nutrients have a specific role to play in the metabolism of plants, the symptom development and its location on the plant parts directly reflect the function and mobility of that particular element within the plant. The deficiency symptoms of various mineral elements (McCauley et al., 2009; Uchida, 2000) are briefly discussed in subsections.

4.5.1 NITROGEN DEFICIENCY

It is available to plants in the form of nitrate and ammonium ions. It is a major constituent of amino acids together with carbon, hydrogen, oxygen, and sulfur. It is a necessary component of enzymes, chlorophyll molecules, and several vitamins. Plants deficient in N show stunted growth. Chlorosis of leaves occurs in older leaves before younger ones. N deficiency causes early crop maturity subsequently reducing the quantity and quality of the yield. Reduction in N reduces the protein content of seeds and other vegetative parts and in extreme cases, flowering is also reduced greatly.

4.5.2 PHOSPHORUS DEFICIENCY

It is available to plants as orthophosphate ions (HPO_4^{2-}, $H_2PO_4^-$). It plays a major role in energy transfer and storage in the form of ADP and ATP. It is an integral part of RNA and DNA. Moreover, it plays a great role in flower initiation, root, fruit, and seed development. The initial overall symptom of P deficiency is delayed, poor, and stunted growth because P is required at high levels during the early phases of cell division. P is a rather mobile substance in plants that can be moved to areas of new development, resulting in symptoms of dark to blue-green color on older leaves in some plants. Purpling of the leaves and stems may appear in cases of extreme shortage. Delay in maturation and poor fruit and seed development might result from a lack of P.

4.5.3 POTASSIUM DEFICIENCY

It is available as K^+ ions. The most typical symptom is chlorosis along the leaf margins (leaf margin scorching). Due to K's high mobility throughout the plant, this first appears in older leaves. K is necessary for protein synthesis and photosynthesis, hence, plants lacking K will develop slowly and stunted. In several crops, weak stems and lodging are common if K levels are low. There is a reduction in the size of fruit and seeds as well as in the quantity of production.

4.5.4 CALCIUM DEFICIENCY

It is available in the form of Ca^{2+} ions to plants. It plays a crucial role in the formation of the cell wall membrane and maintaining its plasticity. It activates certain enzymes involved in protein synthesis and carbohydrate transfer. Due to its immobile nature, its deficiency symptoms first appear on younger leaves. The growing tips of the leaves and roots become brown and die. Newly emerging leaves may stick together at the margins without enough calcium, which is required in the form of calcium pectate to produce solid cell walls. This causes tearing as the leaves grow and unfurl. Additionally, this might weaken the stem structure. In some crops, the terminal bud may be degenerating, and younger leaves may be cupped and crinkled. In some crops, buds and flowers drop off prematurely.

4.5.5 MAGNESIUM DEFICIENCY

It is available as Mg^{2+} ions to plants. Older leaves are the first to display the interveinal chlorosis deficient sign since Mg is a mobile element and a component of the chlorophyll molecule. While the leaf veins stay green, the tissue in between the veins can be yellowish, copper, or reddish. Crops like potatoes, tomatoes, soybeans, and cabbage display an orange-yellow color with green veins, and corn leaves seem yellow-striped with green veins. Severe cases may result in symptoms that develop on younger leaves and premature leaf drop. Symptoms are more common in soils that are acidic, receiving a lot of K fertilizer, or having a lot of calcium.

4.5.6 BORON DEFICIENCY

It is available to plants as a borate. Young leaves that are chlorotic with the main growing point dead are symptoms of B deficiency in plants (terminal bud). In addition to chlorosis, leaves can also acquire uneven, dark-brown lesions that, in severe cases, turn into leaf necrosis. The bases of leaves might also develop whitish-yellow patches. B-deficient plants' leaves and stems will become brittle and twisted, and their leaf tips have a tendency to thicken and curl as a result of abnormalities in cell wall growth. Plants that are affected will develop slowly, look deformed, or stunted, and pollination and seed viability are typically poor in B-deficient plants with short internodes.

4.5.7 COPPER DEFICIENCY

Reduced growth, distorted immature leaves, and possible apical meristem necrosis are symptoms of copper deficiency in plants. In trees there is development of numerous sprouts at their growing points, giving them a bushy appearance. Young leaves start to bleach, and soon twigs start to die back and lose their leaves. Young leaf tips and developing points are first impacted in fodder grasses. The plant is chlorotic and stunted.

4.5.8 IRON DEFICIENCY

Fe deficiency causes immature leaves to produce less chlorophyll and is characterized by interveinal chlorosis, which clearly distinguishes veins from chlorotic patches. The entire leaf will turn whitish-yellow as the deficit progresses, leading to necrosis. Plants can also grow slowly. In Fe-deficient fields, particularly where the subsoil is exposed at the surface, irregularly shaped yellow spots can be seen from a distance.

4.5.9 MANGANESE DEFICIENCY

It is available to plants in the form of Mn^{2+} and Mn^{3+} ions. Chlorosis in developing tissues is the first symptom of Mn deficiency. Mn chlorosis manifests in dicots as little yellow dots, in contrast to Fe chlorosis symptoms. Greenish-gray spots can be seen near the base of new leaves in monocots. The specks could eventually turn yellow to orange. In legumes, necrotic spots called “marsh spots” appear on the cotyledons.

4.5.10 MOLYBDENUM DEFICIENCY

Mo role in the plant is to assimilate N, so its deficiency symptoms match those of N. The margins of older and middle leaves curled inward and became chlorotic. Necrotic patches develop at the leaf margins as a result of nitrate build-up, as opposed to N shortage. Stunted plants may also have a limited ability to develop flowers. Mo deficiency is ubiquitous in legumes that fix nitrogen.

4.5.11 ZINC DEFICIENCY

Interveinal chlorosis, which creates a striped appearance in Zn deficient leaves, is most noticeable halfway between the margin and midrib. Some mottling may also appear. Areas that have chlorosis can be light green, yellow, or even white. The leaves will turn gray-white, drop early, or die from severe Zn shortages. Zn deficiency in plants typically results in severe stunting because Zn is crucial for internode elongation. Infected plants also exhibit poor flowering and seed germination. Alfalfa has smaller leaves, cereal leaves have grey or bronze banding, wheat, and other tiny grains have fewer tillers, and irregular grain formation are all signs peculiar to that crop.

4.5.12 CHLORINE DEFICIENCY

Younger leaves turning chlorotic, and the plant wilting are the symptoms of Cl deficiency. However, its deficiencies are rare because Cl is present in rainfall and the atmosphere.

4.6 CONCLUSION

Both micro- and macro-nutrients are essential for the quality of crops. Even though the techniques utilized by plants to resist many diseases will be strengthened by the application of nutrients. The utilization of nutrients in a balanced method can increase the yield of agricultural crops desired to feed the increasing world population. Due to human activity on farmlands and cultural changes in lifestyle, these areas are becoming less fertile. Thus, coordinated efforts are required to develop plants for increased output as this invariably influences agricultural production and may cause famine and starvation. The most recent technology for precision agriculture (PA), wherein plans are made and directed toward meeting the growing need for food is nanotechnology. Fertilizers provide nutrients to plants, although the majority of them are inefficient at supplying and absorbing nutrients. Therefore, nanofertilizers are designed to be target-oriented and are difficult to lose. By utilizing the features of nanoscale materials, this method uses better materials to enhance agriculture. The agricultural use of nanotechnology is bridging the nutrient loss and crop fortification gaps. In the nano-regime, farmers are using this science to increase the quality and output of agricultural products. The plant-mediated biological methods of synthesis, which make use of raw materials like leftover vegetables, plant extracts, flowers, plant barks, roots, fruit peels, and leather cuttings, should be utilized and developed in order to make this field of study more lucrative and easily applicable to agriculture.

KEYWORDS

- **agricultural products**
- **internode elongation**
- **macronutrients**
- **micronutrients**
- **mobilization**
- **nanofertilizers**
- **nutrients**

REFERENCES

Acosta-Durán, C. M., Ocampo, D., Cedillo, E., & Nava, L. M. (2007). Effect of calcium sulfate and biosolids in crop yield of peanut (*Arachis hypogaea* L.). *Agricultural Research, 4*, 31–38.

Adnan, M., Asif, M., Bilal, H. M., Rehman, B., Adnan, M., Ahmad, T., Rehman, H. A., & Anjum, M. Z. (2020). Organic and inorganic fertilizer: Integral part for crop production. *EC Agriculture, 6*, 1–7.

Arnon, D. I., & Stout, P. R. (1939). The essentiality of certain elements in minute quantity for plants with special reference to copper. *Plant Physiology, 14*, 371–375.

Brady, N. C., & Weil, R. R. (2008). *The Nature and Properties of Soils* (13th ed., pp. 662–710). Pearson Education, Inc.

Bush, D. S. (1995). Calcium regulation in plant cells and its role in signaling. *Annual Review of Plant Physiology and Plant Molecular Biology, 46*, 95–122.

Byrnes, B. H., & Bumb, B. L. (1998). Population growth, food production, and nutrient requirements. *Journal of Crop Production, 1*, 1–27.

Chen, T., Cai, X., Wu, X., Karahara, I., Schreiber, L., & Lin, J. (2011). Casparian strip development and its potential function in salt tolerance. *Plant Signaling & Behavior, 6*, 1499–1502.

Clemens, S., Palmgren, M. G., & Krämer, U. (2002). A long way ahead: Understanding and engineering plant metal accumulation. *Trends in Plant Science, 7*, 309–315.

Conley, D. J., Paerl, H. W., Howarth, R. W., Boesch, D. F., Seitzinger, S. P., Havens, K. E., Lancelot, C., & Likens, G. E. (2009). Controlling eutrophication: Nitrogen and phosphorus. *Science, 323*, 1014–1015.

Curie, C., Panaviene, Z., Loulergue, C., Dellaporta, S. L., Briat, J. F., & Walker, E. L. (2001). Maize yellow stripe1 encodes a membrane protein directly involved in Fe (III) uptake. *Nature, 409*, 346–349.

Drever, J. I., & Vance, G. F. (1994). Role of soil organic acids in mineral weathering processes. In *Organic Acids in Geological Processes* (pp. 138–161). Springer, Heidelberg, Berlin.

Eklund, L., & Eliasson, L. (1990). Effects of calcium concentration on cell wall synthesis. *Journal of Experimental Botany, 41*(7), 863–867.

Fageria, N. K. (2009). *The Use of Nutrients in Crop Plants*. CRC Press.

Fageria, V. D. (2001). Nutrient interactions in crop plants. *Journal of Plant Nutrition, 24*, 1269–1290.

Graham, R. D., Welch, R. M., & Bouis, H. E. (2001). Addressing micronutrient malnutrition through enhancing the nutritional quality of staple foods: Principles, perspectives, and knowledge gaps. *Advances in Agronomy, 70*, 77–142.

Grotz, N., & Guerinot, M. L. (2006). Molecular aspects of Cu, Fe, and Zn homeostasis in plants. *Biochimica et Biophysica Acta, 1763*, 595–608.

Guerinot, M. L. (2000). The ZIP family of metal transporters. *Biochimica et Biophysica Acta, 1465*, 190–198.

Hall, S. M., & Baker, D. A. (1972). The chemical composition of Ricinus phloem exudate. *Planta, 106*, 131–140.

Hanson, J. D., Liebig, M. A., Merrill, S. D., Tanaka, D. L., Krupinsky, J. M., & Stott, D. E. (2007). Dynamic cropping systems: Increasing adaptability amid an uncertain future. *Agronomy Journal, 99*, 939–943.

Hayyat, M. S., Adnan, M., Awais, M., Khan, B., Rahman, H. A., Ahmed, R., Rehman, F., Toor, M. D., & Bilal, H. M. (2020). Effect of heavy metal (Ni) on plants and soil: A review. *International Journal of Applied Research, 6*, 313–318.

Hussain, D., Haydon, M. J., Wang, Y., Wong, E., Sherson, S. M., Young, J., Camakaris, J., Harper, J. F., & Cobbett, C. S. (2004). P-type ATPase heavy metal transporters with roles in essential zinc homeostasis in *Arabidopsis*. *The Plant Cell, 16*(5), 1327–1339.

Ilan, I. (1971). Evidence for hormonal regulation of the selectivity of ion uptake by plant cells. *Physiologia Plantarum, 25*, 230–233.

Jones, C., & Jacobsen, J. (2005). Plant nutrition and soil fertility. *Nutrient Management Module*. Extension Publications MSU, *2*, 44–49.

Kalsoom, M., Rehman, F. U., Shafique, T., Junaid, S., Khalid, N., Adnan, M., Zafar, I., Tariq, M. A., Raza, M. A., Zahra, A., & Ali, H. (2020). Biological importance of microbes in agriculture, food and pharmaceutical industry: A review. *Innovative Journal of Life Sciences, 8*(6), 1–4.

Kochian, L. V., & Lucas, W. J. (1983). Potassium transport in corn roots: II. The significance of the root periphery. *Plant Physiology, 73*, 208–215.

Krämer, U., Talke, I. N., & Hanikenne, M. (2007). Transition metal transport. *FEBS Letters, 581*, 2263–2272.

Landeweert, R., Hoffland, E., Finlay, R. D., Kuyper, T. W., & van Breemen, N. (2001). Linking plants to rocks: Ectomycorrhizal fungi mobilize nutrients from minerals. *Trends in Ecology & Evolution, 16*, 248–254.

Lewandowski, I., Härdtlein, M., & Kaltschmitt, M. (1999). Sustainable crop production: Definition and methodological approach for assessing and implementing sustainability. *Crop Science, 39*, 184–193.

Marschner, H. (1995). *Mineral Nutrition of Higher Plants*. Academic Press.

McCauley, A., Jones, C., & Jacobsen, J. (2009). Plant nutrient functions and deficiency and toxicity symptoms. *Nutrient Management Module, 9*, 1–16.

Nishida, S., Tsuzuki, C., Kato, A., Aisu, A., Yoshida, J., & Mizuno, T. (2011). AtIRT1: The primary iron uptake transporter in the root mediates excess nickel accumulation in *Arabidopsis thaliana*. *Plant Cell Physiology, 52*, 1433–1442.

Palmer, C. M., & Guerinot, M. L. (2009). Facing the challenges of Cu, Fe, and Zn homeostasis in plants. *Nature Chemical Biology, 5*, 333–340.

Pich, A., Manteuffel, R., Hillmer, S., Scholz, G., & Schmidt, W. (2001). Fe homeostasis in plant cells: Does nicotianamine play multiple roles in the regulation of cytoplasmic Fe concentration? *Planta, 213*, 967–976.

Rauser, W. E. (1999). Structure and function of metal chelators produced by plants. *Cell Biochemistry and Biophysics, 31*, 19–48.

Rehman, F., Kalsoom, M., Adnan, M., Toor, M. D., & Zulfiqar, A. (2020b). Plant growth promoting rhizobacteria and their mechanisms involved in agricultural crop production: A review. *SunText Reviews of Biotechnology, 1*(2), 110.

Rehman, F., Kalsoom, M., Nasir, T. A., Adnan, M., Anwar, S., & Zahra, A. (2020a). Chemistry of plant-microbe interactions in rhizosphere and rhizoplane. *Indian Journal of Pure and Applied Biosciences, 8*, 11–19.

Rengel, Z., & Damon, P. M. (2008). Crops and genotypes differ in efficiency of potassium uptake and use. *Physiologia Plantarum, 133*, 624–636.

Robinson, N. J., Procter, C. M., Connolly, E. L., & Guerinot, M. L. (1999). A ferric-chelate reductase for iron uptake from soils. *Nature, 397*, 694–697.

Saeed, M. S., Saeed, A., & Adnan, M. (2020). Production and utilization of single cell proteins—An overview. *Current Research in Agriculture and Farming, 1*, 9–12.

Sancenón, V., Puig, S., Mateu-Andrés, I., Dorcey, E., Thiele, D. J., & Peñarrubia, L. (2004). The Arabidopsis copper transporter COPT1 functions in root elongation and pollen development. *Journal of Biological Chemistry, 279*, 15348–15355.

Sandmann, G., & Boger, P. (1983). The enzymatological function of heavy metals and their role in electron transfer processes of plants. In A. Lauchli & R. L. Bieleski (Eds.), *Encyclopedia of Plant Physiology; New Series, Vol. 15A* (pp. 563–596). Springer-Verlag.

Santi, S., & Schmidt, W. (2009). Dissecting iron deficiency-induced proton extrusion in *Arabidopsis* roots. *New Phytologist, 183*, 1072–1084.

Tian, G., Cox, A. E., Kumar, K., Granato, T. C., O'Connor, G. A., & Elliott, H. A. (2016). Assessment of plant availability and environmental risk of biosolids-phosphorus in a US Midwest corn-belt soil. *Journal of Environmental Management, 172*, 171–176.

Tsai, H. H., & Schmidt, W. (2017). Mobilization of iron by plant-borne coumarins. *Trends in Plant Science, 22*, 538–548.

Uchida, R. (2000). Essential nutrients for plant growth: Nutrient functions and deficiency symptoms. *Plant Nutrition Management: Hawaii's Soils, 4*, 31–55.

Vert, G., Grotz, N., Dédaldéchamp, F., Gaymard, F., Guerinot, M. L., Briat, J. F., & Curie, C. (2002). IRT1: An *Arabidopsis* transporter essential for iron uptake from the soil and for plant growth. *The Plant Cell, 14*, 1223–1233.

Wasaya, A., Affan, M., Yasir, T. A., Sheikh, G. R., Aziz, A., Baloach, A. W., Nawaz, F., & Adnan, M. (2019). Growth and economic return of maize (*Zea mays* L.) with foliar application of potassium sulfate under rainfed conditions. *Journal of Environmental Agriculture, 4*, 268–374.

Wenzel, W. W., Bunkowski, M., Puschenreiter, M., & Horak, O. (2003). Rhizosphere characteristics of indigenously growing nickel hyperaccumulator and excluder plants on serpentine soil. *Environmental Pollution, 123*, 131–138.

White, P. J., & Hammond, J. P. (2009). The sources of phosphorus in the waters of Great Britain. *Journal of Environmental Quality, 38*, 13–26.

Williams, R. R. (1991). Factors determining mineral uptake *in vitro*. *Acta Horticulturae, 289*, 165–166.

Zhai, Z. Y., Gayomba, S. R., Jung, H., Vimalakumari, N. K., Pineros, M., Craft, E., Danku, M. A. R. J., Lahner, B., Punshon, T., Guerinot, M. L., Salt, D. E., Kochian, L. V., & Vatamaniuk, O. K. (2014). OPT3 is a phloem-specific iron transporter that is essential for systemic iron signaling and redistribution of iron and cadmium in *Arabidopsis*. *Plant Cell, 26*, 2249–2264.

CHAPTER 5

Proteomics Applications in Cereal Crops for Food Security Under Increasing Abiotic Stresses

MUHAMMAD IRFAN SOHAIL,[1,2,*] NATASHA ERUM,[3] SHAFIQ UR REHMAN,[1] MUSTANSAR ASLAM,[1] FATIMA AKMAL,[4] MUHAMMAD SHOAIB,[5,6] and UMAIR RIAZ[7]

[1]*Department of Environmental Sciences, Faculty of Life Sciences, University of Okara, Pakistan*

[2]*High Tech Lab-II, Sub Campus Depalpur–Okara, University of Agriculture Faisalabad, Okara, Pakistan*

[3]*Department of Biochemistry, Faculty of Sciences, University of Agriculture, Faisalabad, Pakistan*

[4]*Agriculture Extension Wing, Department of Agriculture Extension, Kamalia, Punjab, Pakistan*

[5]*Institute of Molecular Biology and Biotechnology, Bahauddin Zakariya University, Multan, Pakistan*

[6]*State Key Laboratory of Plant Cell and Chromosome Engineering, Institute of Genetics and Development Biology/Innovative Academy of Seed Design, Chinese Academy of Sciences, Beijing, China*

[7]*Soil and Water Testing Laboratory for Research, Bahawalpur, Agriculture Department, Government of Punjab, Pakistan*

*Corresponding author.

Agricultural Biotechnology: Issues, Challenges, and Recent Developments. Mohd. Sayeed Akhtar and Khalid Rehman Hakeem (Eds.)
© 2025 Apple Academic Press, Inc. Co-published with CRC Press (Taylor & Francis)

ABSTRACT

Proteomics systematically analyzes all expressed proteins in a particular cell, tissue, or organism. It provides a protein profile of an organism at the cellular and tissue levels. Genes are expressed into proteins and implicated in phenotype profiling of crops, ultimately affecting the genomics and breeding scientist's decisions in turning the plant species more resilient and sustainable for crop production. Proteome characterization of cereal crops through proteomic approaches helps confer a novel ability to improve their production and minimize the impact of environmental factors most detrimental to ensuring food security. Proteomics has offered novel molecular markers to assist crop breeders. It also helps agriculture researchers to investigate abiotic stress tolerance mechanisms in various crop species. There has emerged a steady increase in using proteomics techniques in addressing the cereal crops growing under various stressed climatic environments. The increasing world population demands an increase in cereal crop production. Comprehensive proteomic approaches are required to elucidate the mechanism of plant responses to the environment and elevation of crop tolerance towards environmental factors. In this post-genomic era, integrating proteomics with agriculture will speed up crop yield, leading to global food security. This chapter provides a comprehensive overview of proteomics applications in agriculture, focusing on cereal crops for food security. It highlighted the current cereal production scenario across the globe, the state of food security, and also on the most recent advances reported on proteomic successes for the major cereal crops growing in hostile conditions.

5.1 INTRODUCTION

Sustainable development goals (SDGs) by the United Nations (UN) Organization convey a crucial, impressive, and challenging task to scientists and society to emphasize the need to transform the agriculture and the food sector to achieve food security, ecosystem sustainability, economic expansion, and social equity (UN, 2015). Enhanced agricultural production under changing climatic and social factors, food security has become essential to provide sufficient food to the world due to the rise in worldwide population, especially under the increasing hazardous effects of climate change on field crops (FAO, 2022). Among the various threats posed by the current severity of climate change, water scarcity, salinity, high temperature, and radiation are the alarming ones (Farooqi et al., 2020; Hashiguchi et al., 2010; Singh et al., 2022). It is estimated that by 2050, there will be a need to feed 9 billion people; as of now, over 850 million people are suffering from malnutrition (Sohail et al., 2022). Future agriculture necessitates customized solutions that embrace fundamental leaps forward in current knowledge and enabling technology and consider the need to safeguard the environment and meet societal demands (Halder et al., 2022). There is a need to develop crop varieties tolerant of adverse growing conditions, and it demands breakthroughs in plant biotechnology to cope with the abiotic stresses and yield gaps. It requires a complete understanding of the physiological and biochemical processes involved in stress responses (Hashiguchi et al., 2010).

Cereals are the staple food for most of the world's population. After meat and legumes, cereals are the third most important supplier of proteins in the human diet (Agregan et al., 2021), and in some less developed countries, cereals are the only source of protein intake (Sohail et al., 2022). Cereals make up more than 60% of energy intake by the global population. Now, cereal crops are subjected to various biotic and abiotic environmental constraints like drought, salinity, radiations, heavy metal toxicity, high/cold heat, and biotic stresses like pathogens under changing climate. Only the drought has reduced the wheat yield by 32% globally (Zhang et al., 2018), while salinity reduces 40% of cereal yields (Qadir et al., 2014). It is essential to study all proteins of cereal crops to determine their biological function and role in the regulation of crop yield, quality, and response to different stress factors (Rahiminejad et al., 2019). Proteins are directly involved in managing the physiological traits of crops to adapt the phenotype according to environmental factors; hence, played a crucial role in crop sustainability under climate severity (Sinha et al., 2022).

Plant proteomics is a promising approach for cereal crop improvement to enhance yield for the sake of managing food security. Proteomics significantly contributes to identifying functional proteins that participate in crop-environment interactions (Afroz et al., 2011; Sinha et al., 2022). Proteomics is a sub-discipline of the molecular biology of plant sciences become crucial in agricultural research. It is based on the Spatio-temporal analysis of proteins in any biological system (Gallardo et al., 2013; Pandey and Mann, 2000). The term "proteomics" was first introduced in 1995 (Okoli et al., 2021; Wasinger et al., 1995), representing the collation of protein and genome. The role of proteins is crucial in every biological system since proteins are active macromolecules and the center of an understanding of molecular biological mechanisms (Rahiminejad et al., 2019). The biological functions of proteins are determined by some prominent factors, including (i) localization of proteins in the cell; (ii) post-translational modifications (PTMs) of proteins; (iii) protein-protein interactions; and (iv) interaction of proteins with non-proteinaceous compounds. Proteome includes studying the structure, function, modifications, interaction, intracellular location, and quantifying the protein's abundance in a specific biological system (a cell or tissue). The biological system is reflected by its proteome, which is very dynamic and constantly changes in response to abiotic and biotic stresses. Above all, knowing all levels that regulate adaptation mechanisms and agricultural plant resilience in climate change is critical to making significant advances in genomics-driven breeding of main cereal crops for productivity improvement and stress tolerance (Hu et al., 2015). Proteomics is also assisting in the quality of cereal-based foods, that is, protein delivery to humans because cereal grains account for a substantial portion of plant proteins in the human diet.

Researchers have made advancements in developing proteomic technologies for application in agriculture to improve cereal crop yield worldwide. Researchers have generated many proteomic data sets using state-of-the-art technologies, for example, Two-Dimensional Polyacrylamide Gel Electrophoresis (2D PAGE) and mass spectrometry. Analyzing and interpreting the crop proteomic data by implicating the bioinformatics tools and statistical algorithms are equally important. The translation of basic science knowledge for resolving anthropogenic and climatic issues is referred to as "translational plant proteomics," and it focuses on the economic and recreational values of plants, food, and energy (Agregan et al., 2021). On the other hand, post-COVID crises have already irrupted the economic growth globally (Sohail et al., 2021b). Therefore, based on its efficiency and success, the field of proteomics has become a promising strategy in agricultural sciences with a particular focus on cereal crops for food security purposes. This chapter provides a comprehensive overview of proteomics applications in agriculture, focusing on cereal crops for food security. It highlighted the current cereal production scenario across the globe, the state of food security, and also on the most recent advances reported on proteomic successes for the major cereal crops growing in hostile conditions.

5.2 CEREAL PRODUCTION AND STRESS THREATS TO FOOD SECURITY: A GLOBAL PERSPECTIVE

Cereals (Family: Graminae) are those crops that are solely harvested for dry grains, that is, rice, maize, barley, wheat, buckwheat, rye, oats, millet, sorghum, quinoa, fonio, triticale, and canary seed (Wrigley, 2017). On a global scale, the five major cereal crops are wheat, maize, rice, sorghum, and barley, as per area under production, yield, and utility: which collectively account for more than half of the world's food (Li et al., 2019). Wheat and rye hold a distinctive role as they are most suitable for bread-making. Cereals and cereal products are a staple food for 70% world's population and provide an abundance of carbohydrates, proteins, vitamins, minerals, and trace elements (Jespersen et al., 2009; Shah et al., 2016; Shewry and Hay, 2015). Globally 2796 million tons of cereals were harvested in 2021, which is 0.7% higher than the previous year (FAO, 2020). The global wheat, coarse grain, and rice production forecast showed –0.1, +1.2, and +0.7 falls and up from 2020, respectively. Wheat is the most imperative food security crop, with a production of 775.4 million tons in 2021, followed by coarse grains (501 million tons) and rice (519.3 million tons) (FAO, 2022). A global total of about 700 million ha of cropping area is under cereals cultivation, covering almost 50% of the world's caloric intake (Singer et al., 2019).

During current population growth and adverse weather conditions, food security (i.e., adequate food access in quantity plus quality) is projected to be alarming in 44 countries as per the FAO quarterly global report of 2022 (FAO, 2022). Global hunger has shown a gradual upward trend since 2014. We have calculated the food security indices based on FAOSTAT, and data shows that the prevalence of moderate-to-severe food insecure people has encountered a 21.6% increase from 2014 to 2020. It is estimated that 1 in 10 persons is exposed to severe food insecurity, which provides another estimate to measure hunger. If this trend continues, then by 2030, more than 840 million people will be severely affected by hunger (Bongaarts, 2020). Reportedly, 690 million people are currently food insecure in 2019 (FAO, 2020). In low-income countries, cereal intake procures particular relevance as most of the population relies on cereals as the primary food and essential nutrient for total energy intake to meet WHO recommendations for daily consumption. Cereals contribute as a major source of energy and protein, that is, about 56% of global energy and 50% of the protein needs of the human populace either via direct consumption or indirectly via cereal-fed animals and the consumption of their product (Li et al., 2019; Papageorgiou and Skendi, 2018). Our data calculation also agrees with the claims that cereals have contributed to more than 50% of daily caloric energy intake globally. In some developing Asian and African countries like India Sir Lanka and so forth, the importance of cereal consumption gains special where the cereals are the monotonous source of protein supply due to religious limitations (Agregan et al., 2021; Sohail et al., 2022). Proteins are an essential element after carbohydrates, ranging from 7.50 to 16.68% of the grain. Proteins in plants are classed into glutelins, albumins, prolamins, and globulins and are predicated on their solubilization in different solvents (Araus et al., 2002), yet cereals have imbalanced amino acids (Trace-Caba et al., 2015). Cereals are essential in animal nutrition in addition to human usage. Around 60% of the energy intake by the livestock is fulfilled by corn, wheat, and barley. Only 15% of corn grain is utilized in human food directly (Agregan et al., 2021). Nonetheless, cereals are a predominant economic platform for most of the world's countries and a source of sustainable income for livelihood (Oishi, 2021).

On the other hand, climate change, including all associated biotic and abiotic stresses and soil degradation, has become the biggest challenge to food security (Hadebe et al., 2017; Tadesse et al., 2019; UN, 2019). This ever-increasing world population (approximately 60 to 110% by 2050) and current dietary structure are expected to cause a surge in global food demand and make it highly difficult to maintain the equilibrium between cereal consumption and production side by side (FAO, 2020). However, given the significance of cereal in the human diet, the only solution to this problem is to increase cereal production to meet our global food and nutrition security. Furthermore, cereal grain production, including major crops like maize, wheat, rice, barley, rye, and oats, is very prone to abiotic stresses, jeopardizing the sustainable supply of human and animal use (Begcy and Dresselhaus, 2018). Wheat production is already going down in most locations. Simulation studies showed that under the climate change scenario, each 1°C increase in global mean temperature might result in a 6% decrease in the crop yield. Without adaptation, wheat, maize, and rice yields are predicted to decrease in tropical and temperate regions (Reynolds et al., 2016). The yield improvement ascribed to breeding or technical innovation is inversely related to the stress in the growing environment. In Australia, for example, it has been 10 times lower (004 Mg ha^{-1} versus 4 Mg ha^{-1}) than in the UK. Therefore, under varying climatic conditions, yield increments reported as percentage values do not make a complete understanding (Araus et al., 2002).

Globally, crop loss prediction showed that abiotic stresses caused more reduction in yield than biotic stresses (Zamir et al., 2013). The magnitude of the impact of climate change (extremely high temperatures, water scarcity, and unpredictable rainfall patterns) is massive and is the direct challenge amongst all constraints (Muitire et al., 2021). In Africa alone, climate change caused an almost 20% reduction in the production of maize and wheat (Blackie et al., 2016). In Africa, extreme weather conditions curb the possible potential of 2022 cereal production, and the current estimate for

2021 showed a 3.9% reduction in cereal production over 2020 (FAO, 2022). Biotic stresses, that is, pests, diseases, and weeds, are also evolving with climate change; within some time, a new strain of pathogens and diseases arrives for which new crop varieties need to develop (Keno et al., 2018). Some pests and diseases native/non-significant in one region could prove extremely ravaging to crops of other regions. For example, fall armyworm is a minor pest of cereals in northern and southern America but caused 100% yield loss in some instances in Africa, where it only appeared in 2016 (Day et al., 2017). Likewise, diseases could cause significant yield loss if they appear during critical stages of plant development, that is, in a rice crop, a rice blast can lead to 100% yield loss (Fahad et al., 2018).

Over the last 50 years, the yield of cereal crops has increased steadily. An increase in the land under cultivation, canal irrigation, application of chemical fertilizers, and the development of high-yielding cultivars through traditional breeding approaches have resulted in this increase. However, there is evidence that wheat, rice, and maize yields have been holding steady in recent years, with global yearly gains of around 1% since the 1990s, a significant decrease from the 1960s, and a concerning trend in the current climate (FAO, 2017). Our examination of FAO data in suggests that cereals contribute to the gross production index number (GPIN) and total cereal contribution to the per capita gross production index number (TGPCIN). Though there has been a 12% increase in TGCPIN, to feed 9.8 billion people by 2050, the cereal food supply must be increased by 70–100% (Wang et al., 2018).

The demand and supply gap should be decreased to guarantee and secure present and future food security. It is only possible by improving the cereal production system, alleviating the crop losses by biotic and abiotic stresses, and upgrading genetic crop makeup via conventional or molecular breeding techniques and policy changes. Likewise, crop proteomics significantly improves cereal crops by keeping the ever-changing climatic conditions in focus. Today, in the world, very few crop species are used as a staple food, primarily the C_3 (rice, wheat, and barley) and C_4 cereals (maize, pearl millet, and sorghum). However, the richness of genetic resources is mainly overlooked in the C_4 cereals. To relieve the food problem, efforts are being made to engineer the C_3 crop species to have C_4 features, which can significantly increase the crop yield of the C_3 species (Ghatak et al., 2021). Plant proteomics, a model of action/system, responds to the current problem by using different approaches and attempts to provide adequate food supply by guaranteeing increased crop productivity and food security (Agregan et al., 2021).

5.3 SIGNIFICANCE OF PROTEOMICS PROTOCOLS FOR CEREALS OVER OTHER OMICS APPROACHES

Crop yield and abiotic stress tolerance are quantitative traits (He et al., 2018), controlled by the interaction of genetics, environment, and management. Collective utilization of genomics, transcriptomics, and proteomics approaches can determine the molecular mechanisms and pathways involved in crop stress tolerance. Proteomics is the comprehensive characterization of all the proteins in living tissue (Hondius et al., 2018). Identifying protein concoction, amount, isoforms, localization, and mutual interactions under physiological conditions and at a particular time is part of the proteomics study. Protein profiles represent variations in tissue and its cellular compartment's protein expression. Proteomics has a distinct benefit over typical omics approaches in that it may show PTMs and link the functional variation with crop production. To date, greater than 300 PTMs have been revealed (Rolland et al., 2019). Proteo-genomics is a relatively new field that merges genomics and proteomics as a multi-omics setting by employing mass spectrometry (MS) and high-throughput sequencing (HTS) technologies to aid in the annotation of genes and deciphering proteome complexity. A general proteomics workflow consists of three main steps: (i) protein extraction; (ii) identification and separation; and (iii) data integration.

Previously, the researcher focused on only genomics. Carrying out genome-wide investigations and the reverse genetic techniques in model plants, *Arabidopsis*, rice *Thellungiella salsuginea*, aided in exploring the underlying molecular pathways for stress tolerance (Dresselhaus and Huekelhoven, 2018; He et al., 2018). Even though genomics and transcriptomics have yielded a plethora of data about plant response to stress treatment, insignificant associations between the mRNA transcript levels and protein abundance are of serious concern (Pan et al., 2019) also reported zero correlation between the transcription and translation of differentially abundant proteins (DAPs), highlighting the need to research stress-responsive alterations at the proteome level at a specific time. Additionally, an integrated transcriptome and proteomic investigation found that 30% of proteins transcribed under drought stress were not in the root tissue (Wang et al., 2021). At the proteome level, systematic stress-responsive patterns in root and shoot tissues were observed, but the leaf and root transcriptomes displayed distinct responsive patterns under salinity and drought stress (Kumar et al., 2022). The genomic data alone cannot predict the spatio-temporal metabolic level (Kaur et al., 2014). Therefore, proteomics as a modern technology has benefits over other omics techniques (genomics or metabolomics; Halder et al., 2022).

5.4 ADVANCES IN PROTEOMICS PROTOCOLS FOR CEREAL IMPROVEMENT

Based on proteins expressed under stress conditions, proteomics can be categorized into three main classes: expression proteomics, structural proteomics, and functional proteomics (Okoli et al., 2021). In expression proteomics, both qualitative and quantitative expressions of proteins are studied under two variable conditions. For example, in agriculture, crops' normal and stressed/abnormal cells are studied to study the proteins responsible for the stress state. In other words, expression proteomics studies protein expression patterns in abnormal cells and is referred to as translational proteomics (Mirzaei et al., 2016; Pandey et al., 2000). "Bottom-up" approach is regarded as the most common approach in proteomic workflows in which the protein(s) under study is/are digested by proteases (e.g., trypsin) into smaller peptides which are further analyzed by mass spectroscopy (Aebersold and Mann, 2003). Another approach followed in proteomics workflow is the "top-down proteomics approach" in which the protein(s) digestion step is skipped, and proteins are fragmentized directly by ionization in a mass-spectrometer. It has become possible by the availability of advanced high-resolution mass spectrometers (Yates and Kelleher, 2013). In agriculture proteomics, bioanalytical, and bioinformatics tools have made their way to structural and functional studies of proteins. Analytical techniques, that is, 2D gel electrophoresis, matrix-assisted laser desorption/ionization (MALDI), mass spectrometry (MS), time-of-flight mass spectrometry (MALDI-TOF); and many bioinformatics tools are indispensable for proteomics study of cereal crops (Aebersold and Mann, 2003; Okoli et al., 2021).

Over the years in proteomics (the 2D-PAGE), technique has proven to be efficient and dependable in separating protein molecules ranging from several 100 to a few 1000 (Konig, 2020). It is a conventional method of separating proteins in biological samples and is used to compare two similar samples to study specific protein differences. It is the most widely adopted method in proteomics because of its high efficiency in resolving protein on the simplest apparatus. The history of proteomics is associated with the development of 2D gel electrophoresis (2-DGE) in the 1970s, which was the first method to display thousands of proteins simultaneously on a single gel (Klose, 1975; O'Farrell, 1975). The 2-DGE can separate complex samples in a single run and analyze hundreds of proteins in the same gel (Goez et al., 2018). In 2D-PAGE, proteins are separated based on their molecular mass and isoelectric point (pI), followed by visualizing after the staining with reagent (Miller et al., 2006). In the first step, called isoelectric focusing, the protein mixture is resolved based on its net charge. In the following step, proteins are separated based on differences in their molecular masses. The separated proteins are then visualized as spots excised, hydrolyzed by proteases to specific peptides, and ultimately identified by mass spectrometry.

Mass spectrometry is a state-of-the-art method of choice in proteomics for identifying and characterizing proteins in any biological system. In this technique, mass spectra of atoms of molecules making up the sample material are produced, subjected to the determination of the elemental or isotopic signature of that sample, followed by elucidation of the chemical structure of molecules (e.g., peptides). MS provides a protein/peptide map which is then used to search a sequence database for an unidentified protein under study. The removal of the spot and subsequent examination by MS helps to identify the unique proteins making up the spot (Schulze and Usadel, 2010). Before the MS analysis, proteolytic enzymes (endoproteases/endopeptidases) are used to digest the spot. This phase is critical in the bottom-up strategy and differs from the top-down technique in that it allows for broadening the scope of the proteome investigation. Trypsin is the most used protease due to its low cost, ability to produce peptides suited for MS, high specificity, and accounting for 96% of research studies using a bottom-up strategy. Lys C is the second most commonly used protease, followed by pepsin, chymotrypsin, and Glu-C (Wu et al., 2018). Because of short peptides containing Arg or Lys at the C-terminus and quick trypsin digestion facilitated separation by contemporary liquid chromatographic procedures (LC), fragmentation, and identification using search algorithms (Tsiatsiani and Heck, 2015). However, covering the complete proteome is difficult because of the tiny size of the trypsin zed proteins (5 and 20 aa; Dupree et al., 2020). The peptidesionnization is accomplished by electrospray ionization (ESI) and MALDI. MALDI-TOF/MS technique uses a laser beam for ionizing protein samples. MALDI faces fewer interferences induced by popular proteomics buffers (e.g., urea or tris), and thus the production of charged ionic species increases, making MS spectra easier to interpret. On the other hand, ESI encourages the formation of multiple charged ions (Cunsolo et al., 2014). Electron-capture dissociation (ECD), Electron-transfer dissociation (ETD), and collision-induced dissociation (CID) can all be used to dissociate peptides (Schulze and Usadel, 2010).

Liquid phase separation by capillary electrophoresis (CE) or liquid chromatography (LC) is advantageous over the 2-DEG separation with greater dynamic range, higher sensitivity, speed, and ease of automation. The strongly basic/acidic protein of low-high molecular weight can be isolated by reserves phase, ion exclusion, and size exclusion (O'Donnell et al., 2004). For the peptide separation through the one-phase LC, reverse phasing is the most widely used mechanism. The strong cation exchange (SCX) and reverse phase (RP) columns are the recommended multidimensional separation combinations (O'Donnell et al., 2004). The mass analyzers include time-of-flight (TOF), quadrupole (Q), Fourier transform ion cyclotron resonance (FTICR), and ion trap (IT). Furthermore, mass-to-charge separation by utilizing a differential ion motion can be achieved through tandem (MS/MS). After enzymatic digestion of proteins, coupling MS/MS with LC technology (LC-MS/MS) is an HTS technique extensively employed for complicated samples (Weisser and Choudhary, 2017). Shotgun proteomics is a new approach replacing the 2-DGE with a mass spectrometer for proteome analysis (Gao et al., 2012; Marcus et al., 2020). In the shotgun approach for the relative quantitative analysis, two strategies are available that is, label-free quantification (LFQ) and labeled (Bubis et al., 2017). In the label-free strategy, tandem mass tags (TMT) stable isotope labeling with amino acids in cell culture (SILAC), dimethyl labeling, isotope affinity tagging (ICAT), and isobaric tags for relative absolute quantification (iTRAQ), can be used to modify the isotope containing samples. The approaches for label-free quantification rely on the ion intensity and spectral counts (Bubis et al., 2017).

Targeted proteomics, a new proteomic strategy focused on identifying selected proteins, gives the help of dealing with a reduced portion of the proteome rather than quantifying several proteins with no preceding information (Borràs and Sabidó, 2017). This method differs from the shotgun method. Instead of searching for candidate biomarkers to employ in detecting an event, that is, any stress – a pre-recognized biomarker is deployed to detect such an event. Instead of unbiasedly profiling and comparing those to determine the abundance of differential proteins, targeted proteomics aims at quantifying a single or a cluster of desired proteins. As a result, we can assess these protein biomarkers with great precision and repeatability in many samples (Arora and Somasundaram, 2019). Three forms of focused proteomic approaches include selected reaction monitoring (SRM), multiple reaction monitoring (MRM), parallel reaction monitoring (PRM), and data-independent acquisition (DIA), as well as tailored data extraction of the MS/MS spectra (Shi et al., 2016).

5.5 PROTEOME MAPS/DATABASES OF MAJOR CEREAL CROPS

Along with the Arabidopsis database, big proteomics data is also emerging in cereal research. Cereals being the field crop species with a large genomic database, it took a long time to establish the proteomic database by collecting and summarizing the expression data from different crop tissues and various developmental stages and stress conditions. On the other hand, bioinformatics analysis of protein sequence data is a powerful strategy for protein identification and their structural and functional studies. It has significantly improved the standard of proteomics applications in cereals. It has dramatically facilitated the management and interrogation of the vast "omics" technologies generated data such as genome, transcriptome, and proteome (Agarwal et a., 2013). Technological advancements in gel-based (particularly 2DE in conjunction with IPG and DIGE) and gel-free approaches, particularly MS analysis and bioinformatics resources in plant proteomics (Kim et al., 2014), have had a substantial influence on the knowledge of stress tolerance and genetic potential. Further developments to improve proteome databases can be achieved by (i) regular updates; (ii) improvement in data quantity and quality; (iii) integration of different types of data, such as genome, epigenome, proteome, transcriptome, and phenome; and (iv) integration of databases with crops yield-related literature.

Proteome analysis in plant systems has been performed utilizing a variety of tissue samples, including leaves, roots, flowers, and fruits (Feng et al., 2017; Szymanski et al., 2017). Many databases have related to omics output, and their interaction has been published and available. There are a large number of protein databases, that is, Model Organism Protein Expression Database (htttp://moped.proteinspire.org), MaxQB (http://www.uniprot.org/database/DB-0186), Peptide Atlas (http://www.peptideatlas.org/), Global Proteome Machine Database (http://gpmdb.thegpm.org/), and PRIDE database (https://www.ebi.ac.uk/pride/archive/), can be explored to get relevant information (Perez-Riverol et al., 2015). The whole genome of the sequenced crop plants can be accessed from the generalized database that is, NCBI, Viridiplantae, DDBJ, GenBank, and UniProt; moreover, the plant-specific databases are also available that is, PlantGDB and Ensable Plants (https://www.plantgdb.org/).

Compared to the human complement, the model plant *Arabidopsis* has two-fold more protein kinases (Zulawski et al., 2014). Undoubtedly, Arabidopsis-based omics data has been the backbone of cereal proteomics. The number of kinases (1502) and phosphatases (162) (enzymes that catalyze the high-energy, phosphate-donating molecules to specific substrates-phosphorylation) is still growing (Wang et al., 2014), facilitating to identify the functions of signaling molecules in a context-dependent way. Previously, Mitsuda and Ohme-Takagi (2009) have compared the available plant transcription factor databases. The plant TFDB contains the transcriptional factors from more than 80 plant species (Jin et al., 2014), the Arabidopsis Transcriptional Regulatory Map (ATRM) and Arabidopsis Transcription factor database (DATF; Guo et al., 2005) are a subset of to plant TFDB. The Subcellular location of proteins in the Arabidopsis (SUBA) database collects protein localization data from large-scale organellar proteomics and green fluorescence protein investigations (Hooper et al., 2017). The collection of plant-specific activities is defined by the *in vivo* interactions between protein scans. Plant-specific activities are defined by the *in vivo* interactions between proteins. The Arabidopsis Interactome Mapping Consortium (AIMC) has cataloged 6200 binary interactions between 2700 proteins in the Molecular Interaction Database (IntAct; Consortium, 2011; Ngara et al., 2021; Orchard et al., 2014).

Maize (*Zea mays* L.) is the new world's most significant agro-economic crop because of its raw material application in manufacturing food, feed, and biofuel for animals and humans (Urrutia et al., 2021). Nevertheless, the post-transcriptional and post-translation modifications (Luo et al., 2018) result in a near zero correlation between the mRNA levels and the protein (Zenda et al., 2018). Although various genome databases are available for maize genomics, there are fewer publicly available proteome databases for maize. The protein-protein interaction database (PPIM; http://comp-sysbio.org/ppim), was created for maize (Zhu et al., 2011, 2016), PPDB (plant proteomics database) (http://ppdb.tc.cornell.edu; Sun et al., 2009), ProMEX (http://ProMEX.pph.univie.ac.at/ProMEX/; Wienkoop et al., 2012), ProFITS (http://bioinfo.cau.edu.cn/ProFITS; Ling et al., 2010) and PlantMWpIDB (https://plantmwpidb.com/; Mohanta et al., 2022). PPIM is a user-friendly, freely available database covering 2,762,560 interactions among 14,000 different proteins of maize (Zhu et al., 2016). PPDB, launched in 2004, is an integrated database for experimentally identified proteins in maize and *Arabidopsis* (Sun et al., 2009). ProMEX is a reference database for plant proteomics (including maize proteome) comprising mass spectra generated by mass spectrometry (Wienkoop et al., 2012). PlantMWpIDB contains proteome datasets of 342 different plants, including maize.

Similarly, sorghum and barley genomics resources are abundantly available, but proteomics resources are comparatively limited and have been extensively reviewed by Ngara et al. (2021). Few proteomic-based databases are CropPal (https://crop-pal.org) for protein subcellular locations (Hooper et al., 2020), ProtAnnDB (http://www.polebio.lrsv.ups-tlse.fr/ProtAnnDB/) for protein annotation (Clemente et al., 2009), ExPASy (https://www.expasy.org/) a bioinformatics resource for proteins (Artimo et al., 2012) and Uniprot (https://www.uniprot.org) for protein sequence and functional information. Barley ranks 4th among cereals globally and is widely utilized as a feed and as a basic element in making beer. The protein content of whole barley grain ranges between 10 and 17%. Barley is a diploid plant with a 5.1 Gb genome. Barley whole-genomic sequencing has been accomplished (Silvar et al., 2015). Around 26,159 high-confidence genes homologous to other plant genomes and 53,220 low-confidence genes without homology were reported. This indicated that barley has around 26,100 open reading frames, laying the groundwork for barley genomics and proteomics research. Proteome research of the aleurone layer in barley has shown plasma membrane protein identification and targeted investigation of germination-related alterations, including the thioredoxin system (Ngara et al., 2021).

Rice (*Oryza sativa* L.) is one of the most important cereal crops worldwide; almost half of the world's population consumes rice as its primary food source (Mahender et al., 2016). Improving rice yield through rice breeding is crucial not only to feed the rising global population but also because its cultivation is a source of livelihood for many farmers around the globe (Swamy et al., 2016; Tanaka et al., 2016). Rice yield is associated with traits like grain weight, size, and panicle number. Rice breeders and researchers have explored several genes associated with these traits in an extensive collection of rice seeds (Jiang et al., 2018). Many genomics databases of rice have previously been constructed by researchers in different rice genome sequencing projects, such as RGKbase (Wang et al., 2013), RAD (Ito et al., 2005), RMD (Zhang et al., 2006), JCVI (Ouyang et al., 2007), RAP-DB (Rice Annotation Project Database) and CRDC (China Rice Data Center) database. Although these databases are loaded with a vast quantity of data, they lack the integration of genomics with other "omics" of rice, especially proteomics. Integrating genome databases with proteome and other related databases and literature can serve better in facilitating rice researchers and breeders to improve traits for enhanced rice yield. Similarly, the computational system biology database collects plant protein interaction databases such as PPIM. The "*Oryza sativa interacting proteins (DIPOS) database*" is solely devoted to protein interactions in rice. The rice proteome database of functional analysis of rice (https://www.hsls.pitt.edu/obrc/index.php?page=URL1152242630) is based on 2D-PAGE analysis of rice proteins from different tissues and cell organelles. It contains 23 reference maps comprised of 13,129 identified proteins, out of which 5092 proteins are entered with their amino acid sequence (Komatsu, 2005). RicyerDB, a Rice Yield-related Database, focuses on a thorough understanding of rice biology and can be helpful in a comprehensive understanding of improved rice production. It is a combined database based on genomics and proteomics of rice. It also provides a protein-protein interaction network with biological analyses (Jiang et al., 2018).

Numerous rice proteome datasets are not publicly available. For example, PlantPReS (Mousavi et al., 2016) has more than 30 MS (mass spectrometry-based) datasets of rice proteomes, out of which only one is available on PRIDE, a public database (Rahiminejad et al., 2019). The "*ProteomicsID Entifications*" (PRIDE) database is an open, user-friendly proteomics data repository. Data generated by MS-proteome experiments, including raw spectral data, protein identifications, peptides, and associated statistics, can accessed using one centralized web interface that is independent of the hardware or algorithms. Proteomics resources and tools for functional analysis such as PhosphoRice: a meta-predictor of rice-specific phosphorylation sites (http://bioinformatics.fafu.edu.Cn/PhosphoRice) have been developed (Kim et al., 2014), Oryza PG-DB: rice proteome database based on shotgun proteogenomics (http://oryzapg.iab.keio.ac.jp/), and PRIN: a predicted rice interactome network (http://bis.zju.edu.cn/prin/; Que et al., 2012). The advancement in bioinformatics will help identify an individual protein's structure and function, predicting its regulatory network in rice and other crop species.

Wheat is the most widely farmed staple cereal across various soils and climate zones, with a 620 million tons annual production. The wheat genome is complicated and extensive (17 Gb; Shi and Ling, 2018). The first edition of the genomic sequences of wheat was released by the International Wheat Genome Sequencing Consortium (IWGSC; https://www.wheatgenome.org/; Keeble-Gagnère et al., 2018), which delivers an indispensable source of knowledge for wheat researchers; however, the gene annotation and sequence assembly remain challenging tasks. Various algorithms based on defined characteristics and machine learning (ML) have been utilized to predict the subcellular location based explicitly on the N-terminal region of proteins that can contain pre-sequenced targeting information (Emanuelsson et al., 2007). Several other web-based programs are also available to predict chloroplast localization, that is, TargetP (Kamal et al., 2013), ChlroP (Emanuelsson et al., 1999), WolF PSORT (Horton et al., 2007), and PSORT (Nakai, 1999). In

WolF PSORT, the query sequence is aligned to the 14 nearest neighbors, and an integrated link to UniProt and Gene Ontology is provided. The PSORT program was used to identify chloroplast thylakoid space (lumen) because WolF PSORT is not able to predict the sub-organelle location (Kamal et al., 2013).

Wheat is a hexaploid plant with three sub-genomes. The protein sequences from each genome are similar, yet few variations are present. Because of the high proportions of proline and glutamine and low proportions of lysine and arginine, the MS analysis faces problems like cleaving the proteins. Yet, wheat offers isogenic lines, which assist in locating the proteins on a particular chromosome region. Many researchers have analyzed the wheat grain during developmental stages at maturity to discriminate among different wheat cultivars (Clarke et al., 2000; Cornish et al., 2001; Wong et al., 2004). Proteomics has been utilized to investigate the effect of nitrogen and sulfur on gluten proteins (Andon et al., 2002; Grove et al., 2009) explained the characterization of the wheat amyloplast proteome. According to Gobaa (2008), the replacement of wheat chromosome 1BS with rye chromosome 1RS (1BL.1RS translocation) resulted in significant changes in endosperm proteome, also a quantitative and qualitative influence on prolamin production. Akagawa et al. (2007) performed a functional proteome study in wheat, revealing wheat allergens and nine components of low molecular weight glutenins prevalent with IgE-binding proteins. More than 249 transcripts are reported to be highly expressive during the first two weeks of grain development enabling the unraveling mechanism of grain formation (Nadaud et al., 2010).

5.6 RECENT PROTEOMIC DEVELOPMENTS IN CEREAL CROPS GROWING UNDER ABIOTIC STRESSES

The application of proteomics in agriculture is a highly active area of research leading to the identification of various proteins expressed in cereal crops under abiotic stress (drought, salinity, temperature, heavy metals, nanoparticles toxicities, etc.), and the generation of an enormous amount of data and success for stress mitigations strategies (Afroz et al., 2011; Halder et al., 2022; Sohail et al., 2020, 2021a). In the recent past, the cultivable land has been decreasing, thus building pressure on the plant breeders to develop crop varieties with higher yields per unit area even under varying climatic conditions. In order to feed the world population, there is a need to discover efficient pathways for converting sunlight into ATPs under stress conditions (Driever and Kromdijk, 2013). Comparative analyses of the plant chloroplasts at the proteomic level (Zhao et al., 2013) explored the key components that effectively and efficiently carry out the light and CO_2 conversion (Manandhar-Shrestha et al., 2013) and the stress mitigation through antioxidant systems activations. Stressful conditions often lead to several biochemical malfunctions, resulting in decreased crop yield addressed by translational proteomics. The organ-specific proteomic study is another idea to identify the proteins commonly accumulated during abiotic stress (Komatsu and Hossain 2013). The intracellular compartments are critical in plant stress response (Nouri and Komatsu, 2013) reported that several sub-cellular localized proteins like proteins related to signaling, transcriptional regulation, ion/water transporters, and reactive oxygen species (ROS) scavengers are involved in stress tolerance (Kosova et al., 2018; Mitsui et al., 2013). Numerous studies have been reported on post-translational changes during stress in various crops. Yet, upcoming literature presents a general overview of translation changes in major cereals crops (Ghatak et al., 2021).

Despite the abundance of proteomics literature, protein validation still has challenges (Gong et al., 2015; Pan et al., 2019). In rice, for example, protein annotation of only 1% of genes has been experimentally verified (Rhee and Mutwil, 2014). Similarly, the maize genomic sequence has roughly 40,000 genes, however, nothing is known about the function of the majority of them. Furthermore, there is a high level of reputation and/or redundancy particularly in the NCBI. The NCBI database contains many uncharacterized stress-sensitive maize proteins. As a result, experimental validation of stress proteins with a specific role in stress tolerance is critical for bridging the gap between the proteomic discovery of stress proteins and the selection of potential target proteins for future crop enhancements (Wang et al., 2019).

Plants' stationary life has formed a sophisticated gridded antioxidant defense system comprised of several enzyme components, which play an essential role in resisting diverse stress circumstances (Kosova et al., 2018). Several antioxidant enzymes assist plants in dealing with stress, for example, glutathione peroxidase (GPX), superoxide dismutase (SOD), ascorbate peroxidase (APX), peroxidase (POX), glutathione reductase (GR), glutathione S-transferases (GST), monodehydroascorbate reductase (MDHAR), dehydroascorbate reductase (DHAR) and catalase (CAT; Farooqi et al., 2020). During stresses, SOD catalyzes the removal of O_2^- and other ROS species by dismutating them into O_2 and H_2O_2. The CAT disaccociate the H_2O_2 into H_2O and O_2. The POX scavenges H_2O_2 in the extracellular area.

TABLE 5.1 Application of Recent Proteomic Approaches in Major Cereals (Wheat, Maize, Rice, and Soybean) Growing Under Various Abiotic Stressed Environments (Drought, Salinity, and Heat) and Major Advances in Understanding

Crop/Cultivar	Plant Organ	Stress	Technique	Major Results	References
				Wheat	
T. aestivum L. cv. Scepter	Roots	Salinity	LC-MS, Q-TOF	Downregulation of ATP synthase subunits, glycolytic enzymes, TCA enzymes, and translation-related proteins.	Dissanayake et al. (2022)
T. aestivum L. cv. Zhongmai 175	Seedling leaves	Salinity	LC-MS/MS	Upregulation of 194 chloroplast-localized DAPs involved in the Calvin cycle, transcription, translation, and metabolism of amino acids, carbon, and nitrogen.	Zhu et al. (2021)
T. aestivum L. cv. Chinese spring	Seeds (endosperms)	Salinity	LC-ESI-MS/MS	Upregulation of 207 DEPs (differentially expressed proteins) involved in the metabolism of proteins, amino acids, and organic acids.	Yan et al. (2021)
T. aestivum L. cv. BWL4444	Grains	Heat	2-DE, MALDI-TOF-MS/MS	Translational (gliadins, and glutenins) were upregulated while photosynthesis, glycolysis, and high molecular weight glutenins were downregulated.	Chunduri et al. (2021)
T. aestivum L. AABBDD cv. Qingmai	Seedlings	Salinity	iTRAQ (isobaric tags for relative and absolute quantitation).	Upregulation of ribosomal proteins, nucleoside diphosphate kinases, transaldolases, beta-glucosidases, phosphoenolpyruvate carboxylases, superoxide dismutases, and 6-phosphogluconate dehydrogenases.	Ma et al. (2020)
T. aestivum L. cv. Chinese Spring	Seeds	Salinity	SDS-PAGE, LC-ESI-MS/MS	Upregulation of 133 DAPs (differentially accumulated proteins) and downregulation of 264 proteins.	Yan et al. (2020)
T. aestivum L. cv. HD2985 and HD2329	Pooled sample (leaves, stems, spikes)	Heat	SDS-PAGE, iTRAQ, LC-MS/MS	Upregulation of 3600 proteins and downregulation of 5825 proteins; upregulation of carboxylase and α/β-amylases.	Kumar et al. (2019)
T. aestivum L. cv. Bobwhite	Roots	Salinity	LC-MS/MS	Detection of 69 alkaline stress-response proteins; upregulation of five superoxide dismutases, three malate dehydrogenases, three dehydrin proteins, and one V-ATPase protein; downregulation of two malic enzymes and many TCA cycle enzymes.	Han et al. (2019)

TABLE 5.1 *(Continued)*

Crop/Cultivar	Plant Organ	Stress	Technique	Major Results	References
T. aestivum L. cv. Gaocheng 8901	Grains	Heat	iTRAQ, LC-MS/MS	Upregulation of 207 differentially expressed proteins associated with energy metabolism, growth, and ROS.	Zhang et al. (2018a)
T. aestivum L. cv. Triso	Leaves	Heat + elevated CO_2	LC-MS/MS	Significant downregulation of proteins involved in photosynthesis, protein synthesis pathways, and antioxidant activities.	Zhang et al. (2018b)
T. aestivum L. cv. Chinese spring	Grain bearing spikes	Heat	SDS-PAGE, LC-MS	Upregulation of heat and ROS response proteins; downregulation of protein synthesis.	Wang et al. (2018)
T. aestivum L. cv. Yumai34	Seedling leaves	Drought	SDS-PAGE, nano-LC-MS/MS, iTRAQ	Upregulation of carbon metabolism and protein synthesis related proteins and photosynthesis and downregulation of signal transduction proteins.	Ding et al. (2018)
T. aestivum L. cv. Shaanhe 6, Zhengyin 1	Seedling leaves	Drought	SDS-PAGE, LC-MS/MS, Western blotting	Upregulation of differentially expressed LEA (late embryogenesis abundant) proteins.	Li et al. (2018)
T. aestivum L. cv. Kavir, Bahar	Seedling leaves	Drought	2D-PAGE, LC-MS/MS	Downregulation of ADP-glucose pyrophosphate, GST, glyoxalase enzymes, and phosphoribulokinase in Bahar cultivar; downregulation of soluble inorganic phosphatase in both cultivars (Kavir and Bahar).	Michaletti et al. (2018)
			Maize		
Z. mays L. wild type: Vp5	Leaves	Drought	iTRAQ, LC-ESI MS/MS	Na^+/H^+ antiporter phosphoenolpyruvate carboxylase, Arginine serine-rich splicing factor rsp41, Heterogeneous nuclear ribonucleoprotein r-like, and Ubiquitin ligase protein cop one were upregulated.	Hu et al. (2015a)
Z. mays L. cv. Zhengdan 958	Leaves	Drought	iTRAQ, LC-MS/MS	Upregulation of Phosphoenolpyruvate carboxykinase, Phosphoenolpyruvate carboxylase/E9NQE1, Phospholipase c/B4FY17, Probable-trehalose-phosphate synthase/K7V0H0, Serine threonine-kinase wnk4-like/K7TZQ1 and RAB17 protein/A3KLI0.	Hu et al. (2015b)

TABLE 5.1 *(Continued)*

Crop/Cultivar	Plant Organ	Stress	Technique	Major Results	References
Z. mays L. Inbred line B104	Leaves	Drought	Nano-LC-MS/MS	Upregulation of adenosyl homocysteine, Exocyst complex (SEC5A), histidine decarboxylase, GRIP domain-containing, and RANBP2-like protein.	Vu et al. (2016)
Z. mays L. wild type: Vp5	Leaves	Drought	iTRAQ, LC-ESI MS/MS	Upregulation of 4F5 protein family, Peroxidase, Photosystem I-subunit XI, photosystem. II-L, peroxidase, and Ribose-phosphate pyrophosphokinase 4.	Zhao et al. (2016a)
Z. mays L. cv. Zhengdan	Leaves	Drought	iTRAQ, LC-ESI MS/MS	Upregulation of ABA-responsive protein/K7TFB6, asparagine synthetase, dehydrin/C4J477, RAB17 protein/A3KLI0, Stachyose synthase, and photosystem I-subunit V/B4G1K9.	Zhao et al. (2016b)
				Soybean	
Glycine max L. cv. Yunhi	Germinated seeds	Salinity	iTRAQ	Upregulation of 201 DAPs involved in 20 metabolic pathways, including antioxidant proteins and proteins involved in maintaining protein metabolism.	Yin et al. (2018)
Glycine max L. cv. Enrei	Seedling leaves and roots	Drought	Nano-LC-MS/MS	Upregulation of delta-1-pyrroline-5-carboxylate synthase, β-glucosidase 31, and β-amylase 5; downregulation of proteins of TCA (tricarboxylic acid).	Wang et al. (2017)
Glycine max L.	Leaves	Drought	2-DE, LCMS/MS	Upregulation of 174 DEPs; 38.5% involved in photosynthesis and energy metabolism, 36.4% involved in defense, and 25.2% in protein metabolism.	Yu et al. (2016)
Glycine max L. cv. Enrei	Roots	Drought	Nano-LC-MS/MS	Upregulation of peroxidases and aldehyde dehydrogenases.	Khan and Komatsu (2016)
Glycine max L. cv. Surge and Davison	Leaves	Heat	2D-DIGE and MALDI-TOF-TOF/MS	Upregulation of 25 DEPs involved in photosynthesis: upregulation of heat-stress induced EF-Tu protein.	Das et al. (2016)

TABLE 5.1 *(Continued)*

Crop/Cultivar	Plant Organ	Stress	Technique	Major Results	References
				Rice	
O. sativa L. cv. IR29, FL478	Leaves	Salinity	2D-blue native/ SDS-PAGE, MS	Downregulation of proteins involved in photosynthesis electron transport and Calvin cycle in IR29 cultivar.	Hashemi et al. (2016)
O. sativa L. cv. IR29, FL478	Leaves and roots	Salinity	2-DE, MS/MS	Upregulation of ROS (reactive oxygen species) scavenging proteins in both cultivars; up-regulation of proteins involved in amino acid and polyamine biosynthesis in roots of IR29 cultivar.	Hosseini et al. (2015)
O. sativa L. IR64 (sensitive type and tolerant mutant line)	Germinated seedlings	Salinity	2DE, MALDI-TOF/ TOF, MS/MS	Identification of 34 new proteins for photosynthesis, defense to ROS, protein synthesis and signal transduction; upregulation of ferredoxin; downregulation of GST, fructokinase-2, and glutamine synthetase.	Ghaffari et al. (2014)
O. sativa L. cv. IR64, Cheriviruppu	Seedling flag leaves	Salinity	2-DE, MALDI-TOF/ TOF	Identification of 18 proteins involved in plant adaption to salt stress, carbohydrate/energy metabolism, protein synthesis, and assembly, anther wall remodeling and metabolism: upregulation of three isoforms of fructokinase-2 in Cheriviruppu cultivar.	Sarhadi et al. (2012)

Plant GPX catalyzes the H_2O_2 and HO_2^- to H_2O and lipid alcohols. To decrease the glutathione (GSH; monomeric and glutathione GSSG; dimeric), GR catalyzes the reduction process. To convert H_2O_2 to water, APX employs ascorbate as a particular electron donor.

5.7 PROTEOMIC SUCCESS IN ELUCIDATING PROTEIN RESPONSES TO ABIOTIC STRESS MITIGATION

All abiotic stresses are detrimental to plant metabolism producing ROS, and the subsequent activation of ROS mitigation through proteins (enzymatic and non-enzymatic) pathways share a common pathway (Kosova et al., 2018). Protein isoforms and PTMs frequently differ in molecular weight or isoelectric point, distinguishing them by 1-DE and 2-DE. Temperature affects the enzymatic activity of the protein; as a result, under high and low-temperature stress, various isoforms of a specific protein have been discovered through 2-DGE (Kosova et al., 2018). Currently, over 300 different types of protein PTMs have been described by Wu et al. (2016). Mock and Dietz (2016) have explored the possibilities of PTM alteration by reactive oxygen (ROS), nitrogen (RNS), carbonyl (RCS), and sulfur (RSS). The recently adopted techniques and outcomes of proteomic approaches in major cereal crops growing under various abiotic stressed conditions have also been summarized in (Table 5.1).

Zhang et al. (2018) utilized the iTRAQ and found 207 differentially expressed proteins (DEPS) associated with growth and development, energy metabolism, and stress response. Pan et al. (2019) recently reported 4560 proteins identified during hypoxia-stressed wheat growth. Among total proteins, the DEPs were 361, 640, 788 in S and 33, 207, 279 in T in consecutive three days, respectively. These DEPs were 270 common proteins. Some waterlogging stress-related proteins, such as S-adenosylmethionine synthetase, acid phosphatase, and oxidant-protecting enzyme, were considerably exclusive. Similarly, Yan et al. (2020) found a total of 2273 high-confidence proteins, of which 234, 207, and 209 were identified as DEPs under drought, salinity, and submergence stress, respectively. Functional classification revealed that the DEPs were mainly involved in amino acid and organic acid metabolic processes in all stress treatments. Takahashi et al. (2013) investigated heat stress responses. They discovered that the cell membrane plays a crucial role in signal perception and cellular homeostasis, thus considered a critical determinant in regulating the plant response to environmental stress.

Abou-Deif et al. (2019) also found activation of 59 protein spots (4.3%) in maize seedlings growing at 45°C for 2 hr compared to control, while 2-DE analysis of heat shock proteins (HSPs) for treated plants resulted in activation of 104 protein spots (7.7%). Similarly, Luo et al. (2018) also presented those 513 differentially regulated proteins (DRPs) activated in maize seedlings growing under salinity stress. The iTRAQ-based method further revealed that key DRPs, were glucose-6-phosphate 1-dehydrogenase, NADPH-producing dehydrogenase, glutamate synthase, and glutamine synthetase under salt stress conditions.

Salt stress reduces crop yields, yet several crop varieties exhibit remarkable resilience to the detrimental effects of salt. In essential cereal crops, many salt-responsive proteins have been found. Spring barley grew under 30% and 35% soil water capacity and revealed differential protein expression. Most of the proteins were related to energy metabolism and degradation (Vitamvas et al., 2015). Proteomic analysis of the chloroplastic proteins in wheat, maize, rice, and other herbaceous plants has been conducted to screen the transcripts responsive to abiotic stress (Gayen et al., 2019; Kamal et al., 2013, 2016; Xu et al., 2016; Zorb et al., 2009; Zhu et al., 2021). Wang et al. (2014) developed the wheat cytoplasmic male sterile line using a chemical hybridization agent (CHA)-SQ-1. Proteomics analysis revealed that 71 mitochondrial proteins are involved in pollen abortion. To identify variations between mature and germinating maize pollen, a differential gel-based proteomics method was used by Zhu et al. (2011). However, Clark et al. (2014) reported the turnover rates of 508 distinct proteins in barley varied by more than 100-fold. Numerous studies have confirmed and explored the miraculous plethora of proteins in cereal crops (Table 5.1), which can open a new window for the generation of climate-resilient crops.

5.8 CONCLUSION

Proteomics research has progressed to the "next generation," encompassing the post-translational alterations and interaction partners for the distinct functional proteins. Challenges persist, and future advances with molecular techniques and apparatus along with bioinformatics are projected to ultimately allow for the analysis, characterization, and quantification of an entire protein component of a single cell. A complete genetic/genomic database for key grain crops is now accessible. Yet, using MS analysis, 2D gel electrophoresis, and other modern tools, continued development in ongoing proto-genomics investigations is required. Experimental validation of stress proteins with specific roles in stress tolerance is critical for bridging the gap between the proteomic discovery of stress proteins and the selection of potential target proteins for future crop enhancements.. Although proteomics workflows have been successfully implemented in agriculture to improve cereal crop quality and yield, high degrees of maturity are required.

KEYWORDS

- **biotechnology**
- **chloroplastic proteins**
- **crop stress**
- **enzymes**
- **food security**
- **next generation**
- **pollen abortion**
- **proteomics**

REFERENCES

Abou-Deif, M. H., Rashed, M. A. S., Khalil, K. M., & Mahmoud, F. E. S. (2019). Proteomic analysis of heat shock proteins in maize (*Zea mays* L.). *Bulletin of the National Research Centre, 43*, 1–9. https://doi.org/10.1186/s42269-019-0251-2.

Aebersold, R., & Mann, M. (2003). Mass spectrometry-based proteomics. *Nature, 422*, 198–207.

Afroz, A., Ali, G. M., Mir, A., & Komatsu, S. (2011). Application of proteomics to investigate stress-induced proteins for improvement in crop protection. *Plant Cell Reports, 30*, 745–763.

Agrawal, G. K., Sarkar, A., Righetti, P. G., Pedreschi, R., Carpentier, S., Wang, T., Barkla, B. J., Kohli, A., Ndimba, B. K., Bykova, N. V., Rampitsch, C., Zolla, L., Rafudeen, M. S., Cramer, R., Bindschedler, L. V., Tsakirpaloglou, N., Ndimba, R. J., Farrant, J. M., Renaut, J., Job, D., Kikuchi, S., & Rakwal, R. (2013). A decade of plant proteomics and mass spectrometry: Translation of technical advancements to food security and safety issues. *Mass Spectrometry Reviews, 32*, 335–365.

Agregán, R., Echegaray, N., López-Pedrouso, M., Aadil, R. M., Hano, C., Franco, D., & Lorenzo, J. M. (2021). Proteomic advances in cereal and vegetable crops. *Molecules, 26*, 4924. https://doi.org/10.3390/molecules26164924.

Akagawa, M., Handoyo, T., Ishii, T., Kumazawa, S., Morita, N., & Suyama, K. (2007). Proteomic analysis of wheat flour allergens. *Journal of Agricultural and Food Chemistry, 55*, 6863–6870.

Andon, N. L., Hollingworth, S., Koller, A., Greenland, A. J., Yates, J. R., & Haynes, P. A. (2002). Proteomic characterization of wheat amyloplasts using identification of proteins by tandem mass spectrometry. *Proteomics, 2*, 1156–1168.

Araus, J. L., Slafer, G. A., Reynolds, M. P., & Royo, C. (2002). Plant breeding and drought in C3 cereals: What should we breed for? *Annals of Botany, 89*(Suppl. 1), 925–940.

Arora, A., & Somasundaram, K. (2019). Targeted proteomics comes to the benchside and the bedside: Is it ready for us? *BioEssays, 41*, 1800042. https://doi.org/10.1002/bies.201800042.

Artimo, P., Jonnalagedda, M., Arnold, K., Baratin, D., Csardi, G., De Castro, E., Duvaud, S., Flegel, V., Fortier, A., Gasteiger, E., Grosdidier, A., Hernandez, C., Ioannidis, V., Kuznetsov, D., Liechti, R., Moretti, S., Mostaguir, K., Redaschi, N., Rossier, G., Xenarios, I., & Stockinger, H. (2012). ExPASy: SIB bioinformatics resource portal. *Nucleic Acids Research, 40*, 597–603.

Baetens, J. (2015). Paper knowledge: Toward a media history of documents. *Leonardo, 48*, 96–97.

Bateman, A., Martin, M. J., Orchard, S., Magrane, M., Agivetova, R., Ahmad, S., Alpi, E., Bowler-Barnett, E. H., Britto, R., & Bursteinas, B. (2021). UniProt: The universal protein knowledge base in 2021. *Nucleic Acids Research, 49*, D480–D489.

Begcy, K., & Dresselhaus, T. (2018). Epigenetic responses to abiotic stresses during reproductive development in cereals. *Plant Reproduction, 31*, 343–355.

Blackie, M., Reeves, T., Thomas, G., & Ramsay, G. (2016). Save and grow in practice: A guide to sustainable cereal production. Maize, rice, wheat: A guide to sustainable production. *Food Security, 8*, 1207–1208.

Bongaarts, J. (2021). FAO, IFAD, UNICEF, WFP and WHO: The state of food security and nutrition in the world 2020. Transforming food systems for affordable healthy diets. *Population and Development Review, 47*, 558–558. https://doi.org/10.1111/padr.12418.

Borràs, E., & Sabidó, E. (2017). What is targeted proteomics? A concise revision of targeted acquisition and targeted data analysis in mass spectrometry. *Proteomics, 17*, 17–18.

Bubis, J. A., Levitsky, L. I., Ivanov, M. V., Tarasova, I. A., & Gorshkov, M. V. (2017). Comparative evaluation of label-free quantification methods for shotgun proteomics. *Rapid Communications in Mass Spectrometry, 31*, 606–612.

Chen, Z. Y., Brown, R. L., Damann, K. E., & Cleveland, T. E. (2002). Identification of unique or elevated levels of kernel proteins in aflatoxin-resistant maize genotypes through proteome analysis. *Phytopathology, 92*, 1084–1094.

Chunduri, V., Kaur, A., Kaur, S., Kumar, A., Sharma, S., Sharma, N., Singh, P., Kapoor, P., Kaur, S., Kumari, A., Roy, J., Kaur, J., & Garg, M. (2021). Gene expression and proteomics studies suggest an involvement of multiple pathways under day and day-night combined heat stresses during grain filling in wheat. *Frontiers in Plant Science, 12*, 660446. https://doi.org/10.3389/fpls.2021.660446.

Cifuentes, A. (2013). *Foodomics: Advanced Mass Spectrometry in Modern Food Science and Nutrition* (1st ed., pp. 125–152). https://doi.org/10.1002/9781118537282.

Clark, J., Nelson, R. A., Richard, P., Jacoby, A., & Harvey, M. R. (2014). Proteins with high turnover rate in barley leaves estimated by proteome analysis combined with in planta isotope labeling. *Plant Physiology, 166*, 91–108.

Corthals, G. L., & Rose, K. (2007). Quantitation in proteomics. *Proteome Research, 61*, 69–93.

Cunsolo, V., Muccilli, V., Saletti, R., & Foti, S. (2014). Mass spectrometry in food proteomics: A tutorial. *Journal of Mass Spectrometry, 49*, 768–784.

Das, A., Eldakak, M., Paudel, B., Kim, D. W., Hemmati, H., Basu, C., & Rohila, J. S. (2016). Leaf proteome analysis reveals prospective drought and heat stress response mechanisms in soybeans. *Biomedical Research International, 2016*, 6021047. https://doi.org/10.1155/2016/6021047.

Day, R., Abrahams, P., Bateman, M., Beale, T., Clottey, V., Cock, M., Colmenarez, Y., Corniani, N., Early, R., Godwin, J., Gomez, J., Moreno, P. G., Murphy, S. T., Oppong-Mensah, B., Phiri, N., Pratt, C., Silvestri, S., & Witt, A. (2017). Fall armyworm: Impacts and implications for Africa. *Outlooks on Pest Management, 28*, 196–201.

Dhankher, O. P., & Foyer, C. H. (2018). Climate resilient crops for improving global food security and safety. *Plant Cell & Environment, 41*, 877–884.

Ding, H., Han, Q., Ma, D., Hou, J., Huang, X., Wang, C., Xie, Y., Kang, G., & Guo, T. (2018). Characterizing physiological and proteomic analysis of the action of H2S to mitigate drought stress in young seedlings of wheat. *Plant Molecular Biology Reporter, 36*, 45–57.

Dissanayake, B. M., Staudinger, C., Munns, R., Taylor, N. L., & Millar, A. H. (2022). Distinct salinity-induced changes in wheat metabolic machinery in different root tissue types. *Journal of Proteome Research, 256*, 104502. https://doi.org/10.1016/j.jprot.2022.104502.

Dresselhaus, T., & Hückelhoven, R. (2018). Biotic and abiotic stress responses in crop plants. *Agronomy, 8*, 267.

Driever, S. M., & Kromdijk, J. (2013). Will C3 crops enhanced with the C_4 CO_2-concentrating mechanism live up to their full potential (yield)? *Journal of Experimental Botany, 64*, 3925–3935.

Dupree, E. J., Jayathirtha, M., Yorkey, H., Mihasan, M., Petre, B. A., & Darie, C. C. (2020). A critical review of bottom-up proteomics: The good, the bad, and the future of this field. *Proteomes, 8*, 8030014. https://doi.org/10.3390/proteomes8030014.

Emanuelsson, O., Brunak, S., von Heijne, G., & Nielsen, H. (2007). Locating proteins in the cell using TargetP, SignalP, and related tools. *Nature Protocols, 2*, 953–971.

Emanuelsson, O., Nielsen, H., & Heijne, G. V. (1999). ChloroP is a neural network-based method for predicting chloroplast transit peptides and their cleavage sites. *Protein Science, 8*, 978–984.

Fahad, S., Rehman, A., Shahzad, B., Tanveer, M., Saud, S., Kamran, M., Ihtisham, M., Khan, S. U., Turan, V., & ur Rahman, M. H. (2018). Rice responses and tolerance to metal/metalloid toxicity. In *Advances in Rice Research for Abiotic Stress Tolerance* (pp. 299–312). Elsevier. https://doi.org/10.1016/B978-0-12-814332-2.00014-9.

FAO. (2020). *Figure 39. World Exports of Cereals by Main Commodities*. FAO Statistical Yearbook 2020 Datasets. Food and Agriculture Organization of the United Nations. https://doi.org/10.4060/cb1329en-fig39.

FAO. (2020). *Figure 61. World Area Under Organic Agriculture, Main Countries*. FAO Statistical Yearbook 2020 Datasets. Food and Agriculture Organization of the United Nations. https://doi.org/10.4060/cb1329en-fig61.

FAO. (2022). *Crop Prospects and Food Situation #1, March 2022*. FAO. https://doi.org/10.4060/cb8893en.

Farooqi, Z. U. R., Zia Ur Rehman, M., Sohail, M. I., Usman, M., Khalid, H., & Naz, K. (2020). Regulation of drought stress in plants. In *Plant Life Under Changing*

Environment: Responses and Management (pp. 77–104). Elsevier. https://doi.org/10.1016/B978-0-12-818204-8.00004-7.

Feng, D., Wang, Y., Lu, T., Zhang, Z., & Han, X. (2017). Proteomics analysis reveals a dynamic diurnal pattern of photosynthesis-related pathways in maize leaves. *PLoS One, 12*(7), 180670. https://doi.org/10.1371/journal.pone.0180670.

Gao, X., Lukow, O. M., & Grant, C. A. (2012). Grain concentrations of protein, iron, and zinc and bread-making quality in spring wheat as affected by seeding date and nitrogen fertilizer management. *Journal of Geochemical Exploration, 121*, 36–44. https://doi.org/10.1016/j.gexplo.2012.02.005.

Gao, Y., & Yates, J. R. (2019). Protein analysis by shotgun proteomics. In W. A. Tao & Y. Zhang (Eds.), *Mass Spectrometry-Based Chemical Proteomics* (pp. 1–38). John Wiley & Sons, Inc. https://doi.org/10.1002/9781118970195.ch1.

Gayen, D., Barua, P., Lande, N. V., Varshney, S., Sengupta, S., Chakraborty, S., & Chakraborty, N. (2019). Dehydration-responsive alterations in the chloroplast proteome and cell metabolomic profile of rice reveal key stress adaptation responses. *Environmental and Experimental Botany, 160*, 12–24. https://doi.org/10.1016/j.envexpbot.2019.01.003.

Ghatak, A., Chaturvedi, P., Bachmann, G., Valledor, L., Ramšak, Ž., Bazargani, M. M., Bajaj, P., Jegadeesan, S., Li, W., Sun, X., Gruden, K., Varshney, R. K., & Weckwerth, W. (2021). Physiological and proteomic signatures reveal mechanisms of superior drought resilience in pearl millet compared to wheat. *Frontiers in Plant Science, 11*, 1965. https://doi.org/10.3389/fpls.2020.600278.

Gobaa, S., Bancel, E., Branlard, G., Kleijer, G., & Stamp, P. (2008). Proteomic analysis of wheat recombinant inbred lines: Variations in prolamin and dough rheology. *Journal of Cereal Science, 47*(3), 610–619.

Goez, M. M., Torres-Madroñero, M. C., Röthlisberger, S., & Delgado-Trejos, E. (2018). Preprocessing of 2-dimensional gel electrophoresis images applied to proteomic analysis: A review. *Genomics, Proteomics & Bioinformatics, 16*(1), 63–72. https://doi.org/10.1016/j.gpb.2017.10.001.

Gong, F., Hu, X., & Wang, W. (2015). Proteomic analysis of crop plants under abiotic stress conditions: Where to focus our research? *Frontiers in Plant Science, 6*, 418. https://doi.org/10.3389/fpls.2015.00418.

Grove, H., Hollung, K., Moldestad, A., Færgestad, E. M., & Uhlen, A. K. (2009). Proteome changes in wheat subjected to different nitrogen and sulfur fertilization. *Journal of Agricultural and Food Chemistry, 57*(10), 4250–4258.

Hadebe, S. T., Modi, A. T., & Mabhaudhi, T. (2017). Drought tolerance and water use of cereal crops: A focus on sorghum as a food security crop in sub-Saharan Africa. *Journal of Agronomy and Crop Science, 203*(3), 177–191. https://doi.org/10.1111/jac.12191.

Halder, T., Choudhary, M., Liu, H., Chen, Y., Yan, G., & Siddique, K. H. M. (2022). Wheat proteomics for abiotic stress tolerance and root system architecture: Current status and future prospects. *Proteomes, 10*(2), 17. https://doi.org/10.3390/proteomes10020017.

Han, L., Xiao, C., Xiao, B., Wang, M., Liu, J., Bhanbhro, N., Khan, A., Wang, H., Wang, H., & Yang, C. (2019). Proteomic profiling sheds light on the alkali tolerance of common wheat (*Triticum aestivum* L.). *Plant Physiology and Biochemistry, 138*, 58–64. https://doi.org/10.1016/j.plaphy.2019.02.024.

Hashiguchi, A., Ahsan, N., & Komatsu, S. (2010). Proteomics application of crops in the context of climatic changes. *Food Research International, 43*(7), 1803–1813. https://doi.org/10.1016/j.foodres.2009.07.033.

He, M., He, C. Q., & Ding, N. Z. (2018). Abiotic stresses: General defenses of land plants and chances for engineering multistress tolerance. *Frontiers in Plant Science, 871*, 1771. https://doi.org/10.3389/fpls.2018.01771.
Hondius, D. C., Eigenhuis, K. N., Morrema, T. H. J., van der Schors, R. C., van Nierop, P., Bugiani, M., Li, K. W., Hoozemans, J. J. M., Smit, A. B., & Rozemuller, A. J. M. (2018). Proteomics analysis identifies new markers associated with capillary cerebral amyloid angiopathy in Alzheimer's disease. *Acta Neuropathologica Communications, 6*(1), 46. https://doi.org/10.1186/s40478-018-0540-2.
Hooper, C. M., Castleden, I. R., Aryamanesh, N., Black, K., Grasso, S. V., & Millar, A. H. (2020). CropPAL for discovering divergence in protein subcellular location in crops to support strategies for molecular crop breeding. *The Plant Journal, 104*(3), 812–827.
Hooper, C. M., Castleden, I. R., Aryamanesh, N., Jacoby, R. P., & Millar, A. H. (2016). Finding the subcellular location of barley, wheat, rice, and maize proteins: The compendium of crop proteins with annotated locations (CropPAL). *Plant Cell Physiology, 57*(1), e9. https://doi.org/10.1093/pcp/pcv170.
Hooper, C. M., Castleden, I. R., Tanz, S. K., Aryamanesh, N., & Millar, A. H. (2017). SUBA4: The interactive data analysis center for Arabidopsis subcellular protein locations. *Nucleic Acids Research, 45*, D1064–D1074. https://doi.org/10.1093/nar/gkw1041.
Horton, P., Park, K. J., Obayashi, T., Fujita, N., Harada, H., Adams-Collier, C. J., & Nakai, K. (2007). WoLF PSORT: Protein localization predictor. *Nucleic Acids Research, 35*(S2), 585–587. https://doi.org/10.1093/nar/gkm259.
Ito, Y., Arikawa, K., Antonio, B. A., Ohta, I., Naito, S., Mukai, Y., Shimano, A., Masukawa, M., Shibata, M., Yamamoto, M., Ito, Y., Yokoyama, J., Sakai, Y., Sakata, K., Nagamura, Y., Namiki, N., Matsumoto, T., Higo, K., & Sasaki, T. (2005). Rice annotation database (RAD): A contig-oriented database for map-based rice genomics. *Nucleic Acids Research, 33*(S1), D651–D655. https://doi.org/10.1093/nar/gki083.
Jespersen, B. M., & Munck, L. (2009). Cereals and cereal products. In H. D. Belitz, W. Grosch, & P. Schieberle (Eds.), *Infrared Spectroscopy for Food Quality Analysis and Control* (pp. 275–319). Springer Berlin Heidelberg.
Jiang, J., Xing, F., Zeng, X., & Zou, Q. (2018). RicyerDB: A database for collecting rice yield-related genes with biological analysis. *International Journal of Biological Sciences, 14*, 965–970.
Jin, J., Zhang, H., Kong, L., Gao, G., & Luo, J. (2014). PlantTFDB 3.0: A portal for the functional and evolutionary study of plant transcription factors. *Nucleic Acids Research, 42*, 1182–1187.
Kamal, A. H. M., Cho, K., Choi, J. S., Bae, K. H., Komatsu, S., Uozumi, N., & Woo, S. H. (2013). The wheat chloroplastic proteome. *Journal of Proteomics, 93*, 326–342.
Kamal, A. H. M., Cho, K., Kim, D. E., Uozumi, N., Chung, K. Y., Lee, S. Y., Choi, J. S., Cho, S. W., Shin, C. S., & Woo, S. H. (2012). Changes in physiology and protein abundance in salt-stressed wheat chloroplasts. *Molecular Biology Reports, 39*, 9059–9074.
Kaur, B., Sandhu, K. S., Kamal, R., Kaur, K., Singh, J., Röder, M. S., & Muqaddasi, Q. H. (2021). Omics for the improvement of abiotic, biotic, and agronomic traits in major cereal crops: Applications, challenges, and prospects. *Plants, 10*(10), 1989. https://doi.org/10.3390/plants10101989.
Keeble-Gagnère, G., Rigault, P., Tibbits, J., Pasam, R., Hayden, M., Forrest, K., Frenkel, Z., Korol, A., Huang, B. E., Cavanagh, C., Taylor, J., et al. (2018). Optical and physical

mapping with local finishing enables megabase-scale resolution of agronomically important regions in the wheat genome. *Genome Biology, 19*, 112.

Keno, T., Azmach, G., Gissa, D. W., Regasa, M. W., Tadesse, B., Wolde, L., Deressa, T., Abebe, B., Chibsa, T., & Mahabaleswara, S. (2018). Major biotic maize production stresses in Ethiopia and their management through host resistance. *African Journal of Agricultural Research, 13*, 1042–1052.

Khan, M. N., & Komatsu, S. (2016). Proteomic analysis of soybean root including hypocotyl during recovery from drought stress. *Journal of Proteomics, 144*, 39–50.

Kim, S. T., Kim, S. G., Agrawal, G. K., Kikuchi, S., & Rakwal, R. (2014). Rice proteomics: A model system for crop improvement and food security. *Proteomics, 14*, 593–610.

Klose, J. (1975). Protein mapping by combined isoelectric focusing and electrophoresis of mouse tissues: A novel approach to testing for induced point mutations in mammals. *Human Genetics, 26*, 231–243.

Komatsu, S. (2005). Rice proteome database: A step toward functional analysis of the rice genome. *Plant Molecular Biology, 59*, 179–190.

Komatsu, S., & Hossain, Z. (2013). Organ-specific proteome analysis for identification of abiotic stress response mechanism in crop. *Frontiers in Plant Science, 4*, 71. https://doi.org/10.3389/fpls.2013.00071

König, S. (2020). Differential vs. comparative gel electrophoresis: New technology drives standardization and quantification in protein two-dimensional gel electrophoresis. *Trends in Analytical Chemistry, 122*, 115731.

Koondhar, M. A., Aziz, N., Tan, Z., Yang, S., Raza Abbasi, K., & Kong, R. (2021). Green growth of cereal food production under the constraints of agricultural carbon emissions: New insights from ARDL and VECM models. *Sustainable Energy Technologies and Assessments, 47*, 101452. https://doi.org/10.1016/j.seta.2021.101452.

Kosina, P., Reynolds, M., Dixon, J., & Joshi, A. (2007). Stakeholder perception of wheat production constraints, capacity building needs, and research partnerships in developing countries. *Euphytica, 157*, 475–483.

Kosová, K., Vítámvás, P., Urban, M. O., Prášil, I. T., & Renaut, J. (2018). Plant abiotic stress proteomics: The major factors determining alterations in cellular proteome. *Frontiers in Plant Science, 9*, 122. https://doi.org/10.3389/fpls.2018.00122.

Kotsimbos, T., Williams, T. J., & Anderson, G. P. (2012). Update on lung transplantation: Programmes, patients, and prospects. *European Respiratory Review, 21*, 271–305.

Kumar, R. R., Singh, K., Ahuja, S., Tasleem, M., Singh, I., Kumar, S., Grover, M., Mishra, D., Rai, G. K., Goswami, S., Singh, G. P., Chinnusamy, V., Rai, A., & Praveen, S. (2019). Quantitative proteomic analysis reveals novel stress-associated active proteins (SAAPs) and pathways involved in modulating the tolerance of wheat under terminal heat. *Functional & Integrative Genomics, 19*, 329–348.

Li, N., Zhang, S., Liang, Y., Qi, Y., Chen, J., Zhu, W., & Zhang, L. (2018). Label-free quantitative proteomic analysis of drought stress-responsive late embryogenesis abundant proteins in the seedling leaves of two wheat (*Triticum aestivum* L.) genotypes. *Journal of Proteomics, 172*, 122–142.

Li, Y., Guan, K., Schnitkey, G. D., DeLucia, E., & Peng, B. (2019). Excessive rainfall leads to maize yield loss of a comparable magnitude to extreme drought in the United States. *Global Change Biology, 25*, 2325–2337. https://doi.org/10.1111/gcb.14628.

Li, Y., Li, Z., Cui, S., Chang, S. X., Jia, C., & Zhang, Q. (2019). A global synthesis of the effect of water and nitrogen input on maize (*Zea mays*) yield, water productivity, and nitrogen use efficiency. *Agricultural and Forest Meteorology, 268*, 136–145.

Ling, Y., Du, Z., Zhang, Z., & Su, Z. (2010). ProFITS of maize: A database of protein families involved in the transduction of signaling in the maize genome. *BMC Genomics, 11*, 580. https://doi.org/10.1186/1471-2164-11-580.

Luo, M., Zhao, Y., Wang, Y., Shi, Z., Zhang, P., Zhang, Y., Song, W., & Zhao, J. (2018). Comparative proteomics of contrasting maize genotypes provides insights into salt-stress tolerance mechanisms. *Journal of Proteome Research, 17*, 141–153.

Ma, Q., Shi, C., Su, C., & Liu, Y. (2020). Complementary analyses of the transcriptome and ITRAQ proteome revealed mechanisms of ethylene-dependent salt response in bread wheat (*Triticum aestivum* L.). *Food Chemistry, 325*, 126866.

Mahender, A., Anandan, A., Pradhan, S. K., & Pandit, E. (2016). Rice grain nutritional traits and their enhancement using relevant genes and QTLs through advanced approaches. *SpringerPlus, 5*, 2086. https://doi.org/10.1186/s40064-016-3744-6.

Manandhar-Shrestha, K., Tamot, B., Pratt, E. P. S., Saitie, S., Bräutigam, A., & Weber, A. P. M. (2013). Comparative proteomics of chloroplasts envelopes from bundle sheath and mesophyll chloroplasts reveals novel membrane proteins with a possible role in C4-related metabolite fluxes and development. *Frontiers in Plant Science, 4*, 65. https://doi.org/10.3389/fpls.2013.00065.

Marcus, K., Lelong, C., & Rabilloud, T. (2020). What room for two-dimensional gel-based proteomics in a shotgun proteomics world? *Proteomes, 8*, 17.

Michaletti, A., Naghavi, M. R., Toorchi, M., Zolla, L., & Rinalducci, S. (2018). Metabolomics and proteomics reveal drought-stress responses of leaf tissues from spring wheat. *Scientific Reports, 8*, 5710. https://doi.org/10.1038/s41598-018-24012-y.

Miller, I., Crawford, J., & Gianazza, E. (2006). Protein stains for proteomic applications: Which, when, why? *Proteomics, 6*, 5385–5408. https://doi.org/10.1002/pmic.200600323.

Mirzaei, M., Wu, Y., Worden, P., Jerkovic, A., & Atwell, B. J. (2016). How proteomics contributes to our understanding of drought tolerance. In *Agricultural Proteomics Volume 2: Environmental Stresses* (pp. 129–153). https://doi.org/10.1007/978-3-319-43278-6_6.

Mitsuda, N., & Ohme-Takagi, M. (2009). Functional analysis of transcription factors in *Arabidopsis*. *Plant Cell Physiology, 50*, 1232–1248.

Mitsui, T., Shiraya, T., Kaneko, K., & Wada, K. (2013). Proteomics of rice grain under high-temperature stress. *Frontiers in Plant Science, 4*, 36. https://doi.org/10.3389/fpls.2013.00036.

Mock, H. P., & Dietz, K. J. (2016). Redox proteomics for the assessment of redox-related posttranslational regulation in plants. *Biochimica et Biophysica Acta, 1864*, 967–973.

Mohammadi, M., Anoop, V., Gleddie, S., & Harris, L. J. (2011). Proteomic profiling of two maize inbreds during early Gibberella ear rot infection. *Proteomics, 11*, 3675–3684.

Mohanta, T. K., Kamran, M. S., Omar, M., Anwar, W., & Choi, G. S. (2022). PlantMWpIDB: A database for the molecular weight and isoelectric points of the plant proteomes. *Scientific Reports, 12*, 7421. https://doi.org/10.1038/s41598-022-11077-z.

Mousavi, S. A., Pouya, F. M., Ghaffari, M. R., Mirzaei, M., Ghaffari, A., Alikhani, M., Ghareyazie, M., Komatsu, S., Haynes, P. A., & Salekdeh, G. H. (2016). PlantPReS: A database for plant proteome response to stress. *Journal of Proteomics, 143*, 69–72.

Muitire, C., Kamutando, C., & Moyo, M. (2021). Building stress resilience of cereals under future climatic scenarios: The case of maize, wheat, rice, and sorghum. In *Cereal Grains–Volume 1*. IntechOpen. https://doi.org/10.5772/intechopen.96608.

Munkvold, G. P. (2003). Epidemiology of Fusarium diseases and their mycotoxins in maize ears. *European Journal of Plant Pathology, 109*, 705–713.

Nadaud, I., Girousse, C., Debiton, C., Chambon, C., Bouzidi, M. F., Martre, P., & Branlard, G. (2010). Proteomic and morphological analysis of early stages of wheat grain development. *Proteomics, 10*, 2901–2910. https://doi.org/10.1002/pmic.200900792.

Nakai, K., & Horton, P. (1999). PSORT: A program for detecting sorting signals in proteins and predicting their subcellular localization. *Trends in Biochemical Sciences, 24*, 34–35.

Nelson, C. J., Alexova, R., Jacoby, R. P., & Harvey Millar, A. (2014). Proteins with high turnover rate in barley leaves estimated by proteome analysis combined with in planta isotope labeling. *Plant Physiology, 166*, 91–108.

Ngara, R., & Ndimba, B. K. (2014). Understanding the complex nature of salinity and drought-stress response in cereals using proteomics technologies. *Proteomics, 14*, 611–621.

Ngara, R., Goche, T., Swanevelder, D. Z. H., & Chivasa, S. (2021). Sorghum's whole-plant transcriptome and proteome responses to drought stress: A review. *Life, 11*, 1–25. https://doi.org/10.3390/life11070704.

Nouri, M. Z., & Komatsu, S. (2013). Subcellular protein overexpression to develop abiotic stress tolerant plants. *Frontiers in Plant Science, 4*, 2. https://doi.org/10.3389/fpls.2013.00002.

O'Donnell, R., Holland, J. W., Deeth, H. C., & Alewood, P. (2004). Milk proteomics. *International Dairy Journal, 14*, 1013–1023. https://doi.org/10.1016/j.idairyj.2004.04.004.

Oishi, R. (2021). Economic perspectives on the analysis of ensuring cereal production and consumption security. In *Cereal Grains–Vol. 2*. IntechOpen. https://doi.org/10.5772/intechopen.96377.

Okoli, E. E., Nworji, M. J., & Okoronkwo, C. M. (2021). Potentials of proteomics in plant improvement. *International Journal of Science and Advanced Technology, 2*(1), 80–84. https://doi.org/10.51542/ijscia.v2i1.15.

Orchard, S., Ammari, M., Aranda, B., Breuza, L., Briganti, L., Broackes-Carter, F., Campbell, N. H., Chavali, G., Chen, C., Del-Toro, N., Duesbury, M., Dumousseau, M., Galeota, E., Hinz, U., Iannuccelli, M., Jagannathan, S., Jimenez, R., Khadake, J., Lagreid, A., Licata, L., Lovering, R. C., Meldal, B., Melidoni, A. N., Milagros, M., Peluso, D., Perfetto, L., Porras, P., Raghunath, A., Ricard-Blum, S., Roechert, B., Stutz, A., Tognolli, M., Van Roey, K., Cesareni, G., & Hermjakob, H. (2014). The MIntAct project—IntAct is a common curation platform for 11 molecular interaction databases. *Nucleic Acids Research, 42*, 358–363.

Ouyang, S., Zhu, W., Hamilton, J., Lin, H., Campbell, M., Childs, K., Thibaud-Nissen, F., Malek, R. L., Lee, Y., Zheng, L., Orvis, J., Haas, B., Wortman, J., & Buell, R. C. (2007). The TIGR rice genome annotation resource: Improvements and new features. *Nucleic Acids Research, 35*(Suppl 1), D883–D887. https://doi.org/10.1093/nar/gkl976.

Pagoulatos, G. N., & Yaniv, M. (1977). High-resolution two-dimensional electrophoresis of proteins bound to heterogeneous nuclear RNA. *FEBS Letters, 74*, 115–120.

Pan, R., He, D., Xu, L., Zhou, M., Li, C., Wu, C., Xu, Y., & Zhang, W. (2019). Proteomic analysis reveals the response of differential wheat (*Triticum aestivum* L.) genotypes to oxygen deficiency stress. *BMC Genomics, 20*(1), 60. https://doi.org/10.1186/s12864-018-5405-3.

Pandey, A., & Mann, M. (2000). Proteomics to study genes and genomes. *Nature, 405*(6788), 837–846. https://doi.org/10.1038/35015709.

Pandey, A., Fernandez, M. M., Steen, H., Blagoev, B., Nielsen, M. M., Roche, S., Mann, M., & Lodish, H. F. (2000). Identification of a novel immunoreceptor tyrosine-based activation motif-containing molecule, STAM2, by mass spectrometry and its involvement in growth factor and cytokine receptor signaling pathways. *Journal of Biological Chemistry, 275*, 38633–38639. https://doi.org/10.1074/jbc.M007849200.

Papageorgiou, M., & Skendi, A. (2018). Introduction to cereal processing and by-products. In *Sustainable Recovery and Reutilization of Cereal Processing By-Products* (pp. 1–25). Elsevier. https://doi.org/10.1016/B978-0-08-102162-0.00001-0.

Perez-Riverol, Y., Alpi, E., Wang, R., Hermjakob, H., & Vizcaíno, J. A. (2015). Making proteomics data accessible and reusable: Current state of proteomics databases and repositories. *Proteomics, 15*(5–6), 930–950. https://doi.org/10.1002/pmic.201400302.

Qadir, M., Quillérou, E., Nangia, V., Murtaza, G., Singh, M., Thomas, R. J., Drechsel, P., & Noble, A. D. (2014). Economics of salt-induced land degradation and restoration. *Nature Resources Forum, 38*(4), 282–295. https://doi.org/10.1111/1477-8947.12054.

Que, S., Li, K., Chen, M., Wang, Y., Yang, Q., Zhang, W., Zhang, B., Xiong, B., & He, H. (2012). PhosphoRice: A metapredictor of rice-specific phosphorylation sites. *Plant Methods, 8*, 5. https://doi.org/10.1186/1746-4811-8-5.

Rahiminejad, M., Ledari, M. T., Mirzaei, M., Ghorbanzadeh, Z., Kavousi, K., Ghaffari, M. R., Haynes, P. A., Komatsu, S., & Salekdeh, G. H. (2019). The quest for missing proteins in rice. *Molecular Plant, 12*(1), 4–6. https://doi.org/10.1016/j.molp.2018.11.009.

Reddy, B. M. (2021). Pharmacognostical and phytochemical evaluation of an important medicinal plant *Oldenlandia umbellata* L. *Journal of Pharmacognosy and Phytochemistry, 10*(6), 288–291. https://doi.org/10.22271/phyto.2021.v10.i6d.14302.

Reynolds, M. P., Quilligan, E., Aggarwal, P. K., Bansal, K. C., Cavalieri, A. J., Chapman, S. C., Chapotin, S. M., Datta, S. K., Duveiller, E., Gill, K. S., Jagadish, K. S. V., Joshi, A. K., Koehler, A. K., Kosina, P., Krishnan, S., Lafitte, R., Mahala, R. S., Raveendran, M., Paterson, A. H., Prasanna, B. M., Rakshit, S., Rosegrant, M. W., Sharma, I., Singh, R. P., Sivasankar, S., Vadez, V., Valluru, R., Vara Prasad, P. V., & Yadav, O. P. (2016). An integrated approach to maintaining cereal productivity under climate change. *Global Food Security, 8*, 9–18. https://doi.org/10.1016/j.gfs.2016.02.002.

Rhee, S. Y., & Mutwil, M. (2014). Towards revealing the functions of all genes in plants. *Trends in Plant Science, 19*(4), 212–221. https://doi.org/10.1016/j.tplants.2013.10.006.

Rolland, D. C. M., Lim, M. S., & Elenitoba-Johnson, K. S. J. (2019). Mass spectrometry and proteomics in hematology. *Seminars in Hematology, 56*(1), 52–57. https://doi.org/10.1053/j.seminhematol.2018.05.009.

San Clemente, H., Pont-Lezica, R., & Jamet, E. (2009). Bioinformatics as a tool for assessing the quality of sub-cellular proteomic strategies and inferring functions of proteins: Plant cell wall proteomics as a test case. *Bioinformatics and Biology Insights, 3*, 15–28. https://doi.org/10.4137/bbi.s2065.

Schulze, W. X., & Usadel, B. (2010). Quantitation in mass-spectrometry-based proteomics. *Annual Review of Plant Biology, 61*, 491–516.

Shah, T. R., Prasad, K., & Kumar, P. (2016). Maize—A potential source of human nutrition and health: A review. *Cogent Food & Agriculture, 2*(1), 1166995. https://doi.org/10.1080/23311932.2016.1166995.

Shewry, P. R., & Hey, S. J. (2015). The contribution of wheat to human diet and health. *Food Energy Security, 4*(3), 178–202. https://doi.org/10.1002/fes3.64.

Shi, T., Song, E., Nie, S., Rodland, K. D., Liu, T., Qian, W. J., & Smith, R. D. (2016). Advances in targeted proteomics and applications to biomedical research. *Proteomics, 16*(15–16), 2160–2182. https://doi.org/10.1002/pmic.201500449.

Shi, X., & Ling, H. Q. (2018). Current advances in genome sequencing of common wheat and its ancestral species. *The Crop Journal, 6*(1), 15–21.

Silvar, C., Martis, M. M., Nussbaumer, T., Haag, N., Rauser, R., Keilwagen, J., Korzun, V., Mayer, K. F., Ordon, F., & Perovic, D. (2015). Assessing the barley genome zipper and genomic resources for breeding purposes. *The Plant Genome, 8*(3), 45. https://doi.org/10.3835/plantgenome2015.06.0045.

Singer, S. D., Foroud, N. A., & Laurie, J. D. (2018). Molecular improvement of grain: Target traits for a changing world. In *Encyclopedia of Food Security and Sustainability* (pp. 545–555). Elsevier. https://doi.org/10.1016/B978-0-08-100596-5.22439-6.

Singh, P. K., Indoliya, Y., Agrawal, L., Awasthi, S., Deeba, F., Dwivedi, S., Chakrabarty, D., Shirke, P. A., Pandey, V., Singh, N., Dhankher, O. P., Barik, S. K., & Tripathi, R. D. (2022). Genomic and proteomic responses to drought stress and biotechnological interventions for enhanced drought tolerance in plants. *Current Plant Biology, 29*, 100239. https://doi.org/10.1016/j.cpb.2022.100239.

Sinha, R., Bala, M., Ranjan, A., Lal, S. K., Sharma, T. R., Pattanayak, A., & Singh, A. K. (2021). Proteomic approaches to understand plant response to abiotic stresses. In D. K. Srivastava, A. K. Thakur, & P. Kumar (Eds.), *Agricultural Biotechnology: Latest Research and Trends* (pp. 351–383). Springer. https://doi.org/10.1007/978-981-16-2339-4_16.

Sohail, M. I., Ayub, M. A., Zia ur Rehman, M., Azhar, M., Farooqi, Z. U. R., Siddiqui, A., Umar, W., Iftikhar, I., Nadeem, M., & Fatima, H. (2021a). Sufficiency and toxicity limits of metallic oxide nanoparticles in the biosphere. In *Nanomaterials: Synthesis, Characterization, Hazards and Safety* (pp. 145–221). https://doi.org/10.1016/b978-0-12-823823-3.00002-1.

Sohail, M. I., Siddiqui, A., Erum, N., & Kamran, M. (2021b). Phytomedicine and the COVID-19 pandemic. In *Phytomedicine: A Treasure of Pharmacologically Active Products from Plants* (pp. 693–708). Elsevier. https://doi.org/10.1016/B978-0-12-824109-7.00005-4.

Sohail, M. I., Zia ur Rehman, M., Aziz, T., Akmal, F., Azhar, M., Nadeem, F., Aslam, M., Siddiqui, A., & Khalid, M. A. (2022). Iron bio-fortification and heavy metal/(loid)s contamination in cereals: Successes, issues, and challenges. *Crop & Pasture Science, 73*, 877–895. https://doi.org/10.1071/cp21771.

Sohail, M. I., Zia ur Rehman, M., Rizwan, M., Yousaf, B., Ali, S., Anwar ul Haq, M., Anayat, A., & Waris, A. A. (2020). The efficiency of various silicon-rich amendments on growth and cadmium accumulation in field-grown cereals and health risk assessment. *Chemosphere, 244*, 125481. https://doi.org/10.1016/j.chemosphere.2019.125481.

Sun, Q., Zybailov, B., Majeran, W., Friso, G., Olinares, P. D. B., & van Wijk, K. J. (2009). PPDB, the Plant Proteomics Database at Cornell. *Nucleic Acids Research, 37*(S1), 654. https://doi.org/10.1093/nar/gkn654.

Swamy, B. P. M., Rahman, M. A., Inabangan-Asilo, M. A., Amparado, A., Manito, C., Chadha-Mohanty, P., Reinke, R., & Slamet-Loedin, I. H. (2016). Advances in breeding for high grain zinc in rice. *Rice, 9*(1), 49. https://doi.org/10.1186/s12284-016-0122-5.

Szymanski, J., Levin, Y., Savidor, A., Breitel, D., Chappell-Maor, L., Heinig, U., Töpfer, N., & Aharoni, A. (2017). Label-free deep shotgun proteomics reveals protein dynamics during tomato fruit tissue development. *Plant Journal, 90*, 396–417.

Tacer-Caba, Z., Nilufer-Erdil, D., & Ai, Y. (2015). Chemical composition of cereals and their products. In *Handbook of Food Chemistry* (pp. 301–329). Springer Berlin Heidelberg.

Tadesse, W., Bishaw, Z., & Assefa, S. (2019). Wheat production and breeding in sub-Saharan Africa: Challenges and opportunities in the face of climate change. *International Journal of Climate Change Strategies and Management, 11*, 696–715.

Takahashi, D., Li, B., Nakayama, T., Kawamura, Y., & Uemura, M. (2013). Plant plasma membrane proteomics for improving cold tolerance. *Frontiers in Plant Science, 4*, 090. https://doi.org/10.3389/fpls.2013.00090.

Tanaka, J., Hayashi, T., & Iwata, H. (2016). A practical, rapid generation-advancement system for rice breeding using a simplified biotron breeding system. *Breeding Science, 66*, 542–551.

Tsiatsiani, L., & Heck, A. J. R. (2015). Proteomics beyond trypsin. *FEBS Journal, 282*, 2612–2626.

United Nations General Assembly. (2015). Transforming our world: The 2030 agenda for sustainable development (Vol. 25). United Nations.

Urrutia, M., Blein-Nicolas, M., Prigent, S., Bernillon, S., Deborde, C., Balliau, T., Maucourt, M., Jacob, D., Ballias, P., Bénard, C., Sellier, H., Gibon, Y., Giauffret, C., Zivy, M., & Moing, A. (2021). Maize metabolome and proteome responses to controlled cold stress partly mimic early-sowing effects in the field and differ from those of Arabidopsis. *Plant Cell & Environment, 44*, 1504–1521.

Vítámvás, P., Kosová, K., Prášilová, P., & Prášil, I. T. (2010). Accumulation of WCS120 protein in wheat cultivars grown at 9°C or 17°C in relation to their winter survival. *Plant Breeding, 129*, 611–616.

Vítámvás, P., Urban, M. O., Škodácek, Z., Kosová, K., Pitelková, I., Vítámvás, J., Renaut, J., & Prášil, I. T. (2015). Quantitative analysis of proteome extracted from barley crowns grown under different drought conditions. *Frontiers in Plant Science, 6*, 479. https://doi.org/10.3389/fpls.2015.00479.

Wang, D., Xia, Y., Li, X., Hou, L., & Yu, J. (2013). The Rice Genome Knowledgebase (RGKbase): An annotation database for rice comparative genomics and evolutionary biology. *Nucleic Acids Research, 41*, 1225. https://doi.org/10.1093/nar/gks1225.

Wang, J., Vanga, S. K., Saxena, R., Orsat, V., & Raghavan, V. (2018). Effect of climate change on the yield of cereal crops: A review. *Climate, 6*, 6020041. https://doi.org/10.3390/cli6020041.

Wang, X., Chen, J., Ge, J., Huang, M., Cai, J., Zhou, Q., Dai, T., Mur, L. A. J., & Jiang, D. (2021). The different root apex zones contribute to drought priming-induced tolerance to reoccurring drought stress in wheat. *Crop Journal, 9*, 1088–1097.

Wang, X., Hou, L., Lu, Y., Wu, B., Gong, X., Liu, M., Wang, J., Sun, Q., Vierling, E., & Xu, S. (2018). The metabolic adaptation of wheat grain contributes to a stable filling rate under heat stress. *Journal of Experimental Botany, 69*, 5531–5545.

Wang, X., Khodadadi, E., Fakheri, B., & Komatsu, S. (2017). Organ-specific proteomics of soybean seedlings under flooding and drought stresses. *Journal of Proteomics, 162*, 62–72.

Wang, Y., Thilmony, R., Zhao, Y., Chen, G., & Gu, Y. Q. (2014). AIM: A comprehensive Arabidopsis interactome module database and related interologs in plants. *Database, 2014*, 117. https://doi.org/10.1093/database/bau117.

Wasinger, V. C., Cordwell, S. J., Cerpa-Poljak, A., Yan, J. X., Gooley, A. A., Wilkins, M. R., Duncan, M. W., Harris, R., Williams, K. L., & Humphery-Smith, I. (1995). Progress with gene-product mapping of the mollicutes: Mycoplasma genitalium. *Electrophoresis, 16*(1), 1090–1094.

Weisser, H., & Choudhary, J. S. (2017). Targeted feature detection for data-dependent shotgun proteomics. *Journal of Proteome Research, 16*, 2964–2974.

Wienkoop, S., Staudinger, C., Hoehenwarter, W., Weckwerth, W., & Egelhofer, V. (2012). ProMEX – A mass spectral reference database for plant proteomics. *Frontiers in Plant Science, 3*, 00125. https://doi.org/10.3389/fpls.2012.00125.

Wrigley, C. (2017). Cereal-grain morphology and composition. In *Cereal Grains: Assessing and Managing Quality: 2nd Edition* (pp. 55–87). Elsevier. https://doi.org/10.1016/B978-0-08-100719-8.00004-8.

Wu, X., Gong, F., Cao, D., Hu, X., & Wang, W. (2016). Advances in crop proteomics: PTMs of proteins under abiotic stress. *Proteomics, 16*, 847–865.

Wu, Z., Huang, J., Lu, J., & Zhang, X. (2018). Reversible lysine derivatization enabled improved Arg-C digestion, a highly specific Arg-C digestion using trypsin. *Analytical Chemistry, 90*, 1554–1559. https://doi.org/10.1021/acs.analchem.7b04410.

Xie, F., Smith, R. D., & Shen, Y. (2012). Advanced proteomic liquid chromatography. *Journal of Chromatography A, 1261*, 78–90.

Xu, W., Lv, H., Zhao, M., Li, Y., Qi, Y., Peng, Z., Xia, G., & Wang, M. (2016). The proteomic comparison reveals the contribution of chloroplast to salt tolerance of a wheat introgression line. *Scientific Reports, 6*, 32384.

Yan, M., Xue, C., Xiong, Y., Meng, X., Li, B., Shen, R., & Lan, P. (2020). Proteomic dissection of the similar and different responses of wheat to drought, salinity, and submergence during seed germination. *Journal of Proteomics, 220*, 103756. https://doi.org/10.1016/j.jprot.2020.103756.

Yan, M., Zheng, L., Li, B., Shen, R., & Lan, P. (2021). Comparative proteomics reveals new insights into the endosperm responses to drought, salinity, and submergence in germinating wheat seeds. *Plant Molecular Biology, 105*, 287–302.

Yang, F., & Yin, Q. (2016). Comprehensive proteomic analysis of the wheat pathogenic fungus *Zymoseptoria tritici*. *Proteomics, 16*, 98–101.

Yates, J. R., & Kelleher, N. L. (2013). Top-down proteomics. *Analytical Chemistry, 85*, 6151–6151.

Yin, Y., Qi, F., Gao, L., Rao, S., Yang, Z., & Fang, W. (2018). ITRAQ-based quantitative proteomic analysis of dark-germinated soybeans in response to salt stress. *RSC Advances, 8*, 17905–17913.

Yu, X., James, A. T., Yang, A., Jones, A., Mendoza-Porras, O., Bétrix, C. A., Ma, H., & Colgrave, M. L. (2016). A comparative proteomic study of drought-tolerant and drought-sensitive soybean seedlings under drought stress. *Crop & Pasture Science, 67*, 528–540.

Zaidi, P. H., Rafique, S., & Singh, N. N. (2003). Response of maize (*Zea mays* L.) genotypes to excess soil moisture stress: Morpho-physiological effects and basis of tolerance. *European Journal of Agronomy, 19*, 383–399. https://doi.org/10.1016/S1161-0301(02)00090-4.

Zamir, M. S. I., Yasin, G., Javeed, H. M. R., Ahmad, A. U. H., Tanveer, A., & Yaseen, M. (2013). Effect of different sowing techniques and mulches on the growth and yield behavior of spring-planted maize (*Zea mays* L.). *Cercetări Agronomice în Moldova, 46*, 77–82.

Zenda, T., Liu, S., Wang, X., Jin, H., Liu, G., & Duan, H. (2018). Comparative proteomic and physiological analyses of two divergent maize inbred lines provide more insights into drought-stress tolerance mechanisms. *International Journal of Molecular Sciences, 19*, 3225. https://doi.org/10.3390/ijms19103225.

Zhang, J., Li, C., Wu, C., Xiong, L., Chen, G., Zhang, Q., & Wang, S. (2006). RMD: A rice mutant database for functional analysis of the rice genome. *Nucleic Acids Research, 34*, 016. https://doi.org/10.1093/nar/gkj016.

Zhang, J., Zhang, S., Cheng, M., Jiang, H., Zhang, X., Peng, C., Lu, X., Zhang, M., & Jin, J. (2018). Effect of drought on agronomic traits of rice and wheat: A meta-analysis. *International Journal of Environmental Research and Public Health, 15*, 839. https://doi.org/10.3390/ijerph15050839.

Zhang, X., Högy, P., Wu, X., Schmid, I., Wang, X., Schulze, W. X., Jiang, D., & Fangmeier, A. (2018). Physiological and proteomic evidence for the interactive effects of post-anthesis heat stress and elevated CO2 on wheat. *Proteomics, 18*, 262. https://doi.org/10.1002/pmic.201800262.

Zhang, Y. T., Zhang, Y. L., Chen, S. X., Yin, G. H., Yang, Z. Z., Lee, S., & Yang, F. S. (2015). Proteomics of methyl jasmonate induced defense response in maize leaves against Asian corn borer. *BMC Genomics, 16*, 224.

Zhang, Y., Lou, H., Guo, D., Zhang, R., Su, M., Hou, Z., Zhou, H., Liang, R., Xie, C., You, M., & Li, B. (2018). Identifying changes in the wheat kernel proteome under heat stress using ITRAQ. *Crop Journal, 6*, 600–610.

Zhao, Q., Chen, S., & Dai, S. (2013). C4 photosynthetic machinery: Insights from maize chloroplast proteomics. *Frontiers in Plant Science, 4*, 00085. https://doi.org/10.3389/fpls.2013.00085.

Zhu, D., Luo, F., Zou, R., Liu, J., & Yan, Y. (2021). Integrated physiological and chloroplast proteome analysis of wheat seedling leaves under salt and osmotic stresses. *Journal of Proteomics, 234*, 104097. https://doi.org/10.1016/j.jprot.2020.104097.

Zhu, G., Wu, A., Xu, X. J., Xiao, P. P., Lu, L., Liu, J., Cao, Y., Chen, L., Wu, J., & Zhao, X. M. (2016). PPIM: A protein-protein interaction database for maize. *Plant Physiology, 170*, 618–626.

Zhu, Y., Zhao, P., Wu, X., Wang, W., Scali, M., & Cresti, M. (2011). Proteomic identification of differentially expressed proteins in mature and germinated maize pollen. *Acta Physiologiae Plantarum, 33*, 1467–1474.

Zörb, C., Herbst, R., Forreiter, C., & Schubert, S. (2009). Short-term effects of salt exposure on the maize chloroplast protein pattern. *Proteomics, 9*, 4209–4220.

Zulawski, M., Schulze, G., Braginets, R., Hartmann, S., & Schulze, W. X. (2014). The Arabidopsis kinome: Phylogeny and evolutionary insights into functional diversification. *BMC Genomics, 15*, 548.

CHAPTER 6

Signaling and Regulatory Pathways Between Plants and Microbial Communities Towards Environments

MARYAM KHAN,[1] AROOMA JANNAT,[1] ZUHRA QAYYUM,[1] NOSHEEN FATIMA,[1] MOHD. SAYEED AKHTAR,[2] and RABIA AMIR[1,*]

[1]Atta-Ur-Rahman School of Applied Biosciences, National University of Sciences and Technology, Islamabad, Pakistan

[2]Department of Botany, Gandhi Faiz-e-Aam College, Shahjahanpur, Uttar Pradesh, India

*Corresponding author.

ABSTRACT

Plants have a meshwork of coordinated responses to maintain metabolic homeostasis in response to external and internal stimuli. To improve their production, development, and tissue quality, they also constantly interact with their environment. The interactions between plants and their microbiota have drawn a lot of attention because figuring out how they work through different signaling and regulatory pathways can result in the creation of cutting-edge agricultural applications. Since they have a considerable impact on food security and the ecology of natural ecosystems, these interactions must be carefully managed. The manipulation of plant interactions may result from an understanding of these interactions. This could present several chances to enhance crop plants that are crucial to agriculture. The aim of this chapter is to summarize the interactions between plants and microbial communities for their response towards the changing environment through various communication networks in their vicinity.

Agricultural Biotechnology: Issues, Challenges, and Recent Developments. Mohd. Sayeed Akhtar and Khalid Rehman Hakeem (Eds.)
© 2025 Apple Academic Press, Inc. Co-published with CRC Press (Taylor & Francis)

6.1 INTRODUCTION

Plants share a dynamic environment with microbes that can be beneficial, harmful, or symbiotic in nature. Decades of work by molecular and microbial scientists have revealed four basic principles in plant-microbe interaction: (i) specific receptors in plants perceive signals released by microbes and initiate a cascade of defense responses (Jones et al., 2016); (ii) deoxyribonucleic acid (DNA) and effector molecules of the microbes regulate plant cell function (Hwang et al., 2017); (iii) changes in the developmental processes of both microbes and plants that initiate the formation of nutrient-producing organs, such as nodules (Zipfel and Oldroyd, 2017); and (iv) interaction between plants and microbes at both individual and community levels during disease (Hacquard et al., 2017). It has been observed that most interactions between plants and microbes are greatly affected by environmental parameters. This also led to the famous concept of the "disease triangle," which states that a plant can only develop a disease from a microbe if it is susceptible. Moreover, the pathogen must have disease-causing traits, and environmental factors must support the microbe. Similarly, the interaction between plants and symbiotic microbes is also affected by external environmental conditions such as temperature, moisture, and nutrient availability (Cheng et al., 2019). Depending on the nature of the microbe, plants show positive communication towards beneficial microbes and negative communication towards pathogenic microbes (Rodriguez et al., 2019). During this plant-microbe interaction, several compounds are produced by both the plant and microorganism that regulate the pathway, including phytohormones, phytoalexins, secondary metabolites, bacterial volatile compounds, and effector molecules, respectively (Pierik et al., 2006; Ravanbakhsh et al., 2018).

Climate change has affected life on Earth. Due to the rapidly and drastically changing climate, several stresses have emerged such as drought, extreme temperatures, atypical rains, salinity, and heavy metal accumulation. Plant-microbe interactions have co-evolved causing a change in the composition of root exudates thereby altering microbial colonization (Badri et al., 2013). In response to the climatic changes, bacteria have also evolved mechanisms to survive and withstand these adverse environmental conditions, however, some species still have not adapted and would therefore be unable to survive in such conditions. For example, due to the ozone layer depletion flow of ultraviolet radiation (UVR) especially UV-B has increased. The effects of UV have severely damaged the ecosystem in the Polar region. The plants in the Polar region depend on the soil microorganisms for nutrients. The UV-B radiations directly affect the microorganisms in soil and alter the induction of carbon assimilation, growth, and pigment content (Formánek et al., 2014; Sinha et al., 1999). However, plants and microbes have co-evolved mechanisms for UVR tolerance for example, *Erwinia herbicola* and *Deschampsia antarctica* induce high production of secondary metabolites such as carotenoids that confer tolerance against UV-A radiation. Similarly, the bacteria and archaea isolated from Antarctic soil counter UV-A and UV-B radiation through melanin synthesis-protective mechanisms. The aim of this chapter is to summarize the interactions between plants and microbial communities for their response to the changing environment through various communication networks in their vicinity.

6.2 PLANT-MICROBE INTERACTIONS IN A CHANGING ENVIRONMENT

Soil is a mixture of living organisms, minerals, organic matter, liquids, and gases crucial for plant sustainability. The structure and type of soil is dependent on the mineral composition. The mineral composition of the soil has an impact on the soil moisture and porosity in a way that it forms a microenvironment as a feasible niche for microbes (Wilpiszeski et al., 2019). Living organisms such as bacteria, viruses, and fungi actively participate in the nutrient cycle of minerals leading to the bioavailability of essential nutrients crucial for plant growth and development (Uroz et al., 2013). The soil moisture content has a great impact on shaping the microbial community since it allows the mobility of microbes towards the closest part to root exudates (Santoyo et al., 2017). Under extreme climatic conditions such as in the tundra of the Canadian Arctic, Tibetan permafrost, Antarctic, and Polar regions, high soil moisture content is closely associated with microbial abundance especially bacteria like Proteobacteria (Chu et al., 2011; Zhang et al., 2013). The pore dwellings also alter the establishment of microbial communities by influencing carbon mineralization (Ruamps et al., 2013). It has been reported that the establishment of Mycobacterium, *Pseudomonas,* and *Bacillus* communities in the soil promotes the uptake of nitrogen, phosphorus, and potassium in corn plants (Dilfuza Egamberdiyeva, 2007). The soil nutrients also impact the colonization of microbial communities in the rhizosphere. Nutrient imbalance has an effect on the diversity of plants and microbes such that low levels of nitrogen and carbon are a threat to lower plant and microbial diversity (Ahmed et al., 2008; Clark and Tilman,

2008). The application of phosphorus and nitrogen improved the structure of soil and the diversity of bacteria and fungi in the rhizosphere of the plants thus increasing the bioavailability of minerals for plants (Beauregard et al., 2010; Coolon et al., 2013). Furthermore, the presence of hydronium ions $[H_3O]^+$ in the soil determines the pH of the soil. The soil pH alters the establishment of microbial communities and is associated with plant-microbe interaction (Zhalnina et al., 2015). The community of Acidobacteria impacts the nutrient uptake of plants, however, pH has been observed as a limiting factor (Donnarumma et al., 2014). An array of compounds such as H^+ ions, enzymes, mucilage, water, and carbon-containing primary and secondary compounds, collectively called root exudates are synthesized, accumulated, and secreted into the soil by roots and these play a crucial role in shaping the soil microbial community (Baetz and Martinoia, 2014). The physiological effects of microbes and plants are equally important to abiotic stressors. The growth and colonization of microbial communities in soil are directly influenced by the presence of a plant (Santoyo et al., 2017). However, the microbial community in soil benefits the plants by producing a blend of volatile compounds that antagonize phytopathogens, suppress disease-stimulating ISR (induced systemic resistance), and stimulate plant growth (Peñuelas et al., 2014; Pineda et al., 2013; Yi et al., 2013).

6.2.1 COMMUNICATION WITHIN RHIZOSPHERE

The most complex and diverse ecosystem, which is inhabited by arthropod herbivores, fungi, nematodes, and bacteria, is the rhizosphere (Venturi and Keel, 2016). In comparison to bulk soil, the rhizosphere has higher bacterial activity and abundance; however, there is less bacterial diversity because plants affect the colonization of bacteria in the soil through their primary and secondary metabolites (Hacquard et al., 2017; Yang et al., 2017). Commonly, *Pseudomonas, Arthrobacter, Acinetobacter, Bacillus, Paenibacillus,* and *Enterobacter* exist in the soil as plant growth-promoting rhizomicrobes/bacteria (PGPR; Finkel et al., 2017; Sasse et al., 2018; Zhang et al., 2017). The microbes communicate with each other and with plants through chemical messengers. These chemical messengers alter the gene expression in plants, due to which plants subsequently release a blend of primary and secondary metabolites that are crucial for biofilm formation, microbial accumulation and colonization, root-root interaction, pathogen protection, and nutrient availability (Mommer et al., 2016b).

6.2.2 MICROBE-TO-MICROBE COMMUNICATION

The bacterial communities of the soil produce signaling metabolites that alter the gene expression in plants. During metabolic pathways, several volatile organic compounds (VOCs), which are low molecular weight lipophilic compounds, are produced (Kanchiswamy et al., 2015). VOCs play a crucial role in microbial communication. VOCs are generally composed of compounds of sulfur, ketones, alkanes, terpenoids, ketones, and alkenes, however, the complexity of metabolites is highly dependent upon specific metabolic pathways and microbial species (Kai et al., 2016). The colonization of PGPR in roots involves cell-to-cell communication through quorum sensing (QS) and the release of signals that are also called auto-inducers and influence the bacterial density, and inter and intra-bacterial communication among different species (Hassan et al., 2016; Helman and Chernin, 2015).

6.2.3 COMMUNICATION FROM MICROBE TO PLANT

The rhizobacteria form symbiotic associations with plants by emitting or secreting molecules that trigger alterations in plant transcriptome leading to better plant growth and development. In addition to VOCs such as pentadecane, 2-entacapone, and 2-heptanol produced by PGPR, it has been observed that certain species of PGPR also produce phytohormones such as ABA, auxins, JA, SA, gibberellins, and cytokinins which are involved in plant growth modulation, maintenance of soil health and resistance induction (Fahad et al., 2015). Better growth of *Arabidopsis thaliana* was observed when it was co-cultured with two strains of PGPR (*B. amyloliquefaciens* IN937a and *Bacillus subtilis* GB03) due to the release of VOCs such as 2,3 butanediol and 3-hydroxy-2-butanone (acetoin; Ryu et al., 2003). QS signals such as AHLs released by gram-negative bacteria contribute to the establishment of symbiotic relationships between plant and soil bacteria (Schikora et al., 2016).

6.2.4 COMMUNICATION FROM PLANTS TO MICROBES

Plants secrete root exudates which is a chemically complex mixture of metabolites through which plants communicate in the rhizosphere. The chemical complexity of the root exudate is not only dependent on plant species and genotype but also on external factors such as soil conditions, plant size, and photosynthesis activity (Sasse et al., 2018). The chemical composition of root exudate affects the colonization of microbial communities in the rhizosphere. Plants secrete specific compounds to select the microbial community, for example, the release of citric acid in root exudates of the cucumber plant allows the colonization of *B. amyloliquefaciens* SQR9 and leads to biofilm formation. Similarly, the presence of fumaric acid in the root exudates of bananas stimulates *B. subtilis* N11 colonization and biofilm formation (Zhang et al., 2014). Furthermore, it has also been observed that the presence of certain organic acids and sugars in the root exudate of tomato stimulates the antifungal activity of *Pseudomonas* spp. (Kravchenko et al., 2003). The presence of flavonoids in the root exudate induces bacterial node genes stimulating the production of lipo-chitooligosaccharides (LCOs) which lead to nodule formation in roots and facilitate the arbuscular mycorrhizal fungi and plant associations. Moreover, flavonoids mimic QS molecules thus affecting bacterial metabolism and colonization in roots (Rosier et al., 2016). Various low-carbon molecules such as tryptophan, aminocyclopropane-1-carboxylic acid (ACC), and ET precursor are present in roots and serve as precursors for phytohormone biosynthesis by PGPR (El Zahar Haichar et al., 2014).

6.2.5 COMMUNICATION WITHIN PLANTS

Plants release signaling molecules to communicate with each other either above or below the ground. Plants release compounds called phytotoxins with allelopathic properties such as catechin, hydroxamic acids, benzene-derived compounds, terpenes, and phenolics which mediate the plant-plant interaction by inhibiting the growth of other plants thus reducing competition for nutrients (Mommer et al., 2016a). However, the VOCs released from soil microbes act as allelochemicals. The plant-mycorrhizal networks initiate VOC-mediated signaling which confers protection against degradation and stimulates the transmission of compounds from plant to plant (Massalha et al., 2017).

6.2.6 ENVIRONMENT AFFECTING PLANT INTERACTIONS

Plants and microorganisms have coevolved by forming distinct associations. Plants have developed numerous complex mechanisms that may modulate the outcome of plant-microbe interaction (Oldroyd, 2013). Climate change has posed a great threat to the agriculture sector. However, recently it has been demonstrated that plant-microbiome interactions enhance nutrient uptake, stimulate shoot, and root growth, and induce resistance to adverse environmental stressors (Trivedi et al., 2021). Furthermore, the plant-microbiome interaction acts as a biocontrol of phytopathogens (del Carmen Orozco-Mosqueda et al., 2018). There are multiple factors that impact plant-microbiome interactions. Various non-biological components of the ecosystem such as soil structure, soil composition, geographical factors, and cultivation practices affect the plant-microbe interactions. The soil microbiome and plant ecology are altered by abiotic factors thus complicating the plant-microbiome interaction (Santoyo et al., 2017). The roles of different non-biological and biological components of the ecosystem that affect plant-microbe interactions have been summarized in tabular form (Table 6.1).

6.3 SIGNALING PATHWAYS IN PLANT-MICROBE-ENVIRONMENT INTERACTIONS

Plants encounter various microorganisms in their lifespan and these interactions can be beneficial or fatal resulting in mutualistic or pathogenic interactions, respectively (Thrall et al., 2007). Plants respond to these interactions by modulating the innate immune system as triggered by the microbes (Pieterse et al., 2014). While the mutualistic association of plants with microorganisms can induce a response towards another microorganism deleterious microbial association can generate an immune system against themselves (Martínez-Hidalgo et al., 2015). The first line of defense in plants against microbes is the physical barrier such as waxy cuticles, rigid cell walls, and antimicrobial secondary metabolites (Reimer-Michalski and Conrath, 2016). Microbes, pathogens, or damage-associated molecular patterns (MAMP/PAMP/DAMP) are recognized by the plant cell surface receptors known as pattern recognition receptors (PRRs) to initiate a series of responses MAMP-triggered immunity (MTI), PAMP-triggered immunity (PTI) and DAMP-triggered immunity, these are collectively known as pattern triggered immunity (PTI; Figure 6.1; Saijo et al., 2018). Moreover, microbial virulence factors are recognized by PRRs to induce effector-triggered immunity (ETI). All these events collectively result in the suppression of damage caused by pathogen attacks. With the recognition of microbes on the cell surface by PPRs, immunity is induced with the rapid increase in calcium influx, production of reactive oxygen species (ROS), MAPK cascade, alteration in the cell wall, and activation of defense genes (Figure 6.1; De Lorenzo et al., 2018).

TABLE 6.1 Factors Affecting Plant-Microbe Interactions

Abiotic Factor	Plant	Microbiome	Impact on Interaction	References
Soil: Structure, Type, Composition, and pH				
Moisture	Herbaceous species	Proteobacteria, actinobacteria, fungal species	High moisture content aids in microbial motility thus increasing the uptake of carbon and nitrogen by the plant thereby promoting plant development.	Taketani et al. (2017); Zhang et al. (2017)
Soil type and structure	Corn plant	*Mycobacterium*, *Pseudomonas*, and *Bacillus*	Enhanced uptake of nitrogen, phosphorus, and potassium.	Egamberdiyeva (2007)
pH	Alfalfa	Acidobacteria, *Sinorhizobium meliloti*	Enhanced uptake of nutrients.	Donnarumma et al. (2014)
Minerals	Alfalfa	Bacteria and fungi	Enhanced bioavailability of minerals for plants	Beauregard et al. (2010); Coolon et al. (2013)
Climate Change/Abiotic Stress				
UVR	*E. herbicola, D. antarctica*	Bacteria and archaea	Increased production of plant secondary metabolites which alter the composition of root exudate.	Bhattacharyya and Jha (2012); Pineda et al. (2013)
Cold stress	*Triticum aestivum, S. lycopersicum*	*M. phyllosphaerae*, Arthrobacter, flavobacterium, flavimonas, pedobacter, *Pseudomonas*	Improved plant height and root length. Cold stress protection is induced through increased VOC production.	Bradáčová et al. (2016); Subramanian et al. (2016)
Heat stress	*T. aestivum, Sorghum*	*Pseudomonas putida* strain AKMP7, *Pseudomonas* sp. strain AKM-P6	Improved production of metabolites such as prolines, sugars, and organic compounds which confer heat tolerance.	Ali et al. (2011)
Salinity	*S. lycopersicum T. aestivum, A. thaliana, O. sativa*	*P. fluorescens, P. aeruginosa, P. stutzeri. D. natronolimnaea*	VOCs released by soil microbes alter the gene expression in plants and induce salinity tolerance. Accumulation of glycine betaine-like compounds occurs in plants.	Bharti et al. (2016); Mathew et al. (2015)

TABLE 6.1 *(Continued)*

Abiotic Factor	Plant	Microbiome	Impact on Interaction	References
Water stress	*Vigna radiata, S. lycopersicum*	*A. piechaudii, P. fluorescens*	The microbial association enables the production of catalase to confer stress through VOCs.	Mayak et al. (2004); Saravanakumar et al. (2011)
Heavy metal stress	*E. sativa, V. unguiculata sesquipedalis*	*P. putida, P. halotolerans*	Microbial association reduces phytotoxicity via the release of VOCs and stimulating plant genes.	Ho et al. (2013); Kamran et al. (2015)
			Biotic Stress	
Fungal attacks	Gossypium, Musa	*P. fluorescens, P. stutzeri,*	Antibiotic production protects the plant.	Ho et al. (2015)
C. orbiculare, P. cubensis	Cucumber	*B. licheniformis, Achromobacter* sp., *Streptomyces* sp.	Induce systemic resistance in plants through the production of VOCs and phytohormones.	Kloepper and Ryu (2006)

Some of the conserved MAMPs in microbes are flagellin, elongation factor, fungal chitin, and lipopolysaccharides. Receptor-like kinases (RLKs) are plasma membrane-bound proteins with extracellular domains. Moreover, flagellin-insensitive 2 (FLS2) and elongation factor-Tu (EFR) are PRRs that recognize epitope (flg22) and bacterial elongation factor-Tu epitope (elf18), respectively (Bauer et al., 2001; Zipfel et al., 2006). Some plant pathogens secret special enzymes to degrade the cell walls of plants. The degraded fragments of the cell wall act as elicitors to initiate the plasma protein thiolation index (PTI). The bacterial type 2 secretion system (T2SS) exports cell wall degrading enzymes like pectinases, endoglucanases, and cellulases which induce innate immunity by damaging the cell wall and releasing DAMPs (Jha et al., 2007).

Initiation of PTI is linked directly to the accumulation of multiple SERKs (somatic embryogenesis receptor-like kinases) which are the members of the PRRs complex. Apart from that most PRRs also need LRR-RK brassino steroid insensitive-1-associated KINASE/SERK3 (BAK1/SERK3) for their activity. EFR and FLS2 form a complex with BAK1 when treated with EF-Tu and flg22 which causes the phosphorylation of PRR and BAK1. When *Arabidopsis* plants with mutation in *bak1* were observed they showed decreased response to EF-Tu/flg22 treatment hence emphasizing the importance of BAK1 in the phosphorylation cascade (Chinchilla et al., 2007). BIK1 (botrytis-induced kinase 1) is another cytoplasmic protein kinase that has been identified as a regulator of FLS2-BAK1 interaction. Up-regulation of the *bik1* gene as well as rapid phosphorylation of BIK1 upon flagellin perception has been observed. *In vivo*, analysis showed that BIK1 interacts with both FLS2 and BAK1 and trans-phosphorylates both to propagate flagellin-mediated PTI signaling. A reduced response was observed in *Arabidopsis bik1* mutant for flagellin-mediated response and immunity against non-pathogenic bacterial infection emphasizing the importance of *bik1* in triggering PTI (Lu et al., 2010). A recent study showed that flagellin also recruits U-box E3 ubiquitin ligases, PUB12/PUB13, to the FLS2 receptor complex in *Arabidopsis.* BAK1 is required for FLS2-PUB12/PUB13 association. PUB12/PUB13 attenuates the activity of FLS2 by polyubiquitination and subsequent degradation. PUB12/PUB13 mutant *Arabidopsis* plants showed increased resistance to *Pseudomonas* spp. suggesting PUB12/PUB13 as a negative regulator of flagellin-triggered immune responses. The study also revealed that PUB12/PUB13 can be manipulated to enhance disease resistance in plants (Lu et al., 2011). Getting insight into PAMP-triggered immunity can help develop disease resistance in plants against various bacterial pathogens. Chitin elicitor receptor kinase 1 (CERK1) is another example of a PRR that perceives fungal chitin and bacterial peptidoglycan (Miya et al., 2007; Willmann et al., 2011).

Plants also recognize effector molecules released by pathogens by resistance proteins (R proteins) to induce effector-triggered immunity or R gene-mediated resistance (Dodds and Rathjen, 2010). The effector molecules are recognized by the plant cognate resistance (R) gene initiating a hypersensitive response leading to cell death hence limiting the spread of pathogen infection (Chisholm et al., 2006). Although both effectors-triggered immunity (ETI) and PTI give the same kind of immune response, ETI produces a faster and prolonged response to eliminate the pathogen effectively while the pathogen can escape detection from the plant in the absence of elicitor-pathogen interaction (Tsuda and Katagiri, 2010). Resistance genes belong to a group of intracellular proteins having a nucleotide binding and Leucine-rich repeat (LRR) domain (Dangl and Jones, 2001). Some of the identified components of ETI signaling are EDS1 (enhanced disease susceptibility 1) in *Arabidopsis* for recognition by TIR-NB_LRR (Heidrich et al., 2011) and NDR1 (non-race-specific disease resistance 1) required for recognition by CC-NB-LRR mediated signaling (Day et al., 2006). The pattern of recognition of effectors by NB-LRR follows two pathways 1) through direct effector interaction or 2) indirect interaction by an accessory protein part of NB-LRR protein complex. In the indirect path, the effector modifies the accessory proteins as a virulence target and leads to the recognition of protein by NB-LRR protein. A hypothesis indicating the role of guard proteins that is, proteins that interact directly with the effector and then activate the R protein was proposed known as the "guard hypothesis." The guard protein is activated by the pathogen and then induces plant resistance by triggering the R protein (Dangl and Jones, 2001). A study conducted in tomato plants on resistance against *P. syringae* effectors AvrPto and AvrPtoB showed that tomato plant NB-LRR induced resistance against the pathogen while in *Arabidopsis* plants when NB-LRR proteins (RPM1, RPS2, and RPS5) were present resistance developed against the fungal effectors AvrB, AvrRpm1, AvrRpt2 and AvrPphB (Chisholm et al., 2006). During the evolution plants have developed RPS2 (resistance to *Pseudomonas* syringae 2), and RPS5 (resistance to *Pseudomonas* syringae 5), to detect and inhibit the pathogen effectors AvrRpt2. In some cases, where have plants lost the avr gene, alternative pathway R-protein-mediated resistance in the plant to pathogen effector decreases the chances of entry into the plant. The translation of these pathways into crop plants can effectively improve the yield.

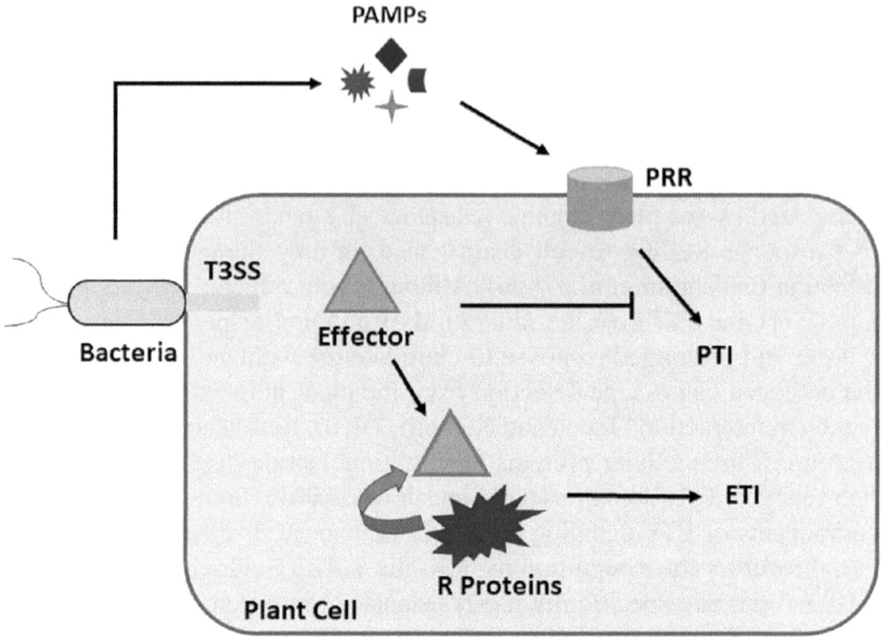

FIGURE 6.1 Schematic representations of signaling for plant immunity.

Note: As the bacterial pathogen attacks pattern recognition protein (PRP) on the surface of plant cells recognizes the PAMP signaling molecules such as flagellin, and EF-Tu and activates the PTI (PAMP-triggered immunity). Bacteria secrete effector proteins through T3SS a specialized structure that makes the condition inside the plant favorable by inhibiting PTI. However, plant activates effector-triggered immunity by recognizing effector molecules through resistance (R) proteins.

Source: Adapted from Pandey and Chaterjee (2013).

6.3.1 MAPK CASCADE

Another line of defense is a group of transferases known as Mitogen-activated protein kinases (MAPK) which catalyzes the phosphorylation of protein substrates on serine or threonine residues (Ligterink and Hirt, 2001). The regulation of MAPK can increase the cell response to various stresses such as high or low temperature, UV irradiation, ROS, water deficiency, heavy metals accumulation, and pathogen attack (Sheikh et al., 2013). The MAPK cascade involves three types of kinases: mitogen-activated protein kinase kinases (MAPKKKs; also known as MAP3Ks or MEKKs), MAP kinase kinases (MKKs; also known as MAP2Ks or MEKs), and MAP kinases (MAPKs; also known as MPKs). MAP kinase kinase kinase kinases (MAPKKKKs) have also been identified in plants (Raja et al., 2017). The MAPK pathway starts with the activation of MAPKKK by stimulation of PRRs. A downstream pathway is activated with the phosphorylation of MAPKK at two serine or threonine residues. After the activation, MAPKK acts as a dual-specificity kinase activating MAPK kinase by phosphorylating on the threonine and tyrosine residues (Cristina et al., 2010). This series of events of phosphorylation of MAPK kinases is known as MAPK cascade. About 110 genes encoding MAPK kinases have been identified in the *Arabidopsis thaliana* genome (Cristina et al., 2010) comparable to the number of MAPK genes in other species such as rice (Xiong et al., 2001), tomato (Mohanta et al., 2015), and maize (Kong et al., 2013). It has been reported that MAPKs regulate JA biosynthesis in plants as well as the expression of JA-dependent genes. The MPAKs in tomato SlMPK6-1 act to positively regulate the JA biosynthesis and signaling pathways (Figure 6.2; Kandoth et al., 2007).

SA actively regulates the expression of AtMPK3 AtEDR1 and ZmMKK6-ZmMPK4-1, induced by ozone stress (Ahlfors et al., 2004). Moreover, MPK3 and MPK6 also increase ethylene production in *Arabidopsis* in response because of induction with *Botrytis cinerea* (Han et al., 2010). The evidence given above clearly indicates the role of MAPKs in plant immunity. There are certain MAPKs such as AtMPK4 and AtMPK6 that are involved in both SA and JA signaling and positively regulate JA/ET responses (Figure 6.2; Brodersen et al., 2006) in contrast AtMEKK1-ArMKK1/2-AtMPK4 negatively regulates plant immunity by modulating SA signaling (Petersen et al., 2000). Tobacco MAPKs NtMPK3 and NtMPK6-1 also positively regulate the JA biosynthesis and repress the synthesis of SA in response to wounding (Seo et al., 2007). Apart from that NtMPK3 and MtMPK6-1 also positively regulate JA signaling and negatively regulate SA biosynthesis to ozone treatment (Gomi et al., 2005). Another study by Zhang et al. (2013) suggests the role of MAPKs (NtMPK16 and NtMPK20) in transcriptional regulation of JA and SA signaling by treatment with MeJA and SA. The role of MAPK in JA and SA signaling has been studied and confirmed in several plants such as *Oryza sativa, Zea mays, and Solanum lycopersicum* with the activation of MAPKs such as OsMPK17-1 because of JA and SA treatment. OsMKK1, OsMPK6, OsMPK17-1, and OsMPK3 are transcriptionally induced in rice when treated with JA and SA. Moreover, overexpression of OsMPK6 causes an accumulation of JA and SA as a result of pathogen treatment which indicates its involvement in OsMKK1-6 in JA and SA-mediated signaling (Cheong et al., 2003). Moreover, plants also have developed another defense response by forming mutualistic associations with microbes. The plant provides shelter to the friendly microbes and in return gets resistance from harmful microorganisms. This is done in order that friendly microbes to produce chemicals or toxins that protect the plant from attacking pathogens (Nishad et al., 2020).

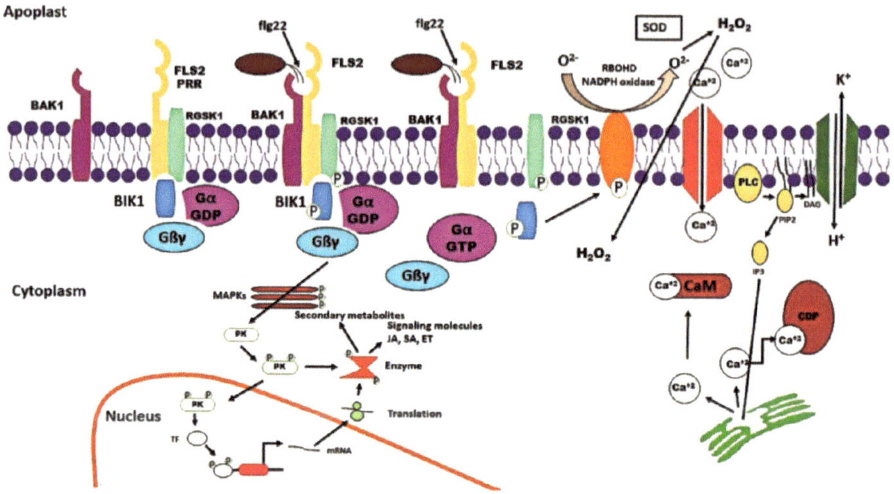

FIGURE 6.2 Schematic representation of MAMP-PR, MAPKs-induced signaling.

Note: PRR proteins recognize receptors (flg22) to form a complex containing FLS2, BAK1, BIK1, and RGSK1 (G-protein). The BIK1-dependent phosphorylation of RGSK1 releases G protein from the complex RGSK1 is released with the conversion of G protein from GDP to GTP. Ca^{2+} influx occurs as the first sign of an MAMP-triggered immune response activating the K^+/H^+ channel. Ca^{2+} activates phospholipase C which releases a secondary messenger IP3 that initiates internal release of Ca^{2+}. The pathway results in the expression of defense genes. MAPK cascade activates the accumulation of JA, SA, and ET and antimicrobial compounds (secondary metabolites) by the phosphorylation of transcription factor (TF).

Source: Adapted from Nishad et al. (2020).

6.4 SIGNIFICANCE OF PLANT INTERACTIONS IN SUSTAINABLE AGRICULTURE AND ENVIRONMENTAL SECURITY

Plants have developed an array of mechanisms to adapt to changing environments for optimal growth, development, and survival. Irrespective of such specialized physiological mechanisms, plants cannot stand alone against a continuously changing environment comprising both biotic and abiotic factors (Nishida, 2014). Plant interactions either antagonistic or symbiotic are equally valuable for its growth, development, survival, ecology, and cultivation (Agrahari et al., 2020).

Modern agriculture is still relying on agrochemicals for enhancing food quality and production resulting in the deterioration of soil quality. Advanced approaches in biotechnology and sustainable agriculture have a promising part in ensuring enhanced food production while keeping the integrity of soil ecology (Sharma et al., 2021).

The use of microorganisms associated with plant growth, development, and resistance/tolerance to stresses along current agriculture practices can revolutionize sustainable agriculture as plant-microbe interactions have global implications for food security and the ecology of natural ecosystems by influencing crop production (George et al., 2016).

In the rhizosphere of a plant, not all the interactions with microbes are beneficial. However, in some cases, the presence of microorganisms is critical for plant survival, community behavior, and success. Among those microbes, PGRP holds a special position as they can affect plant growth and development through hormone production, degradation of harmful chemicals, nutrient acquisition by nitrogen fixation, iron sequestration, solubilization of phosphorus, and inhibiting plant pathogens (Rani et al., 2019). Owning to their higher potential in sustainable agriculture, they can be used as biocontrol agents and alternate for chemical fertilizer, but their inability to colonize the roots of plants makes them ineffective for the use in field. Thus, a detailed study is needed to fully exploit PGRP for its use in sustainable agriculture (Verma et al., 2019).

6.5 CONCLUSIONS

Plants share a dynamic environment with a diverse range of microorganisms. These microbes can be beneficial, harmful, or symbiotic to the plant. As plants are sessile, they need to optimize their internal conditions to deal with the pathogenic microbes while maintaining symbiotic associations with beneficial microbes. This involves the response of individual plants to microbes as well as signaling neighboring plants about the ongoing situation. The interaction between plants and their microbes is greatly influenced by several environmental factors. When it comes to molecular processes our understanding of the impact of environmental conditions on the plant-microbe interaction is still at the beginning stage. This is mainly due to the closed static environment in which the experiments are conducted. Although these experiments give us insight into the plant-microbe interaction they are far from the conditions faced by the plant in nature. In the natural environment change in one factor around the plant is influenced by changes in several other factors. In addition, the insight into the molecular mechanism of plant-microbe interaction and how plants effectively oppose pathogen infection can help develop climate-resilient plants.

As climate change is the leading factor affecting global crop production, the need of the century is that more resilient crops must be developed that are resistant to both abiotic and biotic stresses. Several signaling receptors and molecules involved in plant-microbe interaction have been identified with their roles in these pathways. Such as plants produce ethylene to ward off pathogens, NO production in *Arabidopsis* provides resistance to the plant against *Sclerotina sclerotiorum* (Perchepied et al., 2010), disease resistance against *Botrytis cinerea* in *Nicotiana benthamiana* (Asai and Yoshioka, 2009). Further research into understanding the response of plants to microbes is required to utilize this knowledge to improve plant production.

KEYWORDS

- **ethylene**
- **pathogens**
- **plant immunity**
- **plant production**
- **plant-microbe interaction**
- **signaling pathways**
- **sustainable agriculture**

REFERENCES

Agrahari, R. K., Singh, P., Koyama, H., & Panda, S. K. (2020). Plant-microbe interactions for sustainable agriculture in the postgenomic era. *Current Genomics, 21*, 168–178.

Ahlfors, R., Macioszek, V., Rudd, J., Brosché, M., Schlichting, R., Scheel, D., & Kangasjärvi, J. (2004). Stress hormone-independent activation and nuclear translocation of mitogen-activated protein kinases in *Arabidopsis thaliana* during ozone exposure. *The Plant Journal, 40*, 512–522.

Ahmed, Z., Lim, B. R., Cho, J., Song, K. G., Kim, K. P., & Ahn, K. H. (2008). Biological nitrogen and phosphorus removal and changes in microbial community structure in a membrane bioreactor: Effect of different carbon sources. *Water Research, 42*, 198–210.

Ali, S. Z., Sandhya, V., Grover, M., Linga, V. R., & Bandi, V. (2011). Effect of inoculation with thermotolerant plant growth-promoting Pseudomonas putida strain AKMP7 on the growth of wheat (*Triticum* spp.) under heat stress. *Journal of Plant Interactions, 6*, 239–246.

Asai, S., & Yoshioka, H. (2009). Nitric oxide as a partner of reactive oxygen species participates in disease resistance to necrotrophic pathogen *Botrytis cinerea* in *Nicotiana benthamiana*. *Molecular Plant-Microbe Interactions, 22*, 619–629.

Badri, D. V., Jacqueline, M., Zhang, C. R., Shen, Q., & Vivanco, J. M. (2013). Application of natural blends of phytochemicals derived from the root exudates of *Arabidopsis* to the soil reveals that phenolic-related compounds predominantly modulate the soil microbiome. *Journal of Biological Chemistry, 288*, 4502–4512.

Baetz, U., & Martinoia, E. (2014). Root exudates: The hidden part of plant defense. *Trends in Plant Science, 19*(2), 90–98.

Bauer, Z., Gómez-Gómez, L., Boller, T., & Felix, G. (2001). The sensitivity of different ecotypes and mutants of *Arabidopsis thaliana* toward the bacterial elicitor flagellin correlates with the presence of receptor-binding sites. *Journal of Biological Chemistry, 276*, 45669–45676.

Beauregard, M. S., Hamel, C., & St-Arnaud, M. (2010). Long-term phosphorus fertilization impacts soil fungal and bacterial diversity but not the AM fungal community in alfalfa. *Microbial Ecology, 59*, 379–389.

Bharti, N., Pandey, S. S., Barnawal, D., Patel, V. K., & Kalra, A. (2016). Plant growth-promoting rhizobacteria *Dietzia natronolimnaea* modulates the expression of stress-responsive genes providing protection of wheat from salinity stress. *Scientific Reports, 6*, 34768.

Bhattacharyya, P. N., & Jha, D. K. (2012). Plant growth-promoting rhizobacteria (PGPR): Emergence in agriculture. *World Journal of Microbiology and Biotechnology, 28*, 1327–1350.

Bradáčová, K., Weber, N. F., Morad-Talab, N., Asim, M., Imran, M., Weinmann, M., & Neumann, G. (2016). Micronutrients (Zn/Mn), seaweed extracts, and plant growth-promoting bacteria as cold-stress protectants in maize. *Chemical Biology & Technology Agriculture, 3*, 19.

Brodersen, P., Petersen, M., Bjørn Nielsen, H., Zhu, S., Newman, M. A., Shokat, K. M., Rietz, S., Parker, J., & Mundy, J. (2006). Arabidopsis MAP kinase four regulates salicylic acid- and jasmonic acid/ethylene-dependent responses via EDS1 and PAD4. *The Plant Journal, 47*, 532–546.

Cheng, Y. T., Zhang, L., & He, S. Y. (2019). Plant-microbe interactions face an environmental challenge. *Cell Host & Microbe, 26*, 183–192.

Cheong, Y. H., Moon, B. C., Kim, J. K., Kim, C. Y., Kim, M. C., Kim, I. H., Park, C. Y., Kim, J. C., Park, B. O., & Koo, S. C. (2003). BWMK1, a rice mitogen-activated protein kinase, is located in the nucleus and mediates pathogenesis-related gene expression by activation of a transcription factor. *Plant Physiology, 132*, 1961–1972.

Chinchilla, D., Zipfel, C., Robatzek, S., Kemmerling, B., Nürnberger, T., Jones, J. D., & Boller, T. (2007). A flagellin-induced complex of the receptor FLS2 and BAK1 initiates plant defense. *Nature, 448*, 497–500.

Chisholm, S. T., Coaker, G., Day, B., & Staskawicz, B. J. (2006). Host-microbe interactions: Shaping the evolution of the plant immune response. *Cell, 124*, 803–814.

Chu, H., Neufeld, J. D., Walker, V. K., & Grogan, P. (2011). The influence of vegetation type on the dominant soil bacteria, archaea, and fungi in a low Arctic tundra landscape. *Soil Science Society of America Journal, 75*, 1756–1765.

Clark, C. M., & Tilman, D. (2008). Loss of plant species after chronic low-level nitrogen deposition to prairie grasslands. *Nature, 451*, 712–715.

Coolon, J. D., Jones, K. L., Todd, T. C., Blair, J. M., & Herman, M. A. (2013). Long-term nitrogen amendment alters the diversity and assemblage of soil bacterial communities in tallgrass prairie. *PLoS ONE, 8*, e67884.

Cristina, M. S., Petersen, M., & Mundy, J. (2010). Mitogen-activated protein kinase signaling in plants. *Annual Review of Plant Biology, 61*, 621–649.

Dangl, J. L., & Jones, J. D. (2001). Plant pathogens and integrated defense responses to infection. *Nature, 411*, 826–833.

Day, B., Dahlbeck, D., & Staskawicz, B. J. (2006). NDR1 interaction with RIN4 mediates the differential activation of multiple disease resistance pathways in *Arabidopsis*. *The Plant Cell, 18*, 2782–2791.

De Lorenzo, G., Ferrari, S., Cervone, F., & Okun, E. (2018). Extracellular DAMPs in plants and mammals: Immunity, tissue damage and repair. *Trends in Immunology, 39*, 937–950.

del Carmen Orozco-Mosqueda, M., del Carmen Rocha-Granados, M., Glick, B. R., & Santoyo, G. (2018). Microbiome engineering to improve biocontrol and plant growth-promoting mechanisms. *Microbiological Research, 208*, 25–31.

Dodds, P. N., & Rathjen, J. P. (2010). Plant immunity: Towards an integrated view of plant-pathogen interactions. *Nature Reviews Genetics, 11*, 539–548.

Donnarumma, F., Bazzicalupo, M., Blažinkov, M., Mengoni, A., Sikora, S., & Babić, K. H. (2014). Biogeography of *Sinorhizobium meliloti* nodulating alfalfa in different Croatian regions. *Research in Microbiology, 165*, 508–516.

Egamberdiyeva, D. (2007). The effect of plant growth-promoting bacteria on growth and nutrient uptake of maize in two different soils. *Applied Soil Ecology, 36*, 184–189.

el Zahar Haichar, F., Santaella, C., Heulin, T., & Achouak, W. (2014). Root exudates mediated interactions belowground. *Soil Biology and Biochemistry, 77*, 69–80.

Fahad, S., Hussain, S., Bano, A., Saud, S., Hassan, S., Shan, D., Khan, F. A., Khan, F., Chen, Y., & Wu, C. (2015). Potential role of phytohormones and plant growth-promoting rhizobacteria in abiotic stresses: Consequences for changing environment. *Environmental Science and Pollution Research, 22*, 4907–4921.

Finkel, O. M., Castrillo, G., Paredes, S. H., González, I. S., & Dangl, J. L. (2017). Understanding and exploiting plant-beneficial microbes. *Current Opinion in Plant Biology, 38*, 155–163.

Formánek, P., Rejšek, K., & Vranová, V. (2014). Effect of elevated CO_2, O_3, and UV radiation on soils. *The Scientific World Journal, 2014*, 730149. https://doi.org/10.1155/2014/730149.

George, T. S., Dou, D., & Wang, X. (2016). Plant-microbe interactions: Manipulating signals to enhance agricultural sustainability and environmental security. *Plant Growth Regulation, 80*, 1–3.

Gomi, K., Ogawa, D., Katou, S., Kamada, H., Nakajima, N., Saji, H., Soyano, T., Sasabe, M., Machida, Y., & Mitsuhara, I. (2005). A mitogen-activated protein kinase NtMPK4 activated by SIPKK is required for jasmonic acid signaling and is involved in ozone tolerance via stomatal movement in tobacco. *Plant Cell Physiology, 46*, 1902–1914.

Hacquard, S., Spaepen, S., Garrido-Oter, R., & Schulze-Lefert, P. (2017). The interplay between innate immunity and the plant microbiota. *Annual Review of Phytopathology, 55*, 565–589.

Han, L., Li, G. J., Yang, K. Y., Mao, G., Wang, R., Liu, Y., & Zhang, S. (2010). Mitogen-activated protein kinases 3 and 6 regulate *Botrytis cinerea*-induced ethylene production in *Arabidopsis*. *The Plant Journal, 64*, 114–127.

Heidrich, K., Wirthmueller, L., Tasset, C., Pouzet, C., Deslandes, L., & Parker, J. E. (2011). Arabidopsis EDS1 connects pathogen effector recognition to cell compartment-specific immune responses. *Science, 334*, 1401–1404.

Helman, Y., & Chernin, L. (2015). Silencing the mob: Disrupting quorum sensing as a means to fight plant disease. *Molecular Plant Pathology, 16*, 316–329.

Ho, Y. N., Chiang, H. M., Chao, C. P., Su, C. C., Hsu, H. F., Guo, C. T., Hsieh, J. L., & Huang, C. C. (2015). In planta biocontrol of soilborne Fusarium wilt of banana through a plant endophytic bacterium, *Burkholderia cenocepacia* 869T2. *Plant and Soil, 387*, 295–306.

Ho, Y. N., Hsieh, J. L., & Huang, C. C. (2013). Construction of a plant-microbe phytoremediation system: Combination of vetiver grass with a functional endophytic bacterium, *Achromobacter xylosoxidans* F3B, for aromatic pollutants removal. *Bioresource Technology, 145*, 43–47.

Hwang, H. H., Yu, M., & Lai, E. M. (2017). Agrobacterium-mediated plant transformation: Biology and applications. *The Arabidopsis Book, 15*, e0186. https://doi.org/10.1199/tab.0186.

Jha, G., Rajeshwari, R., & Sonti, R. V. (2007). Functional interplay between two *Xanthomonas oryzae* pv. *oryzae* secretion systems in modulating virulence on rice. *Molecular Plant-Microbe Interactions, 20*, 31–40.

Jones, J. D., Vance, R. E., & Dangl, J. L. (2016). Intracellular innate immune surveillance devices in plants and animals. *Science, 354*, aaf6395. https://doi.org/10.1126/science.aaf6395.

Kai, M., Effmert, U., & Piechulla, B. (2016). Bacterial-plant-interactions: Approaches to unravel the biological function of bacterial volatiles in the rhizosphere. *Frontiers in Microbiology, 7*, 108.

Kamran, M. A., Syed, J. H., Eqani, S. A. M. A. S., Munis, M. F. H., & Chaudhary, H. J. (2015). Effect of plant growth-promoting rhizobacteria inoculation on cadmium (Cd) uptake by *Eruca sativa. Environmental Science and Pollution Research, 22*, 9275–9283.

Kandoth, P. K., Ranf, S., Pancholi, S. S., Jayanty, S., Walla, M. D., Miller, W., Howe, G. A., Lincoln, D. E., & Stratmann, J. W. (2007). Tomato MAPKs LeMPK1, LeMPK2, and LeMPK3 function in the systemin-mediated defense response against herbivorous insects. *Proceedings of the National Academy of Sciences, 104*, 12205–12210.

Kloepper, J. W., & Ryu, C. M. (2006). Bacterial endophytes as elicitors of induced systemic resistance. In B. J. E. Schulz, C. J. C. Boyle, & T. N. Sieber (Eds.), *Microbial Root Endophytes* (pp. 33–52). Springer-Verlag Berlin Heidelberg.

Kong, X., Pan, J., Zhang, D., Jiang, S., Cai, G., Wang, L., & Li, D. (2013). Identification of mitogen-activated protein kinase gene family and MKK–MAPK interaction network in maize. *Biochemical and Biophysical Research Communications, 441*, 964–969.

Kravchenko, L. V., Azarova, T. S., Leonova-Erko, E. I., Shaposhnikov, A. I., Makarova, N. M., & Tikhonovich, I. A. (2003). Root exudates of tomato plants and their effect on the growth and antifungal activity of *Pseudomonas* strains. *Microbiology, 72*, 37–41.

Ligterink, W., & Hirt, H. (2001). Mitogen-activated protein (MAP) kinase pathways in plants: Versatile signaling tools. *International Review of Cytology, 201*, 209–275. https://doi.org/10.1016/s0074-7696(01)01004-x.

Lu, D., Lin, W., Gao, X., Wu, S., Cheng, C., Avila, J., Heese, A., Devarenne, T. P., He, P., & Shan, L. (2011). Direct ubiquitination of pattern recognition receptor FLS2 attenuates plant innate immunity. *Science, 332*, 1439–1442.

Lu, D., Wu, S., Gao, X., Zhang, Y., Shan, L., & He, P. (2010). A receptor-like cytoplasmic kinase, BIK1, associates with a flagellin receptor complex to initiate plant innate immunity. *Proceedings of the National Academy of Sciences, 107*, 496–501.

Martínez-Hidalgo, P., García, J. M., & Pozo, M. J. (2015). Induced systemic resistance against *Botrytis cinerea* by *Micromonospora* strains isolated from root nodules. *Frontiers in Microbiology, 6*, 922.

Massalha, H., Korenblum, E., Tholl, D., & Aharoni, A. (2017). Small molecules below-ground: The role of specialized metabolites in the rhizosphere. *The Plant Journal, 90*, 788–807.

Mathew, D. C., Ho, Y. N., Gicana, R. G., Mathew, G. M., Chien, M. C., & Huang, C. C. (2015). A rhizosphere-associated symbiont, *Photobacterium* spp. strain MELD1, and its targeted synergistic activity for phytoprotection against mercury. *PLoS ONE, 10*, e0121178.

Mayak, S., Tirosh, T., & Glick, B. R. (2004). Plant growth-promoting bacteria that confer resistance to water stress in tomatoes and peppers. *Plant Science, 166*, 525–530.

Miya, A., Albert, P., Shinya, T., Desaki, Y., Ichimura, K., Shirasu, K., Narusaka, Y., Kawakami, N., Kaku, H., & Shibuya, N. (2007). CERK1, a LysM receptor kinase, is essential for chitin elicitor signaling in *Arabidopsis*. *Proceedings of the National Academy of Sciences, 104*, 19613–19618.

Mohanta, T. K., Arora, P. K., Mohanta, N., Parida, P., & Bae, H. (2015). Identification of new members of the MAPK gene family in plants shows diverse conserved domains and novel activation loop variants. *BMC Genomics, 16*, 58.

Mommer, L., Hinsinger, P., Prigent-Combaret, C., & Visser, E. J. W. (2016a). Advances in the rhizosphere: Stretching the interface of life. *Plant and Soil, 407*, 1–8.

Mommer, L., Kirkegaard, J., & van Ruijven, J. (2016b). Root–Root interactions: Towards a rhizosphere framework. *Trends in Plant Science, 21*, 209–217.

Nishad, R., Ahmed, T., Rahman, V. J., & Kareem, A. (2020). Modulation of plant defense system in response to microbial interactions. *Frontiers in Microbiology, 11*, 1298.

Nishida, R. (2014). Chemical ecology of insect-plant interactions: Ecological significance of plant secondary metabolites. *Bioscience, Biotechnology, and Biochemistry, 78*, 1–13.

Oldroyd, G. E. D. (2013). Speak, friend, and enter: Signaling systems that promote beneficial symbiotic associations in plants. *Nature Reviews Microbiology, 11*, 252–263.

Pandey, A., & Chaterjee, S. (2013). Signaling in plant-microbe interactions. *Plant Stress, 7S*, 52–59.

Peñuelas, J., Asensio, D., Tholl, D., Wenke, K., Rosenkranz, M., Piechulla, B., & Schnitzler, J. P. (2014). Biogenic volatile emissions from the soil. *Plant Cell and Environment, 37*, 1866–1891.

Perchepied, L., Balagué, C., Riou, C., Claudel-Renard, C., Rivière, N., Grezes-Besset, B., & Roby, D. (2010). Nitric oxide participates in the complex interplay of defense-related signaling pathways controlling disease resistance to *Sclerotinia sclerotiorum* in *Arabidopsis thaliana*. *Molecular Plant-Microbe Interactions, 23*, 846–860.

Petersen, M., Brodersen, P., Naested, H., Andreasson, E., Lindhart, U., Johansen, B., Nielsen, H. B., Lacy, M., Austin, M. J., & Parker, J. E. (2000). Arabidopsis MAP kinase 4 negatively regulates systemic acquired resistance. *Cell, 103*, 1111–1120.

Pierik, R., Tholen, D., Poorter, H., Visser, E. J. W., & Voesenek, L. A. C. J. (2006). The Janus face of ethylene: Growth inhibition and stimulation. *Trends in Plant Science, 11*(4), 176–183.

Pieterse, C. M., Zamioudis, C., Berendsen, R. L., Weller, D. M., Van Wees, S. C., & Bakker, P. A. (2014). Induced systemic resistance by beneficial microbes. *Annual Review of Phytopathology, 52*, 347–375.

Pineda, A., Dicke, M., Pieterse, C. M. J., & Pozo, M. J. (2013). Beneficial microbes in a changing environment: Are they always helping plants to deal with insects? *Functional Ecology, 27*, 574–586.

Raja, V., Majeed, U., Kang, H., Andrabi, K. I., & John, R. (2017). Abiotic stress: Interplay between ROS, hormones, and MAPKs. *Environmental and Experimental Botany, 137*, 142–157.

Rani, R., Kumar, V., Gupta, P., & Chandra, A. (2019). Application of plant growth promoting rhizobacteria in remediation of pesticides contaminated stressed soil. In H. B. Singh & A. Vaishnav (Eds.), *New and Future Developments in Microbial Biotechnology and Bioengineering* (pp. 341–353). Elsevier.

Ravanbakhsh, M., Sasidharan, R., Voesenek, L. A. C. J., Kowalchuk, G. A., & Jousset, A. (2018). Microbial modulation of plant ethylene signaling: Ecological and evolutionary consequences. *Microbiome, 6*, 52.

Rodriguez, P. A., Rothballer, M., Chowdhury, S. P., Nussbaumer, T., Gutjahr, C., & Falter-Braun, P. (2019). Systems biology of plant-microbiome interactions. *Molecular Plant, 12*, 804–821.

Rosier, A., Bishnoi, U., Lakshmanan, V., Sherrier, D. J., & Bais, H. P. (2016). A perspective on inter-kingdom signaling in plant–Beneficial microbe interactions. *Plant Molecular Biology, 90*, 537–548.

Ruamps, L. S., Nunan, N., Pouteau, V., Leloup, J., Raynaud, X., Roy, V., & Chenu, C. (2013). Regulation of soil organic C mineralization at the pore scale. *FEMS Microbiology Ecology, 86*, 26–35.

Ryu, C. M., Farag, M. A., Hu, C. H., Reddy, M. S., Wei, H. X., Paré, P. W., & Kloepper, J. W. (2003). Bacterial volatiles promote growth in *Arabidopsis*. *Proceedings of the National Academy of Sciences USA, 100*, 4927–4932.

Saijo, Y., Loo, E. P. I., & Yasuda, S. (2018). Pattern recognition receptors and signaling in plant-microbe interactions. *The Plant Journal, 93*, 592–613.

Santoyo, G., Pacheco, C. H., Salmerón, J. H., & León, R. H. (2017). The role of abiotic factors modulating the plant-microbe-soil interactions: Toward sustainable agriculture. *Spanish Journal of Agricultural Research, 15*, 13.

Saravanakumar, D., Kavino, M., Raguchander, T., Subbian, P., & Samiyappan, R. (2011). Plant growth-promoting bacteria enhance water stress resistance in green gram plants. *Acta Physiologiae Plantarum, 33*, 203–209.

Sasse, J., Martinoia, E., & Northen, T. (2018). Feed your friends: Do plant exudates shape the root microbiome? *Trends in Plant Science, 23*, 25–41.

Schikora, A., Schenk, S. T., & Hartmann, A. (2016). Beneficial effects of bacteria-plant communication based on quorum sensing molecules of the N-acyl homoserine lactone group. *Plant Molecular Biology, 90*, 605–612.

Seo, S., Katou, S., Seto, H., Gomi, K., & Ohashi, Y. (2007). The mitogen-activated protein kinases WIPK and SIPK regulate the levels of jasmonic and salicylic acids in wounded tobacco plants. *The Plant Journal, 49*, 899–909.

Sharma, P., Tarun, K., Monika, Y., Sarvajeet, S. G., & Nar, S. C. (2021). Plant-microbe interactions for sustainable agriculture and food security. *Plant Gene, 28*, 100325.

Sheikh, A. H., Raghuram, B., Jalmi, S. K., Wankhede, D. P., Singh, P., & Sinha, A. K. (2013). Interaction between two rice mitogen-activated protein kinases and its possible role in plant defense. *BMC Plant Biology, 13*, 1–11.

Sinha, R. P., Klisch, M., & Häder, D. P. (1999). Induction of a mycosporine-like amino acid (MAA) in the rice-field cyanobacterium *Anabaena* sp. by UV irradiation. *Journal of Photochemistry and Photobiology B: Biology, 52*, 59–64.

Subramanian, P., Kim, K., Krishnamoorthy, R., Mageswari, A., Selvakumar, G., & Sa, T. (2016). Cold stress tolerance in psychrotolerant soil bacteria and their conferred chilling resistance in tomato (*Solanum lycopersicum* Mill.) under low temperatures. *PLoS ONE, 11*, e0161592.

Taketani, R. G., Lançoni, M. D., Kavamura, V. N., Durrer, A., Andreote, F. D., & Melo, I. S. (2017). Dry season constrains bacterial phylogenetic diversity in a semi-arid rhizosphere system. *Microbial Ecology, 73*, 153–161.

Thrall, P. H., Hochberg, M. E., Burdon, J. J., & Bever, J. D. (2007). Coevolution of symbiotic mutualists and parasites in a community context. *Trends in Ecology & Evolution, 22*, 120–126.

Trivedi, P., Leach, J. E., Tringe, S. G., Sa, T., & Singh, B. K. (2021). Plant-microbiome interactions: From community assembly to plant health. *Nature Reviews Microbiology, 19*, 72–72.

Tsuda, K., & Katagiri, F. (2010). Comparing signaling mechanisms engaged in pattern-triggered and effector-triggered immunity. *Current Opinion in Plant Biology, 13*, 459–465.

Uroz, S., Ioannidis, P., Lengelle, J., Cébron, A., Morin, E., Buee, M., & Martin, F. (2013). Functional assays and metagenomic analyses reveal differences between the microbial communities inhabiting the soil horizons of a Norway spruce plantation. *PLoS ONE, 8*, e55929.

Venturi, V., & Keel, C. (2016). Signaling in the rhizosphere. *Trends in Plant Science, 21*, 187–198.

Verma, D. K., Pandey, A. K., Mohapatra, B., Srivastava, S., Kumar, V., Talukdar, D., Yulianto, R., Zuan, A. T. K., Jobanputra, A. H., & Asthir, B. (2019). Plant growth-promoting rhizobacteria: An eco-friendly approach for sustainable agriculture and improved crop production. In D. K. Verma (Ed.), *Microbiology for Sustainable Agriculture, Soil Health, and Environmental Protection* (pp. 3–80). Apple Academic Press.

Willmann, R., Lajunen, H. M., Erbs, G., Newman, M. A., Kolb, D., Tsuda, K., Katagiri, F., Fliegmann, J., Bono, J. J., & Cullimore, J. V. (2011). Arabidopsis Lysin-motif proteins LYM1 LYM3 CERK1 mediate bacterial peptidoglycan sensing and immunity to bacterial infection. *Proceedings of the National Academy of Sciences, 108*, 19824–19829.

Wilpiszeski, R. L., Aufrecht, J. A., Retterer, S. T., Sullivan, M. B., Graham, D. E., Pierce, E. M., Zablocki, O. D., Palumbo, A. V., & Elias, D. A. (2019). Soil aggregates microbial communities: Towards understanding microbiome interactions at biologically relevant scales. *Applied and Environmental Microbiology, 85*, e00324–00319.

Xiong, L., Lee, M. W., Qi, M., & Yang, Y. (2001). Identification of defense-related rice genes by suppression subtractive hybridization and differential screening. *Molecular Plant-Microbe Interactions, 14*, 685–692.

Yang, Y., Wang, N., Guo, X., Zhang, Y., & Ye, B. (2017). Comparative analysis of bacterial community structure in the rhizosphere of maize by high-throughput pyrosequencing. *PLoS ONE, 12*, e0178425.

Yi, H. S., Yang, J. W., & Ryu, C. M. (2013). ISR meets SAR outside the Additive action of the endophyte *Bacillus pumilus* INR7 and the chemical inducer, benzothiadiazole, on induced resistance against bacterial spot in field-grown pepper. *Frontiers in Plant Science, 4*, 122.

Zhalnina, K., Dias, R., de Quadros, P. D., Davis-Richardson, A., Camargo, F. A. O., Clark, I. M., McGrath, S. P., Hirsch, P. R., & Triplett, E. W. (2015). Soil pH determines microbial diversity and composition in the Park Grass Experiment. *Microbial Ecology, 69*, 395–406.

Zhang, N., Wang, D., Liu, Y., Li, S., Shen, Q., & Zhang, R. (2014). Effects of different plant root exudates and their organic acid components on chemotaxis, biofilm formation, and colonization by beneficial rhizosphere-associated bacterial strains. *Plant and Soil, 374*, 689–700.

Zhang, X. F., Zhao, L., Xu, S. J., Liu, Y. Z., Liu, H. Y., & Cheng, G. D. (2013). Soil moisture effect on bacterial and fungal community in Beilu River (Tibetan Plateau) permafrost soils with different vegetation types. *Journal of Applied Microbiology, 114*, 1054–1065.

Zhang, X., Zhang, R., Gao, J., Wang, X., Fan, F., Ma, X., Yin, H., Zhang, C., Feng, K., & Deng, Y. (2017). Thirty-one years of rice-rice-green manure rotations shape the rhizosphere microbial community and enrich beneficial bacteria. *Soil Biology and Biochemistry, 104*, 208–217.

Zipfel, C., & Oldroyd, G. E. D. (2017). Plant signaling in symbiosis and immunity. *Nature, 543*, 328–336.

Zipfel, C., Kunze, G., Chinchilla, D., Caniard, A., Jones, J. D., Boller, T., & Felix, G. (2006). Perception of the bacterial PAMP EF-Tu by the receptor EFR restricts *Agrobacterium*-mediated transformation. *Cell, 125*, 749–760.

CHAPTER 7

Agroecosystem Diversity and Maintenance of Soil and Environment Health

MANEESH KUMAR,[1,*] MOHD. SAYEED AKHTAR,[2,*] DIBYANGANA GANGULY,[3] LOVELY KUMARI,[4] KOMAL KUMARI,[5] and MANSHI AGRAWAL[6]

[1]*Department of Biotechnology, Magadh University, Bodh Gaya, Bihar, India*

[2]*Department of Botany, Gandhi Faiz-e-Aam College, Shahjahanpur, Uttar Pradesh, India*

[3]*Foundation for Innovative New Diagnostics India, New Delhi, India*

[4]*Department of Biotechnology, Jaipur National University, Jaipur, Rajasthan, India*

[5]*Department of Biochemistry, Patna University, Patna, Bihar, India*

[6]*Department of Microbiology, Dr. Harisingh Gour Vishwavidyalaya (A Central University), Sagar, Madhya Pradesh, India*

*Corresponding author.

ABSTRACT

Agroecosystem diversity is one of the significant agrosystem concepts for ecological suitability and sustainability. It is concerned with species diversity and functional groups of species in an agroecosystem. This diversity can be measured and managed in a variety of ways, such as species richness, functional diversity, and complexity. Agroecosystem diversity is key to reducing the effects of pests, diseases, and weed pressure and increasing

Agricultural Biotechnology: Issues, Challenges, and Recent Developments. Mohd. Sayeed Akhtar and Khalid Rehman Hakeem (Eds.)
© 2025 Apple Academic Press, Inc. Co-published with CRC Press (Taylor & Francis)

system stability and resilience. It is also closely linked to biodiversity conservation in agroecosystems, as it is essential to maintain a balance between species and functional groups for optimal ecological performance. It emphasizes biodiversity in agricultural systems that comprise living and non-living elements. Land, soil, water, plants, animals, climate, and humans' diverse ecosystems provide food, energy, water, and air quality. The diversity makes systems adaptable to climate change, disease, and insect outbreaks, promoting sustainability. Farmers can improve economic and social outcomes by diversifying agroecosystems. Pesticide and fertilizer use can be reduced by increasing agroecosystem diversity. Finally, agroecosystem diversity can aid small-scale farmers who lack money or expertise in new technologies by promoting equal access to land, resources, and information. It is essential for providing a balanced environment that can sustain healthy soils and promote a healthy environment. Therefore, the present chapter highlighted the constructive ecological impact of proper utilization of the agroecosystem to maintain soil and environmental health and its socio-economic impact on human beings.

7.1 INTRODUCTION

Agriculture and its productivity are an essential part of the living world. Being a part of well-being is one of the vitals for the survival of humans and animals and, for keeping a healthy ecosystem. It has a significant ecological hoofmark that covers one-third of the global land areas including almost all the biodiverse tropics. In the last two decades, the study of agroecosystem variety has risen in favor of an ecological field of inquiry due to the promise it holds for providing a comprehensive framework for comprehending the intricate workings of these systems. They comprise a wide range of complex systems involving interactions between biotic and abiotic components, and their diversity is critical for preserving ecosystem health and ecological integrity (Gliessman, 2021; Simoni et al., 2021). The investigations concentrated on species variety, genetic diversity, functional diversity, temporal diversity, and landscape diversity. Agroecosystems rely on a wide variety of species to function well, each of which contributes something unique to the whole. However, certain species' contributions to ecosystem services are more crucial than those of others (Gałązka and Grządziel, 2018; Zambrano et al., 2019). Genetic variety is particularly important in agroecosystems because it allows species to adapt to changing environmental conditions and reduces the danger of disease or pest outbreaks. Functional diversity examines the various roles performed by various species and how they interact with one another. The findings can also be useful for land managers in their efforts to preserve healthy, productive agroecosystems, which are crucial for the provision of food, fiber, and other ecosystem services (Rendon et al., 2020). Temporal diversity examines how an agroecosystem's species composition changes over time, while landscape diversity examines how the terrain affects species composition. Each of these areas of investigation is necessary for understanding the many factors that influence agroecosystems and how they can be managed for the benefit of humanity (Williams-Nguyen et al., 2016). Because of their easily manipulable components, simple food webs, and well-defined components, agricultural systems are used to test the ecological perspective. They were previously utilized extensively as ideal model systems for plant physiology, plant-insect interaction, and predator-prey theory.

Recent years have seen a rise in the study of ecological processes in agriculture, with the goal of better manipulating them to increase agroecosystem output while decreasing their negative effects on the environment, in line with the goals of sustainable agriculture (Chakraborty et al., 2017; Kumar and Akhtar, 2019). It is the science of applying ecological principles to the management of agroecosystems, and it is becoming increasingly popular as a way to increase food production while minimizing negative environmental impacts. It focuses on understanding and managing the interactions between plants, animals, and the environment, as well as the relationships between different species within an agroecosystem. It is a holistic approach to farming and food production that takes into account the entire ecosystem and seeks to promote a balance between production and sustainability. It is being used in several different ways. It is being used to develop sustainable farming practices, such as crop rotation and cover cropping, which reduce the need for synthetic fertilizers and pesticides. It is also being used to improve soil fertility, increase water retention, and reduce soil erosion. It actively promotes the growth of beneficial species, such as pollinators, which can increase yields and reduce the need for costly inputs (Hatt et al., 2016; MacLaren et al., 2020).

Agroecology is becoming one of the important in the faces of climate change. As temperatures rise and weather patterns become more unpredictable, agroecology can help farmers adapt their practices to changing conditions and maintain their yields. By promoting sustainable agricultural practices, agroecology encourages farmers to conserve resources, such as water, energy, and soil, which will help them better prepare for and manage increasingly severe weather events. Additionally, agroecology can help reduce the adverse impacts of climate change, such as increasing temperatures, extreme weather events, and drought, on agricultural yields (Aguilera et al., 2020; Kumar et al., 2019a). Agroecology encourages the use of natural resources, such as water conservation and crop rotation, and encourages the use of organic fertilizers and pest control methods. By implementing these practices, farmers can increase their resilience to climate change, helping them maintain their yields and sustain their livelihoods. Therefore, the present chapter highlighted the constructive ecological impact of proper utilization of the agroecosystem to maintain soil and environment health and its socio-economic impact on human beings.

7.2 AGROECOSYSTEM AND ITS ROLE IN THE ENVIRONMENT

The ecological theories of community ecology, ecosystem ecology, physiological ecology, population biology, and landscape ecology are only a few of the many that have been applied to agricultural systems and are thus included in the sub-discipline of "agroecosystem." The term "agroecosystem" refers to a natural ecosystem that has been transformed for agricultural purposes, such as the production of food and fiber. They have intentional as well as unexpected variations. Domesticated plants, animals, and beneficial organisms are grouped in space and time through planned diversity.

Unplanned diversity includes weedy plants, herbivores, predators, microorganisms, and other creatures that remain in the system after it has been altered for agriculture or moved in from the surrounding landscape. Both types of diversity have a big effect on how well an agroecosystem works, how it is stable, how pests are controlled, how the soil works, and how organisms move between farmland and their natural habitats in an agricultural landscape (Hodgson, 2012; Power and Kenmore, 2013). Many researchers and scholars scientifically engage in exploring new methods to develop sustainable crop production and food security without disturbing the socio-economic dimensions and environment for multiple ecosystem services (Schreefel et al., 2020). One of the new concepts for sustainable farming comes out of regenerative agriculture. The idea of "regenerative agriculture" is a way of managing farms that prioritize soil health. Ecosystem services like water filtering, nitrogen cycling, and improved biodiversity are only a few of the many positive outcomes that can result from placing soil quality at the center of agricultural practices (Figure 7.1). Reduced inputs and greater management are common practices in regenerative agriculture, with the goal of adapting both to the needs of the land. A natural ecosystem rarely features a unified crop or plant community (when a single crop is sown in a field rather than multiple crop species).

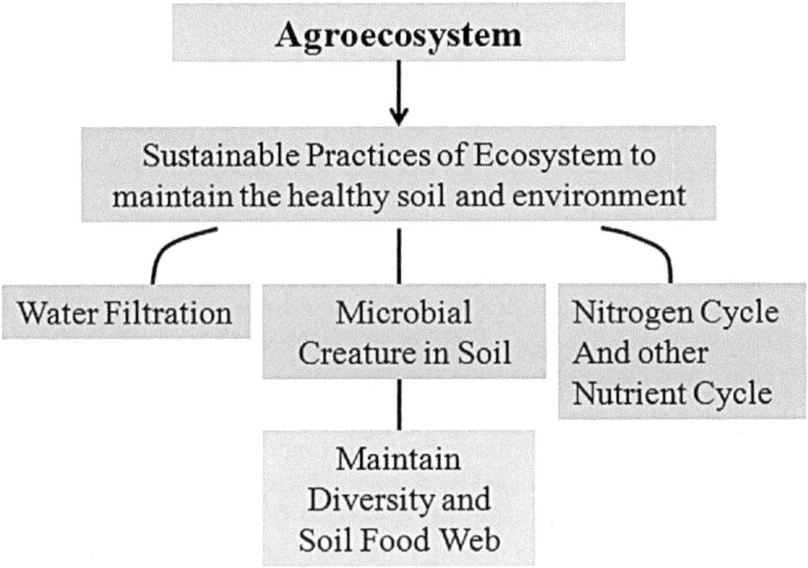

FIGURE 7.1 Significance of agroecosystem.

Nowadays, monocultures are used almost exclusively in farming because they are so easy to maintain. However, the diversity of crop species above ground leads to the diversity of soil creatures below ground, as various plant species associate with various soil organisms. When as many of the interactions between species that are responsible for the different nutrient cycles as possible are present, the soil food web works best. The beauty of natural systems is that they attempt to fix and rebuild themselves automatically (Kumar et al., 2018a). They will do that if they are left alone. Humans, even when well-intentioned, must be managed to prevent this from occurring. In the chapter, we have tried to highlight the significance of regenerative agriculture in the present ecological system. We discussed the rationales for introducing and encouraging the concept of maintaining soil degradation and protecting the ecosystem.

7.3 AGROECOSYSTEM SUSTAINABILITY

Although the agroecosystems share many traits with their natural counterparts, from a toxicological point of view, agricultural ecosystems are distinguished by the use of agrochemicals such as pesticides, fertilizers, and plant growth regulators. The types of crops and livestock being raised will have a major impact on the types and levels of pesticide pollution. The diversity of chemicals will be smaller in monocultures, but the concentrations may be higher, whereas the opposite may be true in agroecosystems supporting multiple crop production. The presence of combustion products from fossil fuels may also offer challenges in highly mechanized agroecosystems (Hodgson, 2012). In addition to agroecosystems, indoor habitats are also considered distinct ecosystems that call for careful consideration. This category includes buildings such as silos and barns that are used for the farming of various animals, most notably chickens and pigs. Toxicologists worry about the potential health effects of inhaling toxicants in an indoor agroecosystem, but this is not an issue in an outdoor agroecosystem. Organic dust, ammonia, hydrogen sulfide (H_2S), nitrogen oxide, and mycotoxins are all potential causes of organic dust toxic syndrome, also known as acute farmer's lung, as well as asthma, bronchitis, and rhinitis. Carbon monoxide poisoning is most often caused by the wrong use of gasoline-powered tools, like pressure washers (Brockerhoff et al., 2017; Kumar et al., 2018b). Such biodiversity is important to maintaining soil health. It increases the number of soil organisms, their activity, and the complexity of the soil environment. Diversity in an agroecosystem can be increased by introducing a variety of plants and animals, such as cover crops, beneficial insects, and livestock. All of these components can contribute to the composition of soil organisms and their activities, which can increase the diversity of the soil environment (Alyokhin et al., 2020; Kumar et al., 2017a). Diversity in an agroecosystem can also help to cycle nutrients, improve water infiltration, reduce soil erosion, and reduce the impacts of weeds and pests. In addition, diverse agroecosystems can provide a variety of ecosystem services, such as pollination, pest control, and soil fertility improvement. All of these benefits can help maintain soil health over time.

Soil health, fertility, and stability can all be preserved and even improved through the use of sustainable agroecosystems. Soil health, fertility, and stability can all be preserved and even improved through the use of sustainable agroecosystems. Agroecosystems provide a number of valuable functions, including nutrient cycling, water filtering, carbon sequestration, and wildlife habitat. The provision of a habitat for a variety of creatures, such as predators and beneficial insects that aid in the management of pests, is another way in which agroecosystems contribute to the promotion of biodiversity (Kumar et al., 2017a). Reducing tillage and growing cover crops, both of which help minimize soil erosion and improve soil structure, are further ways in which agroecosystems help lessen soil degradation. The systems have the potential to cut down on the amount of chemical fertilizer and pesticides that are applied to fields, which in turn lowers the likelihood that water will be contaminated.

7.3.1 RICHNESS OF AGROECOSYSTEM DIVERSITY AND LEGAL SUPPORT

The complexity and intentional diversity of agroecosystems, which could include functional diversity, genetic diversity, species diversity, and structural diversity, vary greatly. The complexity and diversity of agroecosystems depend on the purpose of the system. For example, a small-scale, diversified farming system typically has greater complexity and diversity than a large-scale, monoculture farming system because it requires multiple species, different farming practices, and higher levels of management. Similarly, an agroecosystem designed to provide food, fiber, fuel, and other ecosystem services may have greater complexity and diversity than an agroecosystem that is only used for food production. In terms of functional diversity, agroecosystems may include a variety of soil management practices, such as tillage, crop rotation, cover cropping, and mulching, which help promote soil fertility, water retention, and erosion control. Agroecosystems can also include various types of pest management strategies, such as the use of beneficial organisms, trap crops, and chemical pesticides. Genetic diversity in agroecosystems is typically maintained through the use of crop varieties that are adapted to local environmental conditions. Additionally, genetic diversity can be enhanced through the use of crop breeding practices that promote genetic diversity and the use of heirloom varieties of crops (Alyokhin et al., 2020; Kumar et al., 2017b; Kumari et al., 2021). Species diversity in agroecosystems is typically maintained through the use of multiple crops, the inclusion of beneficial organisms, such as pollinators, and the planting of diverse cover crops. Structural diversity is maintained through the use of different land management practices, such as terracing, contour planting, and alley cropping. Overall, the complexity and diversity of agroecosystems can vary greatly depending on the design and purpose of the system. However, maintaining high levels of complexity and diversity is important for maintaining healthy agroecosystems that can provide multiple ecosystem services (Crews et al., 2018). Low-planned variety and simplified community assemblages characterize the input-intensive monoculture systems that dominate commercial agriculture in most parts of the world. These are two of the system's defining characteristics. Input-intensive monocultures require a large number of inputs, such as fertilizers and pesticides, to grow a single crop or a small number of crops (Yadav et al., 2018). This type of system is characterized by a lack of variety, as only one or a few crops are grown on a large scale. Simplified community assemblages are also a feature of this type of farming, as few other organisms are present other than the crops being grown. This simplification of the environment can have a profound effect on biodiversity and ecosystem health, as well as the availability of resources for farmers. On the other hand, many tropical backyard gardens and conventional farming methods showcase highly organized varieties and assemblages of allied biota. Many studies have been done on the effects of planned diversity on ecological processes like primary production, pest control, decomposition, and nutrient cycling. These studies were made possible by the wide range of planned diversity that has been used in systems that have already been set up (Kumar et al., 2017c; Power and Kenmore, 2013; Seabloom et al., 2017). In such a massive farming land, the agricultural system maintains a balancing act to meet human needs (Figure 7.2). In the upcoming decades, the expectation simultaneously increases with the growing world population. In this, the system must provide some new economic opportunities to the farmers. With this prospect, concerns also have been taken to managing a clean environment with a sustainable and impactful ecosystem (Searchinger et al., 2019). Apart from a healthy ecosystem for the agricultural system, the focus has to be given, to how to minimize food insecurity. The insecurity of the food gap from 2006 to 2050 would be faced by the whole world. Up to 2050 feeding more than 10 billion additional people would lead to an immense burden on the ecosystem. This would only be possible by doing an 11% hike in the total crop production, unexpected agricultural expansion by clearing leftover forests, and, the emission of greenhouse gas (GHG) during agricultural production to hold global warming below 2°C. But the scenario might change with the increase in income, as they start consuming other sources to lower the food gap. They might devour animal-based food times including milk and meat (Ranganathan et al., 2018; Searchinger et al., 2019).

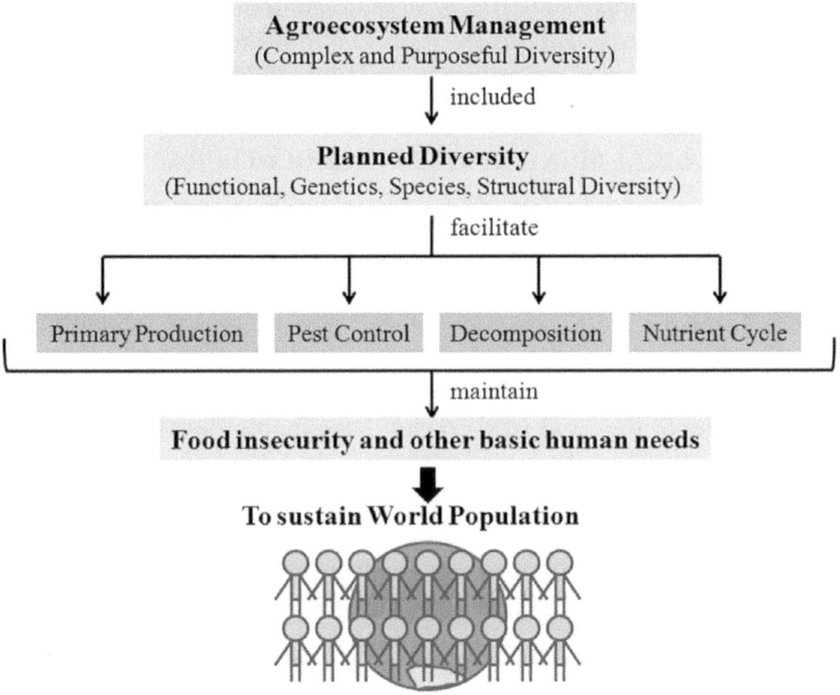

FIGURE 7.2 Essential roles of agroecosystem in maintaining the food availability of the world population.

A non-profit government unit, the Intertribal Agriculture Council (IAC) was made in 1987 with the aim of conserving and developing agricultural resources. With the support of Textile Exchange, they explore the concept of Regenerative Agriculture Landscape Analysis. IAC actively commends the execution of industries in regenerative agriculture. India's participation with IAC is very appreciated in terms of the implementation of infrastructure for research, education, and supporting food for tribal governments and tribal food producers (Duren, 2020; Gordon et al., 2022). Although tribal culture ensures the traditional ways to do their agricultural activities, the modern and indigenous practices in crop production, husbandry, and food preservation always need attention for a stable agriculture system. Proper knowledge and farming practices enable them to do regenerative agriculture against conventional agriculture (Schreefel et al., 2020; Shamon et al., 2022). So, to maintain sustainability in agricultural work, regenerative agriculture was proposed in the late 1980s, by Robert Rodale. He introduced the thought based on various agricultural and ecological practices with minimal soil damage and encouraged the use of organic compost. Many civil societies, non-government organizations (NGOs), and multi-national food companies have strongly promoted regenerative agriculture (Giller et al., 2021).

7.4 ECOLOGICAL SIGNIFICANCE OF AGRICULTURAL LANDSCAPE

For agricultural development to be sustainable, it is vital to raise awareness of agricultural ecological environment protection, promote agricultural ecological environment protection, and maximize the meaning of agriculture's multifunctionality. So, focusing on the development of agricultural variety and focusing on agricultural ecological environmental conservation is essentially the same in various aspects. All of these efforts ultimately serve the goal of fostering sustainable development (Grimm and Wösten, 2018) of the large landscapes. In his article on the classification of landscapes, Morseletto (2020) explained the distinctions between an agricultural and a rural landscape. Accordingly, agriculture (including forestry, livestock husbandry, byproducts, fishing, etc.), a limited population, few transportation options, a dispersed built environment, and primarily low-rise buildings are characteristics of rural areas. The rural landscape shows how the economy, people, society, and environment all work together in rural areas. The term "agricultural landscape" is used to describe the whole scene of farming, including the land, the water system, and the landscape made by foundation ponds, little bridges, flowing water, and people. Landscape theory is the academic study of how people can use, be motivated by, and take enjoyment in well-designed, well-created, and well-maintained outdoor spaces for their benefit. The role that nature performs is very important. The Romans were the first to perfect the art of urban planning, which involved integrating architecture and landscape. It was built as a system of routes, and within that system was a walled city. The notions are the fundamental principles of design-symmetry, organization, repetition, and coherence that give guidelines for arranging or organizing information to achieve a pleasant or beautiful environment (Carus and Dammer, 2018).

Outstanding contributions have been made by the method system and theory of landscape ecological planning (LANDEP), as well as the DLU strategic framework that was proposed by Gontard et al. (2018). The criteria that people have for their living environment are continually growing, and as a result, the government has raised the amount of money it invests in general aesthetic enhancements to the village. In particular, during the time of rural reform and revitalization, the consensus of society is to explore nature, and people expect that, via transformation, rural buildings featuring traditional characteristics can indeed be restored. This is especially true in rural areas (Peng and Pivato, 2019). Garske et al. (2020) created a new image of rural landscapes by closely integrating tourism and agricultural output in rural areas. This led to the creation of a new market for agricultural development. By going back to more traditional and basic forms of geographic landscape civilization, Fytili and Zabaniotou (2018) were able to reach their goal of rural construction.

Since the beginning of the Industrial Age, and especially since the "green revolution" in the middle decades of the 20th century, agricultural practices have been steadily improving. However, it will be harder in the future decades than it has been so far to meet the demand for increased agricultural productivity. Ecologically all these causes are rooted in the natural environment. Many of the natural mechanisms that allow contemporary agriculture to thrive are becoming unstable as a result of global climate change. Still, the sustainability dilemma is in no small part due to modern agricultural farming that needed stability during crop production. Ecological agricultural farming with new policies is leading the trend toward more stability in the human food supply, the preservation of wild ecosystems, and the protection of bees (Carus and Dammer, 2018; Kumar et al., 2019b; Rayne and Aula, 2020). Most ecological agricultural farming methods may be traced back to the days of the small family farm, where a variety of crops were farmed and most of the land, such as wetlands and field margins, was kept untouched. In truth, ecological agricultural farming has been the predominant method of farming throughout human history. Genetically modified (GM) crops, artificial fertilizers, and pesticides are all prohibited in ecological agricultural farming. By avoiding massive monocrop plantations and ensuring ecosystem diversity, ecological agriculture growers prevent pest damage to their crops. Ecological farming restores soil nutrients with natural composting systems. It also keeps the soil from being washed away by wind and water, which happens in many places where industrial farming has taken over (Rayne and Aula, 2020).

7.5 COMPOST BOOST THE AGRICULTURAL BIODIVERSITY

The physical and chemical characteristics of natural compost account for a myriad range of organic fertilizers and can extensively find its usage as soil improvers. The overall physiochemical properties of the soil high in manure content advance to a large extent (Yadav et al., 2019; Rayne and Aula, 2020). The major reason behind this improvement majorly is that organic manure can help in reducing the declination degree of soil pH because of its elevated buffer capacity. Such a reduced soil pH rate may also contribute to a decrease in soil aluminum toxicity. On the contrary, some studies disclose how compost can lead to an increase in the soil pH in the presence of raised amounts of calcium, potassium, magnesium, sodium, organics anions, carbonates, and bicarbonates; thereby raising the exchange of cations and buffer capacities (de la Luz Mora et al., 2017). Manure serves as an excellent source of essential nutrients and minerals, in the form of complex nitrogenous compounds like ammonium releasing inorganic nitrogen into the soil, carbon, phosphorus, sulfur, and metals like copper and zinc (Cavalli et al., 2017). From the sustainability perspective, the abundant availability of organic compost can help farmers beat the expensive usage of chemical fertilizers, indirectly posing a lesser threat to the environment in terms of the release of toxic by-products (Mishra et al., 2019). More importantly, studies show that compost leads to a substantial increase in the soil carbon stocks, thereby stabilizing the carbon/nitrogen ratio and overall organic matter quotient. This contributes to a long-term stable fraction in the soil mineral content – a vital driving force toward soil microbial assortment (Cui et al., 2018; Liu et al., 2020).

Nutrients present in the soil in the form of organic manure enhance the catalytic activity of soil enzymes, as well as microbial activity due to the abundance of soil fauna and biomass. A study shows that enhanced fungal consortium led to minimalization of potatoes from dry rotting, on the application of bovine manure (Gleń-Karolczyk et al., 2018). It is very important to know when compost is judged as liquid and when it is considered to be solid while using it in regenerative agriculture. If the dry mass surpasses 20%, then the manure is supposed to be solid. However, slurry accounts for a dry mass content ranging from 4 to 20%; therefore, manure with a dry mass content of less than 4% is taken as a liquid. In solid manure, nitrogen is strongly bound to organic matter, which results in a slower mineralization process and thus can persist in soils for almost 5 to 10 years after its application (Massé et al., 2014). Urine samples of herbivores and liquid manure are great sources of inorganic nitrogen, as they can be easily taken up by plants as ammonium ions in soil samples with lower nitrogen content. Nonetheless, while practicing regenerative agriculture, farmers and cultivars need to be aware of the fact that too much manure content in soil can prove to be detrimental, in terms of high nitrogen leaching rates. Excessive manure application for longer periods can cause a rise in the levels of potassium and phosphorus percentages, triggering nitrogen surpluses followed by subsequent leaching (Geng et al., 2019).

Certain studies show how intensive techniques of agriculture can have a profound impact on the soil microbiota, and alterations in their activity and abundance rates. Many times, agricultural soils undergo loss of biodiversity due to physiological factors like soil erosion, harsh climatic changes, over-intensified agricultural systems, and even soil contamination (Figure 7.3). Recent observations reveal facts like the application of untreated slurry can lead to the destruction of the soil biome, by incorporating genes with antimicrobial resistance properties, and a wide range of antipathogenic controls-largely decreasing the diverse dung fauna (Chen et al., 2019). When there is an evident decrease in the soil diversity, it reduces the chance for plants to completely absorb nutrients. Intensive systems of agriculture often come with the drawback of low nutrient retention properties. Heavy usage of mineral fertilizers does not contribute to the additional organic carbon content in the applied soil (Liu et al., 2020). Hence, reduced soil biodiversity and compost content are bound within a self-accelerating positive feedback loop with damaging environmental consequences.

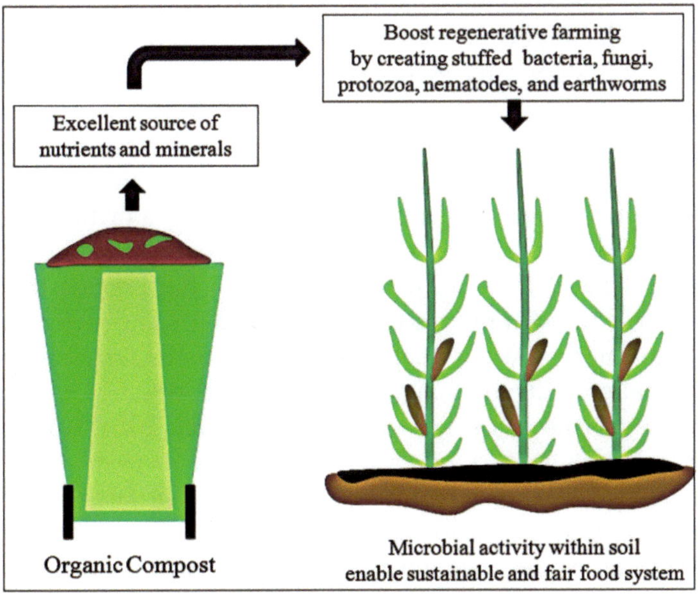

FIGURE 7.3 Role of organic compost in regenerative agriculture.

For their positive effects on the environment, the alternative movements of regenerative farming, organic farming, and agroecology stand out as the most significant and rapidly expanding ones. When it comes to international reach, though, agroecology may be at the top of the list, particularly in underdeveloped nations. It assures the interactions of animals, humans, plants, and the environment while addressing the social issues required for a sustainable and equitable food system, according to the Sustainable Agriculture Network. Family farm agriculture is also the inspiration for this system of farming because it makes use of readily available natural resources, information exchange, scientific inquiry, inexpensive technology, and resourceful independence. The incorporation of diversity is one of the many similarities between agroecology and regenerative agriculture. It is important for agricultural and natural systems to have a certain amount of variability in their processes, living things, elements, and habitats. This helps achieve good interactions, make good use of natural resources, and build resilience so that people can deal with change.

7.6 REGENERATIVE AGRICULTURE ROLE IN ENHANCING BIODIVERSITY

Conventional farming practices today are destroying biodiversity and productive soil. Farmers will not have enough usable topsoil to grow enough food to sustain a growing global population if this continues. There is a potential obligation to make the shift toward more regenerative farming techniques. A set of farming ideas and methods known as "regenerative agriculture" have been shown to boost biodiversity, soil quality, water quality, and ecosystem services. Regenerative agriculture is a method that seeks to revitalize depleted soils by reversing the effects of industrial agriculture. Ecosystems benefit from its practices because the organic matter in the soil is restored through integrative farming and grazing methods, which in turn help to improve water quality. It allows for the restoration of annual and perennial crop diversity in a farm landscape, which is essential for the regeneration of land and water resources (Melvani, 2016; Rodale Institute, 2014). In addition, it takes into account potential environmental and social impacts by reducing reliance on synthetic inputs in favor of site-specific management practices that sustain and improve soil quality over the long term, create new employment opportunities, and assist communities in better planning for and adapting to the effects of climate change.

The prime objective of regenerative agriculture is to improve the physical condition of the soil or bring back extremely degraded soil. This improves the water, plants, and soil qualities. In the end, regenerative agriculture depends on the diversity of the soil, and there may not be any diversity in the soil without regenerative agriculture practices (Johns, 2020; Leitch et al., 2018). They have grown and changed mutually. Regenerative farming techniques allow for the creation of new soils in addition to boosting the organic carbon content of existing ones. This can be accomplished through reducing soil erosion, re-establishing the soil food web, enhancing soil fertility, and utilizing the actions of humans, plants, animals, insects, fungi, and bacteria. Soil degradation, a growing need for food, climate change, and a lack of water are all things that could cause problems in the agricultural sector in the future. We must spread awareness of soil biodiversity and think of soil biology as a long-term solution if we are to reap the great benefits of regenerative agriculture and satisfy the demands of consumers and farmers in agriculture and land management. Similarly, soil biodiversity aids in soil formation and improves the functions, intern production, services, and increased efficiency of regenerative agriculture. Soil biodiversity (the activities of animals, bacteria, fungi, plant roots, insects, etc.) would then be looked at as a natural way to restore the soil food web, improve soil physical condition and fertility, increase agricultural productivity, and lock up carbon from the air (Rhodes, 2017).

To make soil biodiversity visible, education and awareness are essential. To convey and use scientific findings about the benefits of soil biodiversity and steps to restore and conserve it, there is a need to promote communication between researchers, policymakers, and the wider public. Agricultural soils are under threat (El-Ramady et al., 2021). While it has been largely ignored in global, regional, and national policies dealing with forest management, food production, climate variability, loss of ecosystem biodiversity, and desertification, these organizations, along with Elizabeth et al. asserted that soil biodiversity is crucial for soil functioning and plant production (Bach et al., 2020). Most significantly, agriculture requires healthy, full people to feed the world's rising population. To use soil biodiversity to regenerate the ecosystem and agriculture, the impact on human health must be considered. In the past, soil organisms have been utilized to develop a variety of novel drugs and vaccines, such as the well-known antibiotic bleomycin, penicillin, which is used to treat cancer, amphotericin, which treats fungal infections, and therapeutic approaches to managing diseases (Luis et al., 2018). In light of the similarities between soil and human systems, it makes sense that soil management should be based on biology to some extent than industrial chemistry. There are incredibly rich and varied communities of organisms in soils; these species range in size from bacteria measuring only a few micrometers in length to earthworms measuring several meters in length. Bacteria and fungi make up the bulk of the soil's microbial population, contributing somewhere between 0.6% and 1.1% of the organic soil (Fierer et al., 2009; Luis et al., 2018) and reflecting a biodiversity pool including an estimated number of species per gram of soil in the tens of thousands (Kumar, 2021; Roesch et al., 2007).

7.7 CONCLUSION

Regenerative agribusiness is one of the critical agrosystem originations for biological appropriateness. The idea resolves bunches of worldwide issues, including the support of organic variety, further developing stock chains, reestablishing biodiversity, evenhandedly circulating area value and access, and empowering ranchers, organizations, and networks to succeed. This enables precision in farming and combines high food production with precise management of rural assets. Analysts and researchers logically investigate new strategies to foster practical yield creation and food security without upsetting the financial aspects and climate of various biological system administrations. One of the new ideas for sustainable cultivation emerges from regenerative agribusiness. The concept of "regenerative horticulture" is an approach to overseeing ranches that focuses on soil wellbeing. Biological system administrations like water separating, nitrogen cycling, and further developed biodiversity are a couple of the numerous positive results that can come about because of setting soil quality at the focal point of rural practices. The types and levels of pesticide contamination will be greatly influenced by the types of yields and animals raised. The variety of synthetics will be limited in monocultures; however, the concentrations may be higher, though the inverse may be true in agroecosystems that support multiple yield creation. The presence of burning items derived from petroleum may likewise offer difficulties in exceptionally motorized agroecosystems. Aside from agroecosystems, indoor environments are regarded as distinct biological systems that require careful consideration. This classification incorporates structures, for example, storehouses and stables, which are utilized for the cultivation of different creatures, most prominently chickens and pigs.

Toxicologists stress the potential well-being impacts of breathing in poisons in an indoor agroecosystem, yet this is not an issue in an open-air agroecosystem. Natural residue, smelling salts, H_2S, nitrogen oxide, and mycotoxins are some of the potential causes of natural residue harmful disorder, also known as "intense rancher's lung," as well as asthma, bronchitis, and rhinitis. Carbon monoxide harm is most frequently brought about by some unacceptable utilization of gas-controlled devices, similar to pressure washers. Agroecosystems typically change as a result of diverse plans and cognizant grouping, which may include significant combinations, acquired variety, species collection, and central arrangement. The system recycles these harmful wastes into valuable material under the influence of microbes present in the soil and environment. This structure leads to monoculture systems, which may increase agricultural production. The information concentrated in these monoculture structures later helped in the standardization of businesses that developed in various regions of the planet. Then again, different tropical grass nurseries and standard-creating philosophies flaunt fundamentally arranged assortments and assortments of bound-together biota. Various appraisals have been completed on the impacts of coordinated combinations on regular cycles like major creation, bug control, decay, and supplement cycling. These evaluations were made possible by the extensive arrangement of coordinated collection that has been used in recently established frameworks. In the future, the assumption will rise in tandem with the growing total population. In this regard, the framework should provide ranchers with a few new income opportunities. Concerns have also been raised about the possibility of dealing with the perfect climate in a practical and substantial setting. The entire world would be watching the delicacy of the food opening. Dealing with over 10 billion additional individuals until 2050 would overburden the natural framework. This would be possible by increasing crop production by 11%, unforeseen cultivating development by clearing more boondocks, and decreasing ozone-depleting substance (GHG) emissions during green items to keep global temperature change under 2°C. Nonetheless, with an expansion in compensation, the situation might change when individuals start devouring from different sources to close the food hole. Economically many prospects have to be focused on the implementation of an agrosystem approach, to advance regenerative agriculture. It would be an integrative methodology that increases the adaptability of a changing environment, reverses biodiversity loss, and improves soil health.

KEYWORDS

- **agroecosystem**
- **biodiversity**
- **nitrogen cycling**
- **regenerative agriculture**
- **soil degradation**
- **sustainability**
- **tropical grass**

REFERENCES

Aguilera, E., Diaz-Gaona, C., Garcia-Laureano, R., Reyes-Palomo, C., Guzmán, G. I., Ortolani, L., Sanchez-Rodriguez, M., & Rodriguez-Estevez, V. (2020). Agroecology for adaptation to climate change and resource depletion in the Mediterranean region: A review. *Agricultural Systems, 181*, 102809.

Alyokhin, A., Nault, B., & Brown, B. (2020). Soil conservation practices for insect pest management in highly disturbed agroecosystems—A review. *Entomologia Experimentalis et Applicata, 168*(1), 7–27.

Bach, E. M., Ramirez, K. S., Fraser, T. D., & Wall, D. H. (2020). Soil biodiversity integrates solutions for a sustainable future. *Sustainability, 12*, 2662. https://doi.org/10.3390/su12072662.

Brockerhoff, E. G., Barbaro, L., Castagneyrol, B., Forrester, D. I., Gardiner, B., González-Olabarria, J. R., Lyver, P. O. B., Meurisse, N., Oxbrough, A., Taki, H., & Thompson, I. D. (2017). Forest biodiversity, ecosystem functioning, and the provision of ecosystem services. *Biodiversity and Conservation, 26*, 3005–3035.

Carus, M., & Dammer, L. (2018). The circular bioeconomy—Concepts, opportunities, and limitations. *Indian Biotechnologist, 14*, 83–91.

Cavalli, D., Corti, M., Baronchelli, D., Bechini, L., & Gallina, P. M. (2017). CO2 emissions and mineral nitrogen dynamics following application to soil of undigested liquid cattle manure and digestates. *Geoderma, 308*, 26–35.

Chakraborty, D., Saha, S., Singh, R. K., Sethy, B. K., Kumar, A., Saikia, U. S., Das, S. K., Makdoh, B., Borah, T. R., Nomita Chanu, A., & Walling, I. (2017). Trend analysis and change point detection of mean air temperature: A spatio-temporal perspective of North-Eastern India. *Environmental Processes, 4*, 937–957.

Chen, D., Xing, W., Lan, Z., Saleem, M., Wu, Y., Hu, S., & Bai, Y. (2019). Direct and indirect effects of nitrogen enrichment on soil organisms and carbon and nitrogen mineralization in a semi-arid grassland. *Functional Ecology, 33*, 175–187.

Crews, T. E., Carton, W., & Olsson, L. (2018). Is the future of agriculture perennial? Imperatives and opportunities to reinvent agriculture by shifting from annual monocultures to perennial polycultures. *Global Sustainability, 1*, e11.

Cui, S., Liang, S., Zhang, X., Li, Y., Liang, W., Sun, L., Wang, J., Martijn Bezemer, T., & Li, Q. (2018). Long-term fertilization management affects the C utilization from crop residues by the soil micro-food web. *Plant and Soil, 429*, 335–348.

de la Luz Mora, M., Demanet, R., Acuña, J. J., Viscardi, S., Jorquera, M., Rengel, Z., & Durán, P. (2017). Aluminum-tolerant bacteria improve plant growth and phosphorus content in ryegrass grown in volcanic soil amended with cattle dung manure. *Applied Soil Ecology, 115*, 19–26.

Duren, C. D. (2020). The Native Farm Bill Coalition and the 2018 Farm Bill: Building a strong, sustained voice on food and agriculture issues in Indian country. *Renewable Agriculture and Food Systems, 35*, 463–464.

Elevitch, C. R., Mazaroli, D. N., & Ragone, D. (2018). Agroforestry standards for regenerative agriculture. *Sustainability, 10*, 3337.

El-Ramady, H. R., Singh, A., Rajput, V. D., Amer, M. M., Omara, A. E. D., Elsakhawy, T., Elbehiry, F., Elbasiouny, H., & Abdalla, N. (2021). Environment, biodiversity, and soil security: A new dimension in the era of COVID-19. *Environment, Biodiversity and Soil Security, 5*, 1–14.

Fierer, N., Strickland, M. S., Liptzin, D., Bradford, M. A., & Cleveland, C. C. (2009). Global patterns in below-ground communities. *Ecology Letters, 12*, 1238–1249.

Fytili, D., & Zabaniotou, A. (2018). Circular economy synergistic opportunities of decentralized thermochemical systems for bioenergy and biochar production fueled with agro-industrial wastes with environmental sustainability and social acceptance: A review. *Current Sustainable Renewable Energy Reports, 5*, 150–155.

Gałązka, A., & Grządziel, J. (2018). Fungal genetics and functional diversity of microbial communities in the soil under long-term monoculture of maize using different cultivation techniques. *Frontiers in Microbiology, 9*, 76.

Garske, B., Stubenrauch, J., & Ekardt, F. (2020). Sustainable phosphorus management in European agricultural and environmental law. *Review of European, Comparative & International Environmental Law, 29*, 107–117.

Geng, Y., Cao, G., Wang, L., & Wang, S. (2019). Effects of equal chemical fertilizer substitutions with organic manure on yield, dry matter, and nitrogen uptake of spring maize and soil nitrogen distribution. *PLoS ONE, 14*, e0219512.

Giller, K. E., Hijbeek, R., Andersson, J. A., & Sumberg, J. (2021). Regenerative agriculture: An agronomic perspective. *Outlook on Agriculture, 50*, 13–25.

Gleń-Karolczyk, K., Boligłowa, E., & Antonkiewicz, J. (2018). Organic fertilization shapes the biodiversity of fungal communities associated with potato dry rot. *Applied Soil Ecology, 129*, 43–51.

Gliessman, S. R. (2021). *Package Price Agroecology: The Ecology of Sustainable Food Systems*. CRC Press.

Gontard, N., Sonesson, U., Birkved, M., Majone, M., Bolzonella, D., Celli, A., Angellier-Coussy, H., Jang, G. W., Verniquet, A., Broeze, J., & Schaer, B. (2018). A research challenge vision regarding the management of agricultural waste in a circular bio-based economy. *Critical Reviews in Environmental Science and Technology, 48*, 614–654.

Gordon, E., Davila, F., & Riedy, C. (2022). Transforming landscapes and mindscapes through regenerative agriculture. *Agriculture and Human Values, 39*, 809–826.

Grimm, D., & Wösten, H. A. (2018). Mushroom cultivation in the circular economy. *Applied Microbiology and Biotechnology, 102*, 7795–7803.

Hatt, S., Artru, S., Brédart, D., Lassois, L., Francis, F., Haubruge, E., Garré, S., Stassart, P. M., Dufrêne, M., Monty, A., & Boeraeve, F. (2016). Towards sustainable food systems: The concept of agroecology and how it questions current research practices. *Biotechnology, Agronomy, Society and Environment, 20*(S1), 215–224.

Hodgson, E. (2012). Human environments: Definition, scope, and the role of toxicology. *Progress in Molecular Biology and Translational Science, 112*, 1–10.

Johns, C. (2020). Growing our future—How regenerative agriculture can achieve economies of scale. *Strategic Analysis Paper*. Future Directions International Pty Ltd. Hampden Road, Australia.

Kumar, H., Kumar, M., Kumar, B., Bharti, S., Kumar, P., Kumar, M., & Kumar, J. (2017b). Effect of acetone extract of *Zea mays* L. leaves on the growth of *Aspergillus flavus* and aflatoxin production. *International Journal of Advanced Research, 5*(8), 1667–1671.

Kumar, M. (2021). *In silico* efficacy of [S]-8-gingerol (a derivative) with 6-gingerol against PT-domain of polyketide synthase A (PksA). *International Journal of Medical Microbiology and Tropical Diseases, 7*, 62–64.

Kumar, M., & Akhtar, M. S. (2019). Application of microbial biotechnology in improving salt stress and crop productivity. In M. S. Akhtar (Ed.), *Salt Stress, Microbes, and Plant Interactions: Mechanisms and Molecular Approaches* (Vol. 2, pp. 133–159). Springer.

Kumar, M., Kumar, H., Topno, R. K., & Kumar, J. (2019b). Analysis of the impact of the anaerobic condition on the aflatoxin production in *Aspergillus parasiticus* Speare. *Agricultural Sciences Digest, 39*(1), 75–78.

Kumar, M., Rana, S., Kumar, H., Kumar, J., Mansuri, R., & Sahoo, G. C. (2017a). Inhibition of product template (PT) domain of aflatoxin producing polyketide synthase enzyme of *Aspergillus parasiticus*. *Letters in Drug Design & Discovery, 14*(7), 811–818.

Kumar, M., Rana, S., Kumar, H., Kumar, P., Dikhit, M. R., Mansuri, R., Kumar, J., & Sahoo, G. C. (2017). Computational, structural, and functional aspects of hypothetical protein of *Aspergillus flavus* pheromone receptor Pre-A (PRP-A). *Journal of Applied Pharmaceutical Science, 7*(7), 89–97.

Kumar, M., Sahoo, G., Topno, R., Pandey, K., Ravidas, V., Dikhit, M., Kumar, H., & Das, P. (2018b). Molecular interaction of antifungal compounds with fungal type I polyketide synthase and preventing hepatocarcinogenesis. *International Journal of Infectious Diseases, 73*, 285.

Kumar, M., Topno, R. K., Kumar, H., & Kumar, J. (2018a). Biological conversion of carcinogenic aflatoxin B1 to AFB1-8, 9-exo-epoxide, and its potential prevention. *International Journal of Research and Analytical Reviews, 5*(4), 935–946.

Kumar, P., Mishra, R., Topno, R. K., Kumar, M., Dinesh, D. S., & Singh, D. K. (2019b). Seasonal prevalence of Japanese encephalitis (JE) in Patna district of Bihar, India. *Journal of Communicable Diseases, 51*(4), 58–61.

Kumari, A., Kumar, M., Sahoo, G. C., & Kumar, J. (2021). *In silico* prediction of glabridin potency against human tyrosinase in hyperpigmentation condition. *International Journal of Research and Analytical Reviews, 6*(2), 332–341.

Liu, S., Wang, J., Pu, S., Blagodatskaya, E., Kuzyakov, Y., & Razavi, B. S. (2020). Impact of manure on soil biochemical properties: A global synthesis. *Science of The Total Environment, 745*, 141003.

Luis, A. D., Kuenzi, A. J., & Mills, J. N. (2018). Species diversity concurrently dilutes and amplifies transmission in a zoonotic host-pathogen system through competing mechanisms. *Proceedings of the National Academy of Sciences, 115*, 7979–7984.

MacLaren, C., Storkey, J., Menegat, A., Metcalfe, H., & Dehnen-Schmutz, K. (2020). An ecological future for weed science to sustain crop production and the environment: A review. *Agronomy for Sustainable Development, 40*, 1–29.

Massé, D. I., Cata Saady, N. M., & Gilbert, Y. (2014). Potential of biological processes to eliminate antibiotics in livestock manure: An overview. *Animals, 4*, 146–163.

Melvani, K. (2016). *Handbook for Regenerative Agriculture*. USAID, From the American People.

Mishra, J., Kumar, M., Sharan, A. K., & Kumar, J. (2019). Detoxification assessment of aflatoxin in *Aspergillus flavus* under the effect of *Calendula officinalis* Linn's methanolic extract. *Agricultural Sciences Digest, 39*(1), 21–25.

Morseletto, P. (2020). Restorative and regenerative: Exploring the concepts in the circular economy. *Journal of Indian Ecology, 24*, 763–773.

Peng, W., & Pivato, A. (2019). Sustainable management of digestate from the organic fraction of municipal solid waste and food waste under the concepts of back-to-earth alternatives and circular economy. *Waste and Biomass Valorization, 10*, 465–481.

Power, A. G., & Kenmore, P. (2013). Exploiting interactions between planned and unplanned diversity in agroecosystems: What do we need to know? *Agroecology and Sustainable Food Systems, 37*(3), 252–261.

Ranganathan, J., Waite, R., Searchinger, T., & Hanson, C. (2018). How to sustainably feed 10 billion people by 2050, in 21 charts. *World Resources Institute*. https://www.wri.org/insights/how-sustainably-feed-10-billion-people-2050-21-charts (accessed on 12 September 2024).

Rayne, N., & Aula, L. (2020). Livestock manure and the impacts on soil health: A review. *Soil Systems, 4*, 64.

Rendon, P., Steinhoff-Knopp, B., & Saggau, P. (2020). Assessment of the relationships between agroecosystem condition and the ecosystem service soil erosion regulation in Northern Germany. *PLoS ONE, 15*(12), e0234288.

Rhodes, C. J. (2017). The imperative for regenerative agriculture. *Science Progress, 100*, 80–129.

Rodale Institute. (2014). *Regenerative Organic Agriculture and Climate Change: A Down-To-Earth Solution to Global Warming*.

Roesch, L. F., Fulthorpe, R. R., Riva, A., Casella, G., Hadwin, A. K., Kent, A. D., Daroub, S. H., Camargo, F. A., Farmerie, W. G., & Triplett, E. W. (2007). Pyrosequencing enumerates and contrasts soil microbial diversity. *The ISME Journal, 1*, 283–290.

Schreefel, L., Schulte, R. P., De Boer, I. J., Schrijver, A. P., & Van Zanten, H. H. (2020). Regenerative agriculture—The soil is the base. *Global Food Security, 26*, 100404.

Seabloom, E. W., Borer, E. T., Gross, K., Kendig, A. E., Lacroix, C., Mitchell, C. E., Mordecai, E. A., & Power, A. G. (2015). The community ecology of pathogens: Coinfection, coexistence, and community composition. *Ecology Letters, 18*, 401–415.

Searchinger, T., Waite, R., Hanson, C., Ranganathan, J., Dumas, P., Matthews, E., et al. (2019). *Creating A Sustainable Food Future: A Menu of Solutions to Feed Nearly 10 billion People by 2050*. Washington, DC: World Resources Institute.

Shamon, H., Cosby, O. G., Andersen, C. L., Augare, H., BearCub, S. J., Bresnan, C. E., Brock, B. L., Carlson, E., Deichmann, J. L., Epps, A., Guernsey, N., Hartway, C., Jørgensen, D., Kipp, W., Kinsey, D., Komatsu, K. J., Kunkel, K., Magnan, R., Martin, J. M., Maxwell, B. D., McShea, W. J., Mormorunni, C., Olimb, S., Rattling Hawk, M., Ready, R., Smith, R., Songer, M., Speakthunder, B., Stafne, G., Weatherwax, M., & Akre, T. S. (2022). The potential of bison restoration as an ecological approach to future tribal food sovereignty on the Northern Great Plains. *Frontiers in Ecology and Evolution, 10*, 826282.

Simoni, S., Caruso, G., Vignozzi, N., Gucci, R., Valboa, G., Pellegrini, S., Palai, G., Goggioli, D., & Gagnarli, E. (2021). Effect of long-term soil management practices on tree growth, yield, and soil biodiversity in a high-density olive agroecosystem. *Agronomy, 11*(6), 1036.

Williams-Nguyen, J., Sallach, J. B., Bartelt-Hunt, S., Boxall, A. B., Durso, L. M., McLain, J. E., Singer, R. S., Snow, D. D., & Zilles, J. L. (2016). Antibiotics and antibiotic resistance in agroecosystems: State of the science. *Journal of Environmental Quality, 45*(2), 394–406.

Yadav, D. P., Kumar, M., & Jamal, M. A. (2018). Larvicidal and juvenomimetic property of *Withania somnifera* extracts against larvae of *Tribolium confusum* Duval. *International Journal of Research and Analytical Reviews, 6*(1), 559–563.

Yadav, D. P., Kumar, M., & Jamal, M. A. (2019). Larvicidal and juvenomimetic effect of different plant extracts against *Tribolium confusum* Duval (Coleoptera: Tenebrionidae). *International Journal of Research and Analytical Reviews, 6*(1), 420–432.

Zambrano, J., Garzon-Lopez, C. X., Yeager, L., Fortunel, C., Cordeiro, N. J., & Beckman, N. G. (2019). The effects of habitat loss and fragmentation on plant functional traits and functional diversity: What do we know so far? *Oecologia, 191*, 505–518.

CHAPTER 8

Recent Advances in the Use of Bionanotechnology for Bioremediation

MUNIR OZTURK,[1,*] SARAH FATIMA,[2] SHAFIA MARYAM,[2] SIDRA UROOJ,[2] FAREEHA IQBAL,[2] LAILA-TUL-QADAR,[2] SYEDA GUL NISA,[2] MUHAMMAD RAZA KALEEM,[2] ALVINA GUL,[2,*] VOLKAN ALTAY,[3] BENGU TURKYILMAZ UNAL,[4] and FAZILET VARDAR SUKAN[5]

[1]*Ege University, Department of Botany and Center for Environmental Studies, Izmir, Türkiye*

[2]*Atta-Ur-Rahman School of Applied Biosciences, National University of Sciences and Technology, Islamabad, Pakistan*

[3]*Hatay Mustafa Kemal University, Faculty of Arts and Sciences, Department of Biology, Hatay, Türkiye*

[4]*Nigde Omer Halisdemir University, Faculty of Science and Arts, Department of Biotechnology, Niğde, Türkiye*

[5]*Faculty of Engineering and Natural Sciences; Director of SUNUM Nanotechnology Research and Application Center, Sabanci University, Istanbul, Türkiye*

*Corresponding author.

ABSTRACT

Environmental problems are the major reason for the destruction of agricultural land, animals, and human diseases on earth. It causes pollution in the air, water, and soil that is difficult to control. Bio-nanotechnologies play a significant role in environmental protection and sustainability by adopting novel techniques to clear, mitigate, and monitor environmental pollutants.

Agricultural Biotechnology: Issues, Challenges, and Recent Developments. Mohd. Sayeed Akhtar and Khalid Rehman Hakeem (Eds.)
© 2025 Apple Academic Press, Inc. Co-published with CRC Press (Taylor & Francis)

Nanotechnology helps to reduce energy consumption, cost, and resource consumption through more efficient methods. So, nanomaterials with large surface area and small size have the unique capacity to solve complex problems. They require less activation energy and have high reactivity; these additional benefits make them ideal candidates for sustained bioremediation. Bio-nano-technology techniques have a positive impact on the environment and its sustainability by reducing environmental contaminants. The present chapter discusses the role of bio-nanotechnology in solving the persisting urgent environmental issues such as soil contamination, air, and water pollution, and climate change and also focuses on the effectiveness, safety, and scalability issues as well as any potential advantages and difficulties.

8.1 INTRODUCTION

Environmental pollution is one of humanity's most prominent issues in the 21st century. The sustainability of our world and its well-being are threatened by many intricate, interrelated concerns, such as resource depletion, pollution, and climate change (Ashraf et al., 2010; Ghori et al., 2019; Imanberdieva et al., 2018; Ozturk and Altay, 2021; Ozturk et al., 2015, 2022a, b, 2023a; Yalcin and Altay, 2023). These environmental issues affect every part of our lives, including the air we breathe, the water we drink, the food we eat, and the places we live in (Fakruddin et al., 2012). The seriousness of the issue demands increased awareness and efforts to solve it. As a result, it is becoming increasingly apparent that creative and revolutionary methods are required to address these issues and build a sustainable future. This is where the exciting new possibilities of bio-nanotechnology are employed, providing solutions to environmental issues that are effective, efficient, and sustainable (Gardner and Stern, 1996). Conventional approaches to environmental concerns have mitigated certain issues. For instance, the concentration of atmospheric CO_2 continues to grow at alarming rates despite decades of international attempts to cut greenhouse gas (GHG) emissions and address climate change (Imanberdieva et al., 2018; Ozturk and Altay, 2021; Ozturk et al., 2022). According to the World Meteorological Organization, the average worldwide CO_2 concentration in the atmosphere in 2020 hit a record high of 413.2 parts per million (ppm), 148% over pre-industrial levels. Despite attempts to minimize pollution, air quality remains a serious problem in many regions. According to the World Health Organization (WHO), air pollution causes more than 4.2 million fatalities yearly, accounting for 7 million premature deaths. The plastic waste dumped into the ocean causes disruption in the oceanic ecosystem. It is estimated that 8 million metric tons of plastic enter the oceans annually. Given the scope and severity of these environmental concerns, new and revolutionary ideas that go beyond traditional procedures are necessary. By combining the creative potential of biological systems with nano-scale materials, bio-nanotechnology opens fresh avenues for tackling these problems. Bio-nanotechnology offers the potential to provide effective, efficient, and long-lasting solutions to various environmental issues by using the unique qualities of biological molecules, such as enzymes and proteins. The opportunities and challenges presented are opportunities to evaluate the potential of bio-nanotechnology (Mäler, 1990; Ozturk et al., 2023).

8.1.1 THE EMERGENCE OF BIO-NANOTECHNOLOGY

Bio-nano-technology development may be dated to the 1950s when researchers first started to investigate the distinctive characteristics of materials at the nanoscale (Singh et al., 2013). However, the area did not take off until the 1990s, when new methods and instruments for working with and analyzing nano-scale materials were created. Since then, bio-nanotechnology has expanded quickly as researchers from several fields collaborate to develop novel materials and systems with previously unheard-of control and accuracy. The creation of deoxyribonucleic acid (DNA) nanotechnology in the 1980s and 1990s was one of the early achievements in bio-nanotechnology. This entailed utilizing DNA molecules to build exact shapes and sizes in nano-scale structures, opening the door for new uses in fields including medication delivery, sensing, and nano-electronics. The 1991 discovery of carbon nanotubes, which are extraordinarily strong and light and might be used in fields like energy storage and water purification, was another critical development. Since biotechnology has only been around for a few generations and nanotechnology has only been around since the late 20th century, its scope is still being defined. Biotechnology integrates scientific knowledge and equipment with cellular, molecular, and genetic phenomena to create products or commodities used in various sectors, from agriculture to medicine (Kulkarni et al., 2022; Ozturk et al., 2023b). The area of nanotechnology known as "bio-nanotechnology" includes atom-level engineering and manufacturing affected by biological antecedents. It is also closely related to biotechnology but offers the opportunity to shape and modify the atomic-scale details of the creatures produced. Bio-nano-technology equipment is created at the atomic level and completes 3D molecular tasks and processes for individual control.

8.1.2 BIO-NANOTECHNOLOGY AND ITS SCOPE

Bio-nanotechnology is an interdisciplinary discipline combining nanotechnology with biological sciences to create innovative materials and systems at the nanoscale. Biological molecules like proteins, DNA, and enzymes are combined with nano-scale materials like nanoparticles, nanotubes, and nanofibers to design, produce, and manipulate materials and devices. Bio-nanotechnology has several uses and many applications (Thakur et al., 2018). Bio-nanotechnology has made it possible to create tailored drug delivery systems (DDS) in the medical profession that can administer medications to specific cells or tissues while minimizing toxicity and adverse effects (González et al., 2016). For instance, medication carriers have been utilized to treat cancer and other disorders using liposomes, nano-sized vesicles made of lipids. Bio-nanotechnology has aided in creating biosensors with high sensitivity and specificity for detecting disease biomarkers. Bio-nanotechnology has created brand-new solar cells, fuel cells, and battery materials in the energy sector. For instance, through improving light absorption and charge separation, titanium dioxide and zinc oxide nanoparticles have been employed to raise the efficiency of solar cells. Enzymes have also been utilized to catalyze chemical processes in fuel cells, producing very efficient energy. Bio-nanotechnology can solve problems with pollution, water treatment, and waste management in environmental remediation. For instance, heavy metals, organic pollutants, and other contaminants have been removed from water and soil using iron nanoparticles and other metals. Bio-nanotechnology has aided in creating nano-sensors that can track air and water quality, allowing for early identification and response to environmental threats. In agriculture, bio-nanotechnology has resulted in the creation of nano-fertilizers and nano-pesticides that can increase crop yields while using fewer chemicals, lessening the impact of agriculture on the environment (Rohela et al., 2019). Bio-nanotechnology has aided in the creation of solar cells, batteries, and fuel cells in the energy sector that are more effective, affordable, and environmentally friendly than conventional energy technologies. The growing number of articles and patents in bio-nanotechnology illustrates its potential. The worldwide bio-nano-technology market was estimated at USD 45.6 billion in 2020. It is projected to increase at a compound yearly growth rate of 9.2% from 2021 to 2028, according to research by Grand View Research.

8.1.3 ROLE OF BIO-NANOTECHNOLOGY IN ADDRESSING ENVIRONMENTAL PROBLEMS

Some of the most urgent environmental issues facing the world today may be solved by bio-nanotechnology. Air pollution is one of the main areas where bio-nanotechnology may substantially influence. Nitrogen oxides and other dangerous air pollutants, such as particulate matter, can be eliminated using nanoparticles. Additionally, real-time air quality monitoring using nano-sensors enables the early identification and mitigation of environmental threats. Bio-nanotechnology may be applied to water treatment to eliminate infections, organic pollutants, and heavy metals. Iron and other metal nanoparticles have effectively and inexpensively eliminated harmful pollutants. Nano-sensors can monitor water quality to help with remediation efforts by allowing for early contamination detection. Bio-nanotechnology may be applied to agriculture to create nano-fertilizers that can supply nutrients to plants more effectively, cutting down on waste and increasing crop yields (Primožič et al., 2021). To optimize nutrient utilization and lessen environmental effects, farmers can employ nano-sensors to monitor soil quality. Bio-nanotechnology has the potential to transform medication delivery and diagnostics in the realm of healthcare. Nanoparticles may target specific cells or tissues, lowering their toxicity and increasing their effectiveness. High sensitivity and specificity biosensors may be used to identify illness biomarkers, allowing for early diagnosis and customized care.

Comparing bio-nanotechnology to more conventional methods of resolving environmental issues, there are several significant benefits. Firstly, this enables accurate targeting of pollutants or toxins, reducing the influence on species or places that are not targeted. This accuracy can be especially crucial in delicate environments, like aquatic ecosystems, where conventional methods might have unintended effects. Secondly, compared to conventional methods, bio-nanotechnology is frequently more productive. For instance, nanoparticles may be made with a high surface area to volume ratio to increase their reactivity and, consequently, their capacity to remove contaminants. Like nano-fertilizers, which may be made to release nutrients gradually, they can be tailored to use less fertilizer overall and produce less runoff. Thirdly, bio-nanotechnology may be very scalable in the future. Nanoparticle-based technologies are appropriate for industrial-scale applications since they can be generated in enormous volumes. Nano-sensors may also be used to monitor environmental conditions across wide regions, giving important information for resource management and decision-making.

According to studies, bio-nanotechnology can be used to solve a variety of environmental issues. For instance, iron nanoparticles had a clearance effectiveness of up to 99% when utilized in one research to remove heavy metals from polluted water (Ng et al., 2015). In another research, real-time information on pollutant levels was obtained by monitoring the city's air quality with nano-sensors. Research has revealed that bio-nanotechnology may be used to solve agricultural and health problems. For instance, in one study, scientists created a medication delivery system based on nanoparticles that increased the efficiency of cancer therapies. In another study, scientists created a nano-fertilizer that upped rice plants' growth rates by 30%. The present chapter discusses the role of bio-nanotechnology in solving the persisting urgent environmental issues such as soil contamination, air, and water pollution, and climate change and also focuses on the effectiveness, safety, and scalability issues as well as any potential advantages and difficulties.

8.2 ROLE OF BIO-NANOTECHNOLOGY IN MANAGEMENT OF ENVIRONMENTAL POLLUTANTS

The contamination of the environment is a global concern leading to catastrophic side effects. Air consists of different pollutants such as hydrocarbons, chlorofluorocarbons, carbon monoxide, nitrogen oxide, and volatile organic compounds (VOCs). Soil and water contaminated with inorganic and organic materials as industrial effluents, pesticides, oil spills fertilizers and sewage water (Altay and Ozturk, 2012; Darwash et al., 2023; Hameed et al., 2017; Haq et al., 2021; Mohamed, 2017; Ozturk et al., 2017a, 2019a; Sardar et al., 2018; Yalcin and Altay, 2023). Environmental nanotechnology has a significant role in shaping recent environmental science and engineering. Nanotechnology introduces new and cost-effective approaches for the remediation of pollutants, pollution detection, and monitoring of pollution (Savage et al., 2007). The materials made up of nanostructure are used for the evaluation and determination of compounds (Vaseashta et al., 2007). The nanomaterials have the property of adsorbent high reactivity and big surface area. The particle size is not in the nano-range, it is usually 10–100 um (Bhawana and Fulekar, 2012). Nanoparticles can be modified chemically (Bhawana and Fulekar, 2012)., TiO_2 functions with ethylenediamine and it is tested for the capability to clean contaminated groundwater from anion metals or pollutants (Mattigod et al., 2005). Traditional technologies are used to clean toxic waste by chemical oxidation, adsorption, and bio-oxidation.

Nanotechnology also gained popularity due to nano-materials' potential to clean and monitor three different parts of the environment gas, liquid, and solid (Figure 8.1; Mohamed, 2017).

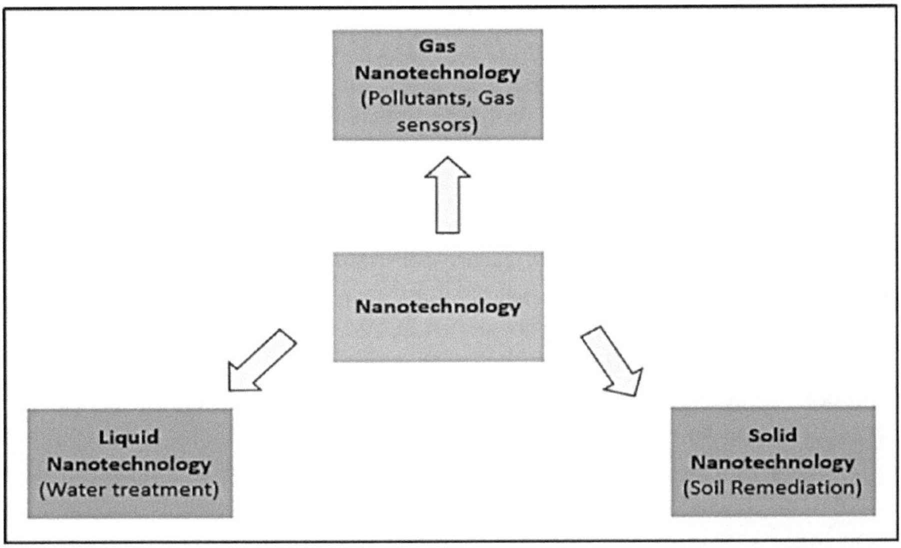

FIGURE 8.1 Applications of nanotechnology in three different parts of the environment.

8.2.1 ROLE OF BIO-NANOTECHNOLOGY IN AIR POLLUTANTS

The contamination in the air is most challenging due to the modification in atmospheric composition introduced by various pollutants such as carbon monoxide, chlorofluorocarbons, hydrocarbons, organic chemicals, heavy metals, nitrogen oxides, biological substances, and sand particles, and so forth (Araújo et al., 2014; Darwash et al., 2023; Ghori et al., 2019). Human health and the ecosystem are severely affected by air pollution which causes different types of diseases such as respiratory, cardiovascular diseases, and cancer. The WHO stated in 2014 that 7 million deaths in one year were due to heavy air pollution (Ngô and Van de Voorde, 2014). There are different types of air contaminants such as gas contaminants, particulate contaminants, dust, and biological contaminants, , and so forth (Mohamed, 2017). The nano-materials function as sensory devices due to their small size and big surface-to-volume ratio. They are designed in a way that reacts with contaminants to degrade them to toxic-free products and they also replace toxic materials with safety materials. Moreover, it also has a self-cleanup feature and nano-technology coating that is resistant to pollutants (Das et al., 2022). Nanotechnology has a lot of advantages in environmental control, and it is divided into three phases detection and sensing, remediation and treatment, and pollution prevention (Yadav et al., 2017). There are three ways to treat and remediation control air pollution degradation by nano-catalysis, adsorption by nano-absorptive materials, and separation/filtration by using nano-fillers (Mohamed, 2017). In adsorption by nano-absorptive materials, there is an approach that uses nano-structured membranes that have little pores to separate methane and carbon dioxide (CO_2) from the exhaust. The CNT (carbon nanotubes) trap gases at a much faster rate as compared to other conventional approaches and separate large volumes of gases effectively (Bhawana and Fulekar, 2012). Innovative nano-technology techniques are adopted to reduce air pollutants through different processes such as reducing GHG emissions, enhancing manufacturing processes, lessen the use of synthetic plastics and hazardous chemicals (Figure 8.2). Nano-technology applications help to form an environmentally friendly material or substance, replacing toxic substances. The method has reduced cost, increased efficiency, and low impact to create a clean environment (Yadav et al., 2017). The new types of nano-made materials that are less toxic, bio-degradable plastics are known to have structure to degrade and non-toxic crystalline substance in rechargeable battery electrodes made up of lithium-graphite. Other examples include environment-friendly self-clean glasses made of TiO_2 and nanotubes which perform better function as compared to other conventional tubes (Yadav et al., 2017).

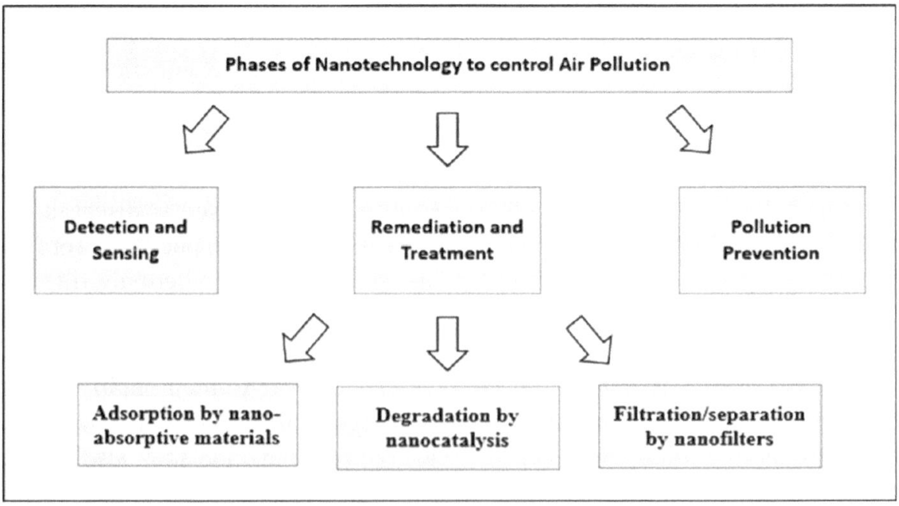

FIGURE 8.2 Different phases of nanotechnology to control air pollution.

8.2.2 ROLE OF BIO-NANOTECHNOLOGY IN SOIL REMEDIATION

Soil is the most essential supporting system of life and a necessary part of the terrestrial ecosystem as the soil ensures life on land (Abhilash et al., 2012; Ozturk et al., 2017b). Soil health and quality management are imperative not only to protect food but also to maintain the balance in the ecosystem (Lal, 2015; Ozturk et al., 2012). Approximately 30% of land is degraded across the world. Lands contaminated with heavy metals and pesticides are unsuitable for crop production (Bakshi and Abhilash, 2020; Mushtaq et al., 2020). The nano-devices or materials are used for remediation due to their cost-effectiveness and ease of use as compared to other methods (Li et al., 2019). Nowadays, nanomaterials or nanoparticles for soil remediation to treat the contaminated soil with chemicals such as pesticides, persistent organic pollutants (POPs), heavy metals, and so forth, are widely used. They are widely distributed and non-degradable in nature. Different nano-technology-based approaches are used for soil remediation (i) POP degradation and nano-materials for pesticides; (ii) nano-materials to convert heavy metals to less toxic form; (iii) pesticide residual detection in soil by nano-based sensors; and (iv) bio-remediation or phytoremediation of contaminated soil by the help of nano-material (Bakshi and Abhilash, 2020). Heavy metals such as Cr(VI), Pb(II), As(III), and Cd can be remediated using nanoparticles. Pesticides such as hexachlorobenzene, metolachlor, malathion, dicofol, and cypermethrin. POPs include dioxins, PCBs (polychlorinated biphenyls), DDT (dichlorodiphenyltrichloroethane), and heptachlor (Bakshi and Abhilash, 2020). Nanoparticles commonly employed for soil remediation are MWCNTs (multiwalled carbon nanotubes), ZnO (zinc oxide), TiO_2 (titanium dioxide), nZVI (nanoscale zerovalent iron), stabilized nanoparticles, and bimetallic nanoparticles. Nano-remediation uses nanomaterials in contaminated soil to remove pollutants or for its transformation (Yan et al., 2013). The nano-technology technique is used to remove heavy metals because of their ability to absorb and immobilize metal ions. The nanoparticles high ability as adsorbent increases their application to remove heavy metals from soil (Cai et al., 2019). The nano-material's excellent properties allow them to function as catalysts that reduce chemicals and pollutants. The nanomaterials remediation of contaminated soil to remove pollutants such as PAHs (polycyclic aromatic hydrocarbons), chlorinated organic solvents, PCBs (polychlorinated biphenyls), and organochlorine pesticides (Figure 8.3). The soil remediation techniques can be *in situ* and *ex-situ*, but mostly it is processed on-site without including transportation and pumping of soil making nano-remediation a cheaper and easily available method (Cai et al., 2019; Jiang et al., 2018).

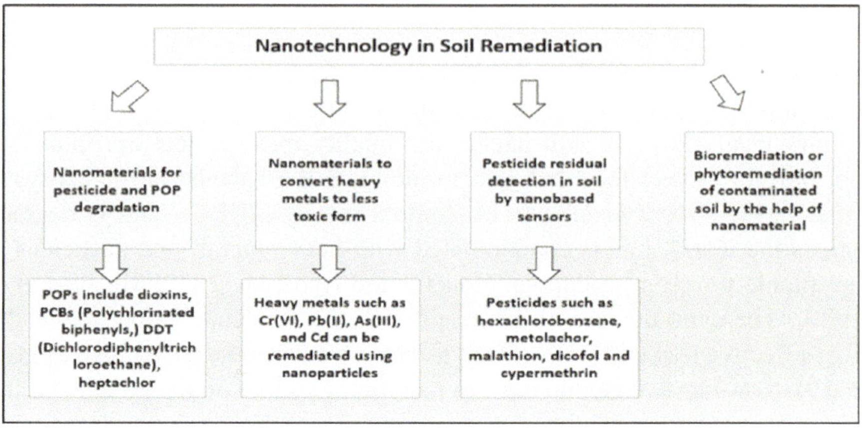

FIGURE 8.3 Nanotechnology in soil remediation.

8.2.3 ROLE OF BIO-NANOTECHNOLOGY IN WASTEWATER TREATMENT

Water is one of the most valuable and scarce resources on earth, continues to be recycled by means of the water cycle. The primary contributors to water pollution include a variety of pollutants such as organic and inorganic materials, microorganisms, industrial waste, and many other toxic substances (Figure 8.4). Water is a valuable and necessary component of life, and its role in the global economy is significant. The world is currently facing severe issues with freshwater lack due to excessive utilization and water pollution (Haq et al., 2021; Rafique et al., 2019; Yalcin and Altay, 2023).

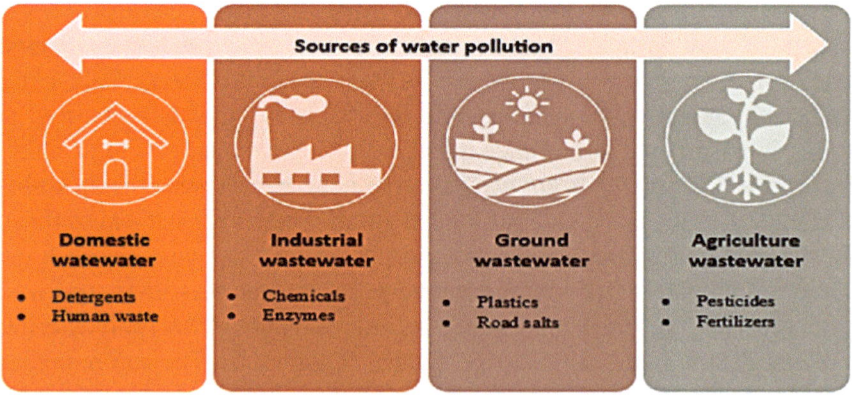

FIGURE 8.4 The image depicts sources of water pollution. It includes domestic wastewater, industrial wastewater, agricultural runoff, and groundwater contamination.

Agriculture water gets polluted mainly by pesticides and fertilizers. Ground wastewater includes animal waste, sludge, plastics, polythene bags, and road salts. Domestic waste is also a major source of water pollution and includes detergents, human waste, and chemicals (Rafique et al., 2019). Clean water is essential for the survival of life on Earth. Various Conventional and advanced approaches are under consideration for the removal of these contaminants, but the implications of these technologies have their advantages and shortcomings (Baruah et al., 2019). To address this scarcity, numerous techniques have been developed globally for cleaning polluted water before distributing it into the natural environment. Physical and chemical approaches are the two broad categories into which frequently employed water purification techniques can be categorized. Conventional techniques of water purification, like filtration with the help of sand, sedimentation, flocculation, coagulation, chlorination, and adsorption on activated carbon, do not show promising results for the elimination of hazardous metal ions and dispersed organic compounds (Baruah et al., 2019). Other options for cleaning up water include using ultraviolet light, ozone treatment, and incineration but these methods are not economical for getting rid of trace contaminants. The scientific world has begun investigating the exceptionally exciting area of research because of the ground-breaking developments in nanoscience and nanotechnology. Here, the special and beneficial features of unique nano-structured objects could be used to provide more effective and long-lasting solutions to the existing challenges associated with water (Lu et al., 2016). Nanomaterials typically range in size from 1 to 100 nm. They contain significantly fewer atoms due to their smaller dimensions, which results in distinct characteristics in them when compared with bulk substances. Nanomaterials have a large surface area to volume ratio due to their small size which causes greater surface-related characteristics. Nanomaterials were discovered to be outstanding adsorbents for an array of water pollutants due to their useful physicochemical properties (Tyagi et al., 2017). Several distinct types of nanomaterials include iron nanoparticles, metal nanoparticles, polymer nanoparticles, zeolites, carbon-based nanomaterials, self-assembled monolayers on mesoporous supports (SAMMS), biopolymers and nano-scale semiconductor photocatalysts.

The full recovery of nutrients in particular nitrogen, potassium, and phosphorus, and energy neutrality is the goal of the next generation of residential wastewater treatment plants (DWWTP). Additionally, there are increased forces pushing for the recovery of valuable materials from industrial wastewater. Heavy, valuable, or radioactive metals, as well as new contaminants like pharmaceuticals, enzymes, hormones, fertilizers, biofuels, biogas, and bioplastics, can be extracted using biological and nano-technology protocols. Despite laws of resource dispersion domestic wastewater alone does not satisfy the essential energy demands of industrialized communities (Puyol et al., 2017). It is advised to use a functionalized PSf/SiO_2 nano-composite membrane to separate oil from water emulsions (Ahmad et al., 2011). For phosphorus recovery from wastewater, Hybrid ion exchange nanotechnology (HIX-Nano) is a particular technique that may enable the production of a useful nutrient solution after the desorption of phosphorus from saturated media. HIX-Nano-demonstrates promising and inexpensive solutions to the phosphorus issue in wastewater treatment. The recovered phosphorus can be implied as a useful fertilizer (Ownby et al., 2021). There have been many adsorbents generated to eliminate metal ions and dyes from water but most of these are not commercially viable solutions due to low absorbance ability. The preparation of versatile materials such as organic-inorganic hybrid polymers aids in this regard by providing greater capacity for adsorption, higher thermal resistance, and higher recyclability (Lofrano et al., 2016). Heavy metal ions (Cd, Zn, Pb, and Cu) were removed from water solutions at various pH levels using Fe_3O_4 magnetic nanoparticles with modification of 3-aminopropyltriethoxysilane and copolymers of acrylic acid and crotonic acid. Even with an acidic pH, these magnetic polymer nanocomposites showed extremely effective metal ion adsorption.

Nanomaterials have been used in the development of novel membrane materials. Graphene oxide (GO) coating on the membrane's surface functioned as a multipurpose layer of protection to improve the membrane antifouling and chlorine resistance while preserving its separation efficiency. Recently, the functionalized carbon nanotubes in the composite have drawn attention. As demonstrated addition of zwitterion-modified carbon nanotubes to the polyamide membrane can prominently improve the water flow and salt denial ratio. These positive characteristics of the composite membranes are attributable to the water-repellent nano-channels made by the CNTs and polyamide that are in active layers (Baruah et al., 2019).

Nanostructures are used for the disinfection of water polluted with pathogenic microbes. TiO_2, ZnO, and silver nanoparticles are used in water filters and nano-sized membranes for killing microbes in water. These nanoparticles mainly release ROS in their mechanism of action. Another category of nanomaterials with strong antimicrobial properties is carbon nanotubes (CNTs), which can be used in a wide range of anti-microbial possibilities (Kudhier et al., 2018). CNTs can filter out microbes due to their brittle structure by depth and size exclusion purification. Pathogenic microbes like *E. coli, S. aureus,* and *K. pneumoniae* can be killed by using TiO_2 nanoparticles of different diameters. Titanium dioxide nanoparticles can produce reactive oxygen species (ROS) in cells when exposed to light. It has been discovered that their activity is size-dependent, with size as a major contributing factor to product efficiency. Strong anti-bacterial properties against *E. coli* have been shown by Sol-gel thin films and silver-doped titania nanofibers. Zinc oxide nanoparticles are being used in commercial water purification devices and antibacterial drugs because it is documented to have low cytotoxicity. Other metal oxide nanoparticles (copper, magnesium, and iron oxide) have also been used as antibacterial agents (Kudhier et al., 2018). Another highly effective method for wastewater treatment is the degradation of organic contaminants by photocatalysis with the help of nano-scale catalysts but it also has several flaws that prevent its widespread application. Nanotechnology also offers a wide range of opportunities for detecting and tracking water pollutants by providing water-sensing capabilities (Kumar et al., 2013). Although the side effects of nanoparticles on human health are unknown. Procedures such as clearing freshwater of organic colors and metal ions by antimicrobial nano-agents or nano-adsorbents to remove organic colors and metal ions might have negative effects on the aquatic environment (Baruah et al., 2019). Nanoparticles progressively endure bio-concentration after ingestion. The toxic effects are primarily caused by build-up in the body. The use of nanotechnology on a larger scale is hampered by these significant obstacles. It is essential to develop nano-technology-based equipment with active nanomaterials not compromised and renewed via affordable and energy-efficient approaches after a period of usage (Baruah et al., 2019).

8.3 ENVIRONMENTAL QUALITY MONITORING THROUGH BIOSENSORS

The advent of civilization has raised the level of more serious environmentally harmful pollutants such as oxides of nitrogen, sulfur, carbon, heavy metals, VOCs, air microbes, and dioxins. Therefore, it is not possible to use some of the most cost-effective methods for handling these substances by means of Spectroscopic and chromatography techniques since they are limited with a time-consuming method, but others have potentially risky consequences (Mohamed and Awad, 2022). To address this specific problem, it is therefore necessary to provide a novel and cost-effective solution. The assessment of environmental risk advanced by bio-sensing devices has been developed to meet the needs of rapid, accurate, precise, and real-time pollution detection and screening. At a reasonable cost, carefully combined techniques of analysis with biotechnology are mainly based on biosensors (Gavrilaş et al., 2022; Mohamed and Awad, 2022). Biosensors are widely used as fast, affordable, accurate, sensitive, and instant detection methods of pollution. The basic mechanism is based on integrating the target chemicals such as pesticides and tiny organic molecules with detection molecules such as enzymes, antibodies, or DNA sequences as analytical tools to predict substances represented in Figure 8.5. The biosensor uses a biological analyte rather than the sensor transducer and signal processing via transducer sensor technology (Mohamed and Awad, 2022).

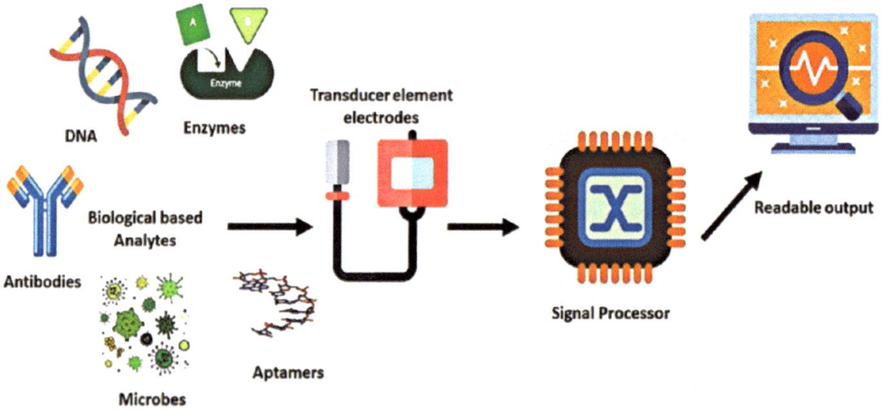

FIGURE 8.5 The image depicts the basic working principle of biosensors.

For this specific purpose, several biosensors have been designed. Based on the type of biotic signaling transduction, biosensors may be classified. The various categories of biosensors include fluorescence, electrochemical, mass change, optical, and thermal biosensors, depending on the basic transduction mechanisms utilized to integrate the reactions of the target materials and the sensing (Gavrilaş et al., 2022). Nanocomposites are techniques that have the potential to improve the sensibility and diversity of complicated environmental sample analysis. Nanomaterials such as tubing and nanoparticles are used to modify the properties of biological sensors to achieve this goal. To provide a high degree of analysis with constant, programmed, and instant data, Nigam and Shukla (2015) observe that biosensors still need to be improved for the environmental area in order to achieve such an outcome. However, accuracy should be taken into consideration as the primary concern. To determine the quality of the environment, these biosensors shall be used (Gavrilaş et al., 2022).

8.3.1 ENZYME-BASED BIOSENSORS

Enzymes function as biological catalysts by being three-dimensional, complex macromolecules. The biosensor-employed enzymes serve as the organic detecting component, which is linked to transducers to convert enzymatic response output into quantifiable responses proportional to the analyte concentration. An example of an enzyme response signal is energy release, proton density changes, oxygenation or absorption, illumination, and absorption. Transducers are electrical, electrotherapeutics, optical, heat, and piezoelectric devices that convert this signal into a voltage, temperature exchange, absorption, or other properties that can be monitored in a variety of ways. Several advantages have emerged for researchers in enzyme biosensors over the last few decades, including the exceptional sensitivity and precision of enzyme-mediated reactions. A broad spectrum of detectable substances with the capacity to detect diverse highly pure accessible enzymes (Mohamed and Awad, 2022). The fundamental operating principles for enzymatic biosensors and mechanisms of action include metabolizing the target analyte by the enzyme and triggering impediment, or modification of the enzyme by the substance. The ability of enzymes to be immobilized on the transducer surface and activated after being immobilized are key requirement for enzyme biosensors. Immobilizing enzymes have greater stability. Cross-linking, adsorption, nano-capsulation, entrapment, and covalent bonding are the most commonly used techniques for the immobilization of enzymes (Gavrilaş et al., 2022; Naresh and Lee, 2021).

8.3.2 MICROBIAL BIOSENSORS

Naturally or biologically adapted microbes, fungi, algae, protozoans, and viruses capable of controlling a range of substances allow the specific transducer to identify its signal to form cell biosensors. Electrical sensors have been combined with microorganisms called microbiological sensors have a higher sensitivity to environmental signals than their natural counterpart. There are benefits to this technique, such as minimal detection limits, good selectivity, and high sensitivity. As a result of these characteristics, whole-cell receptors may be applied in different applications (Mohamed and Awad, 2022). Due to the close similarity of detection methods between bacteria and enzymes, microbiological sensors can be considered an advanced form of biosensors. They must be immobilized by an immobilization technique to ensure that the biological materials are contained within transducers and supporting matrices. Physical adsorption or chemical techniques combined with covalent binding and cross-linking ability can bind and restrict microorganisms, such as enzymes (Naresh and Lee, 2021). In addition, the immobilization method to be applied has to provide for chemical resistance, cell viability, safety handling, and prolonged storage. Compared to conventional methods, bacterial bioassays encounter issues, which include their high sensitivity, and simultaneous detection of multiple chemicals as well as significant potential drawbacks with regard to site monitoring and economic efficiency. They are limited by a variety of parameters, including extensive reaction periods, rapid detection of external heat, acidity, and alkalinity. Any factor may influence cell survival over time. Recent research has used terrestrial aquatic biota to identify contaminants from the environment such as heavy metals, insecticides, Cadmium, Arsenic, copper, mercury, lead, or phenols among others (Gavrilaş et al., 2022).

8.3.3 ANTIBODY-BASED BIOSENSORS

An antigen or an immunoglobulin is a widespread category of glycoprotein, assembled by specific compartments in the internal organism and useful for detecting dangerous substances such as antibiotics, and so forth. The antibodies can identify and bind antigens, which results in antigen complexes that remain stable for an extended period. Antibodies can be monoclonal or polyclonal form. An antibody-based biosensor is a small devices that use a transducer to detect and measure unique interactions of immunoglobulins and antigens. Electrolytes, potentiometric, impedimetric, colorimetric, light, and microgravimetric are defined as immunoassays depending on their transduced function. They are also known as labeling or labeling sensors. For labeling, the target analyte or bioreceptor is to be affixed with a sensitivity-detectable marker. During the analysis, these tags' activities are monitored (Mohamed and Awad, 2022). These tags can be classified as enzymes, fluorescent dyes, electrolytes, and nanoparticles based on the type of compound. An immunosensor with a nonlabelled tag is designed to estimate changes in physical characteristics that result from its development for direct determination of the antigen-antibody complex. Immunological instruments have an advantage over conventional analytical methods in terms of selectivity and sensitivity (Gavrilaş et al., 2022). At the same time, it is possible to observe an evolution of immunoreactions on detector surfaces in real-time. The limitations to the use of antibody biosensors, including pH and temperature sensitivity constraints as well as extensive time spent developing specialized reagents for each compound which is related to their use should also be considered (Naresh and Lee, 2021).

8.3.4 APTAMER-BASED BIOSENSORS

Aptamers are linear oligonucleotide sequences with a length of 15 to 80 base pairs which may be attached to certain molecules of the target. Aptamers target all types of small molecules as well as proteins, carbohydrates, and metal ions in cells, viruses, or bacteria. To create aptamers *in vitro*, it is possible to use a technique known as sodium laureth sulfate (SELEX) that allows ligands to be systematically evolved by exponential enrichment (Naresh and Lee, 2021). SELEX will be created by creating a huge collection of DNA sequences from various alignments whereby targeted compounds will be exposed for a period. Unfolded molecules and target-bound oligonucleotides should be screened either by heating or washing at the end of incubation. The polymerase chain reaction (PCR) amplifies the attached aptamer molecules to generate the input for the successive screening steps. During the entire process, 5 to 15 selection and amplification cycles are conducted. Various biosensors have been developed in recent years to detect environmental toxins integrating aptamers as bio-receptors (McConnell et al., 2020).

8.3.5 DNA-BASED BIOSENSORS

In DNA biosensors single-stranded DNA is employed as reference components. Their functionality depends on the following two processes: (i) the phenomenon of hybridization between the desired genetic material and its complementary thread confined on a detecting region via instinctive hydrogen bonding among nitrogenous bases; and (ii) the modification of the single stranded-DNA structure by the selected substrate molecules. This mechanism, which can be converted to an accurate response by suitable transducers, for example, electrolysis or Oscillators, is made up of the combination of different chemical changes resulting in an individual signal being generated. In order to develop DNA-based biophysicists, immobilization of nucleic acid electrode fragments onto the surface is a critical step (Mohamed and Awad, 2022). Nonetheless of the adsorption, covalent bond, or avidin-biotin reaction methods used to adsorb them, these fragments have to be immobilized so that they remain active and can maintain their stability and access to target molecules (Gavrilaş et al., 2022). DNA biosensors are used for a wide variety of applications, for example, prediction, medical diagnostics, nutrition monitoring, and ecological monitoring owing to their recompense which includes specificity, sensitivity, biocompatibility, and cost-effectiveness. A number of studies have shown that DNA biosensors are capable of detecting the presence in the environment of large amounts of metal. A working notion is built in this example based on the inclination of specific ions of heavy metals to build durable double helical structures containing specified nitrogenous bases. The mercuric ions prefer to interact with nitrogenous bases called thymine, resulting in a thermally stable T–Hg^{2+}–T structure. Similarly, silver ions interact with two cytosines to form C–Ag^{+}–C base combinations, which strengthen the nucleotide strand. When the presence of specific metal ions allows precise detection of metals using proper transducers, thymine-rich or cytosine-rich single-stranded DNA can form strong structures for this purpose (Naresh and Lee, 2021). Examples of biosensors are mentioned in Table 8.1.

TABLE 8.1 Examples of Biological-based Biosensors Targeting Different Environmental Pollutants

Substance (Analyte)	Substrate	Detection Method	Target	Type of Sensor	References
Hg^{2+}, Cu^{2+}, Cd_2	Urease	Entanglement in sol-gel matrix	River water	Enzyme-based sensor	Kurbanoglu et al. (2020)
Cu^{2+}	*S. cerevisiae*	Entanglement in alginate beads	Water	Microbial-based biosensor	Vopálenská et al. (2015)
Okadaic acid	Graphene	Impedimetric	Sea water	Antibody-based biosensor	Antunes et al. (2018)
Cd^{2+}	Aptamer	SYBR green I as an indicator	Tap water. River water	Aptamer-based biosensor	Fang et al. (2018)
Ostreopsis cf. *ovata*	DNA	Colorimetric	Plankton, bentonite	DNA-based biosensor	Ravikumar et al. (2017)

8.4 ROLE OF BIO-NANOTECHNOLOGY IN AGRICULTURE

The forecast that more than 9 billion individuals will populate the world by 2050 is well established. This exponential growth will threaten food and water security, particularly in underdeveloped countries. Producing more food requires more natural resources, water supply, land consumption, and energy (Glenn et al., 2016). To fulfill this elevated food, demand crop productivity needs to be multiplied through efficient use of modern technology (Manjunatha et al., 2016). International agencies like the United Nations (UN) Food and Agriculture Organizations (FAO) and the World Bank are urging trailblazing solutions to cope with this multiplex problem. Nano-biotechnology is a promptly emerging field that ensures sustainable food production and assures resistance to crop pathogens (Marchiol, 2018).

Bio-nanotechnology comprises nanoparticles that are smaller than 100 nm in size. They can be manipulated at the molecular or atomic level. Its agricultural applications are drastically being explored these days as a well-structured strategy to promote plant growth resulting in an increase in crop yield (Giraldo et al., 2019). The nanoscopic size of nanoparticles enables them to make their way across biological barriers, diffusing into vascular tissues of plants after foliar or root applications, thus opening new avenues for nutrients and pesticide delivery. The surface chemistry and structure of these nanoparticles can be reformed to direct them to the required plant tissues (Lowry et al., 2019). Gainful applications of bio-nanotechnology in agriculture are as in subsections (Figure 8.6).

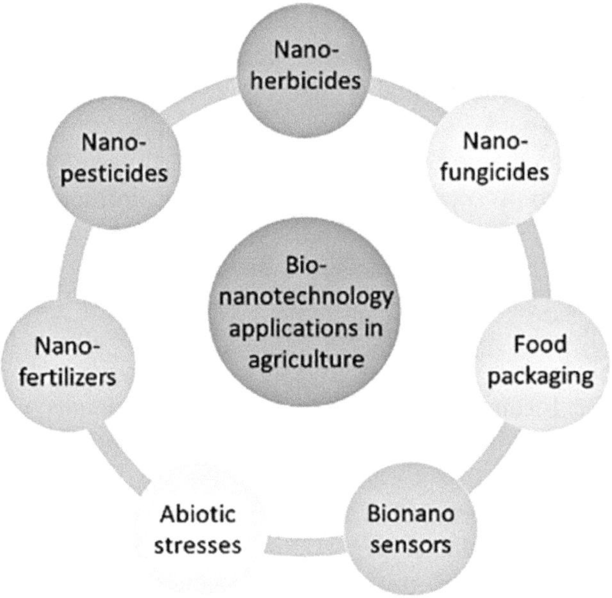

FIGURE 8.6 Applications of bio-nanotechnology in agriculture.

8.4.1 NANO-FERTILIZERS

Unsystematic use of chemical-based fertilizers in agricultural systems to keep pace with escalating global food demand has adverse effects on the ecosystem. A large proportion of these fertilizers added to the soil are lost and become inaccessible to plants. To redeem the lost fertilizer, more fertilizer is added to the soil, which disturbs the nutrient balance in the soil (El-Saadony et al., 2021). To overcome this issue, fertilizer particles are coated on nano-membranes that will aid slow release of nutrients. These nano-fertilizers showed up as great alternatives to inorganic fertilizers as they permit plants to grab the majority of nutrients without losses (Guo et al., 2018). Higher surface areas of nano-fertilizers as compared to traditional fertilizers permit nano-sized fertilizers to grab more nutrients which they release in accordance with plant needs without any harmful consequences. Nano-fertilizers can be supplemented to plants by either roots or foliage depending upon the properties of the particles. These nano-sized particles act as a smart nutrient delivery system to the plants, thus elevating crop productivity (Babu et al., 2022).

8.4.2 NANO-PESTICIDES

Crops are attacked by a substantial number of insects and pests, which deteriorates the quantity, quality, and commercial value of crop products. To combat pest attacks, diverse chemicals are applied that either kill them or hinder their reproduction or feeding habits. The overreliance on pesticides has led to several problems such as pesticide poisoning and toxicity (Dangi and Verma, 2021; Mushtaq et al., 2020). Nanotechnology has been employed to protect crops from pest attacks in a more efficient way. A high surface area to volume ratio enables nano-pesticides to get into the cell membranes of the insects and damage them. Nano-pesticides are preferably synthesized from Nobel metals such as silver and gold that are non-toxic to humans (Connor et al., 2005). To attain a greener and more eco-friendly route, scientists have begun to use plant extract containing reducing agents such as dehydrogenases, flavonoids, and reductases for the synthesis of nano-pesticides (Vadlapudi and Kaladhar, 2014). Thus, nano-pesticides offer greater stability and permeability to enhance pest control efficiently for longer durations of time. Nano-pesticides are "magic bullets" loaded with pesticides that hit the target site with great precision (Dangi and Verma, 2021).

8.4.3 NANO-HERBICIDES

Weeds are hazardous for agriculture as they lessen crop yield to a substantial extent (Rehman et al., 2020; Yarcı et al., 2007; Yarci and Altay, 2016). Therefore, eradication of weeds is crucial for sustainable agriculture. The approach to control multiple species with a single herbicide is rather causing herbicide resistance even to chemicals (Chinnamuthu and Boopathi, 2009). Employing nano-herbicide is a promising and eco-friendly approach to clearing out weeds in a more efficient way. Their nano-size allows them to blend with the soil in a better way to deal with weed species that have become herbicide-resistant over time (Yadav et al., 2015). Herbicides available commercially are only able to deal with the above-ground part of the weeds. None of the available herbicides can counter the growth of underground more viable parts of the weeds like tubers or rhizomes. Nano-herbicides being more target specific would aim for specific receptors in the roots of target weeds. These nano-substances would make their route through the root system to the site where glycolysis of reserved food in roots takes place to inhibit it. Resultantly, targeted weed plants would starve to death (Chinnamuthu and Boopathi, 2009).

8.4.4 NANO-FUNGICIDES

Synthetic fungicides remain to be the most efficient and cost-effective way to control fungal diseases to date. This great asset of plant protection against fungal infection is challenged by a number of constraints such as environmental pollution, development of disease resistance, and side effects in non-target organisms (Malandrakis et al., 2022). Due to the finite number of available fungicides and their infinite applications on plant surfaces, pathogens have acquired resistance against these chemical-based fungicides (Alghuthaymi et al., 2015). In recent times, nano-biotechnology has proved its incredible potential in combating fungal diseases both as an alternative to conventional fungicides and/or as a partner to them. Nano-fungicides can be synthesized by chemical, physical, or biological (by green synthesis employing microorganisms or plants or their metabolites) to counter plant diseases in a more efficient way (Nayak et al., 2011). The positive charge of metal NPs adheres to the negatively charged cell membrane of microbes because of electrostatic attractions. This affinity between opposite charges destroys membrane integrity and ultimately leads to microbe death (Malandrakis et al., 2022).

8.4.5 FOOD PACKAGING

The elevating demand for premium food quality and enhanced awareness of customers about healthy lifestyles opens new avenues for researchers to refine food quality without compromising on its nutritional value. As bio-nanotechnology incorporates biological molecules into nanostructures, NPs are also being implemented in packaging food. Employing NPs in food packaging offers longer shelf life of food products along with better food quality (Primožič et al., 2021). The majority of these nanomaterials are composed of essential elements, which are non-toxic and possess high stability at elevated pressures and temperatures (Mohamadian et al., 2019). Nanotechnology embodies NPs into a film called "biofilm," which enhances the permeability of certain gases to decrease the concentration of damaging gases. These include oxygen (O_2) and CO_2 which negatively affect the shelf life of food. Furthermore, these NPs also act as a barrier to intercept food spoilage by microbes (Sonkaria et al., 2012). The most universally used materials for food packaging are plastic polymers, which are non-biodegradable thus posing a serious threat to human health. Edible coats and films of bio-nanocomposites can be potential alternatives to plastics as packaging materials. NPs engineered with antimicrobial particles not only prolong the shelf life of vegetables and fruits but also improve the storage quality of foods (Xing et al., 2019).

8.4.6 BIO-NANO-SENSORS

Conventional farming approaches damage crops due to biotic and abiotic factors. Here more innovative technology, that is, nanobiotechnology has been employed to overcome these shortcomings along with promising food security (Mittal et al., 2020). Unchecked use of fertilizers, heavy metals, and pesticides can be checked with the help of nano-sensors. The health condition of the soil and possible disease incidence could be checked regularly with the help of emerging nanotechnology. Linking modern Information and Communication Technologies (ICTs) with nano-sensors provides online and novel ways to detect each component present with precision (Maryam and Gul, 2023; Omanović-Mikličanina and Maksimović, 2016). Nanotechnology has enhanced the performance of biosensors specifically in the field of agriculture. Nano-sensors have proved their potential in many fields of agriculture such as pathogen detection, crop cultivation and harvesting, and soil parameters (nutrients and pH). Nanoparticles have unique thermal and electrical characteristics, enhanced sensitivities, distinct surface chemistry, and greater detection limits that establish them as refined innovative sensing systems (Yao et al., 2014).

8.4.7 ABIOTIC STRESSES

Plants show a number of complex responses against abiotic stresses and change their metabolism, physiology, and morphology accordingly (Garcia-Caparros et al., 2020; Ozturk et al., 2019b; Shahrasbi et al., 2021). Various strategies including genetic engineering are being implemented to add to plant's tolerance against abiotic stresses (Das and Das, 2019). Bio-nano-technology being a versatile field offers solutions to combat abiotic stresses. These stresses are major and apparent factors affecting agriculture. These abiotic stresses include drought, salinity, heavy metal stresses, water logging, and metals and minerals deficiency/toxicity (Ayub et al., 2021; Ghori et al., 2019; Hameed et al., 2017; Mushtaq et al., 2020; Ozturk et al., 2019a; Raza et al., 2019). Investigative studies on NPs proved that they are better able to cope with abiotic stresses, as they are nano-sized, so they have high mobility. They can be transported in all parts of plants (Tariq et al., 2021). NPs respond to water stress by increasing roots' hydraulic conductance and enhanced uptake of water by plants' roots. In addition, proteins involved in ROS detoxification, oxidation-reduction, hormonal pathways, and stress signaling are produced in varied concentrations (Das and Das, 2019).

8.5 ROLE OF BIO-NANO-TECHNOLOGY IN HUMAN DISEASE MANAGEMENT

Disease management is a vast term to use that includes the whole management including the knowledge related to the disease, diagnosis, prevention, treatment, and improvising the life after disease, but most importantly disease management relies on two important points that are diagnosis and treatment. Nanotechnology has been proven technology in the management of human diseases, especially in their diagnosis, prevention, and treatment. The word "nano-diagnostics" refers to the application of nanotechnology's ideas and approaches to diagnostics. To address the demand for clinical diagnostics, nano-diagnostics is a growing use of nano-scale technology. It involves identifying the pathophysiology of the ailment and the organisms that are responsible for it, as well as determining the disease status and any susceptibility to it. Nano-scale diagnostics made possible by nanotechnology have sparked a trend toward the usage of readily available, marketable handheld gadgets. Different kinds of nano-based diagnostics options such as nano-tubes, biorobots, quantum dots, nano-biosensors, and nano-bio-chips, and so forth. Carbon NTs, fullerene NTs, and silicon nanowires were successfully designed for the diagnosis of oral cancer, Lung cancer, and Stomach cancer (Shehada et al., 2015).

Nanoparticles can be classified into several types as follows; Quantum dots emit fluorescence in response to low-energy light. These dots of various sizes can be implanted in a microbead and agitated to produce varied colors. Such straightforward excitation is advantageous for genotype identification, image-guided surgery, and biological diagnostics due to its high sensitivity and broad excitation spectrum. Diagnostics and therapies are combined with quantum dots. Quantum dots with detectable luminescence encapsulated in carbohydrates are useful in cancer imaging, and citrate-capped silver nanoparticles (Citrate-AgNPs) modified graphene quantum dots (GQDs) nano-ink can be used to detect prostate-specific antigen (PSA). Cys-AuNPs were utilized to amplify the signal. It immobilizes PSA biotinylated antibodies with a large exterior surface. First, cellulose paper (ivory sheet and photographic paper) was directly printed with citrate-AgNPs-GQDs nano-ink before being altered by Cys-AuNPs. Ab1 and PSA (antigen) completed the immune device construction. DPVs were employed to detect PSA in immunosensors. The calibration plot's linear range was 0.07 to 60 μg/L, and the first immune device's LLOQ was 0.07 μg/L. The suggested immunoassay successfully monitored PSA glycoprotein in unprocessed human plasma. Findings suggest that the immunosensor can be used as a portable bio-device to analyze human plasma PSA (Farshchi et al., 2020). Nanofibers are ultrafine fibers, manufactured from polymer solutions and have nano-meter-scale diameters (Islam et al., 2019). Another study developed fluorescent molecularly imprinted conjugate polythiophenes nanofibers (FMICP NFs) to detect cancer biomarkers serum alpha-fetoprotein (AFP) and carcinoembryonic antigen (CEA). Researchers sought to address enzyme-linked immune sorbent assay (ELISA's) complex enzyme labeling and pricey antibodies. Borate produces enzyme-affine cavities in the device. Molecularly imprinted polymers (MIPs) are cheaper than antibodies, stable in severe circumstances, easy to make, and selective but they do have drawbacks. They have heterogenous bindings, low potential binding sites, and inaccessible recognition sites. These drawbacks limit their use for real sample analysis and the analytes they can detect (Tawfik et al., 2020).

Nanocrystals are a class of crystalline materials characterized by size-dependent thermodynamic and electrical properties and at least one dimension less than 1 μm. At 10 nm, nano-porous nano-crystals function as efficient semiconductors. Protein is adsorbed when silica molecules alter the pores (Montanarella and Kovalenko, 2022). Biochips or lab chips are also novel and very effective tools for diagnosis. It is a simple glass or silicon gadget that performs numerous analyses. The device specifically interacts with cellular components like metabolites, proteins, and DNA. It has been an effective tool in managing chronic diseases like diabetes and kidney disorders (Wu et al., 2018). Nano-bio-robots or nano-bots are a very advanced science intervention and are still in their early stages, but they seem quite cost-effective and promising in terms of output. Scientists in China developed Fluorescence magnetic spore-based micro-robots (FMSMs) to detect the Real time *Clostridium difficile* toxins in patients' stool (Zhang et al., 2019). Different targeted drug delivery options have become feasible with nano-bots. Nano-machines are also a type of nano-bio-robots used for targeted therapy. Different MEMS (micro-electromechanical systems) devices are currently available and are a major application of nanotechnology in healthcare systems. They are being used as labs on chips, biosensors, and also a method for targeted drug delivery. With the development of science and advances in nanotechnology, there have been multiple hybrids and conjugated nanotechnologies that are used as a biosensor or diagnostic tools for different diseases. Chitosan is an especially important material, and it has a lot of beneficial properties it is biocompatible, biodegradable, non-toxic, and has the potential as a Nano-carriers encapsulating active compounds and allows a controlled release procedure, chitosan has several applications. It is used in cancer diagnostics, glucose detection sensors, wound dressings, bone tissue engineering, histamine biosensors, and dentistry (Jafernik et al., 2023). It was observed in diabetic rabbits, that chitosan-PEG hydrogel with silver nanoparticles helps wounds heal faster (Masood et al., 2019).

8.5.1 ROLE OF NANOTECHNOLOGY IN CARDIOVASCULAR DISEASES

For disorders like atherosclerosis, Heart Attack methylisothiazolinone (MI), and hypertension, exclusive micro- or nanoparticles are frequently injected into the targeted blood vessel in an effort to avoid problems related to conventional drug delivery, such as adverse systemic side effects. In addition, Novel nano-drug transporters that enter the bloodstream may be selectively absorbed by immune system cells to modulate inflammatory responses and migrate to plaques for therapeutic payload delivery. Indeed, novel nanoparticle composition, formulation, and functionalization have revolutionized the field as an approach toward therapeutic success in a wide range of cardiovascular conditions. Different kinds of nano-technological integrations and inclusion like implants, wearable devices, stents, nanoparticles coated surfaces have made a huge mark in combating cardiovascular diseases. Nanocomposites like PLGA and liposomes have been a very good option for drug delivery to a specific area in blood vessels (Wang et al., 2021; Zheng et al., 2022).

8.5.2 ROLE OF NANOTECHNOLOGY IN NEUROLOGICAL DISORDERS

Treating nervous system disorders has always been a challenge due to the BBB. Nanotechnology is utilized according to reports Pgp-mediated phenytoin resistance is suppressed by NPs of PBCA in rodents. Additionally, a recent study demonstrates that andrographolide (a neuroprotective medication) is more permeable to the BBB when it is enclosed in SL NPs than when it is free (Graverini et al., 2018). Nanoparticles lower the toxicity of drugs and make them more biologically available in the brain. Several neurological illnesses and infections, as well as brain tumors, have been successfully analyzed and characterized using MI approaches (Bocan et al., 2019). Transporting targetable probes across the BBB using PS-80-coated PBCA dextran polymeric NPs has been proven to be effective in an Alzheimer's disease model (Ji et al., 2013). Sulfated dextran-coated IO NPs have been shown in a recent study to enhance bio-imaging of activated microglia-induced brain inflammation. This is accomplished through the binding of NPs to highly expressed class A scavenger receptors (Tang et al., 2018). After binding to integrin V3, RE-doped NPs have been found to emit short-wave infrared light, making them useful for fluorescence imaging. Different fluorescent NPs have proven to be efficient in bio-markers discovery for Alzheimer's disease that can easily distinguish Amyloid Beta plaques, inflammatory cytokines, and Tau protein NFTs (Fernández-Cabada and Ramos-Gómez, 2019; Sun et al., 2021; Zeng et al., 2018).

A study in mice reveals that memory deficits caused by *Trypanosoma evansi* infection may be effectively treated with nerolidol-loaded NPs, which also normalizes ROS level and regulates the activity of Na-K ATPase and acetylcholine esterase (Baldissera et al., 2016). Researchers have reported that treating mice infected with *Schistosoma mansoni* with Au and zinc oxide (ZnO) NPs significantly improves oxidant/antioxidant status and neural damage via modulating several genes (Bauomi, 2020). NP are helpful in Ischemic stroke, to lower the levels of ROS and also, they have been implicated in amnesia. It is observed that memory deficits in amnesia animal mode can be corrected by treated with galantamine-loaded thiolate CSNPs (Singh and Mishra, 2019). Multiple Sclerosis is a disease with currently no available effective cure, the glucocorticoid NPs have shown therapeutic effects in MS (Montes-Cobos et al., 2017) also in a mouse model of multiple sclerosis, chondroitinase ABC 1 encapsulated in porous silicon NPs reduces the neuronal damage and promotes remyelination (Rezaei et al., 2020). Recent research studies demonstrate the expressions of TNF-, cooxygenase-2/inducible nitric oxide synthase, and intercellular adhesive molecule one are considerably suppressed in migraine patients treated with omega-3 fatty acids and curcumin NPs, leading to a considerable reduction in inflammation (Abdolahi et al., 2018).

8.5.3 ROLE OF NANOTECHNOLOGY IN CANCER

Utilizing nanotechnology to perform molecular tests for the diagnosis of cancer is made possible through the utilization of minuscule substances such as gold nanoparticles, quantum dots, and other nanomaterials. The detection of cancer-associated biomarkers has been simplified owing to advances in assisted molecular diagnostics. In particular, despite its tiny size, a nano-biosensor is capable of rapidly detecting a high number of protein biomarkers (Tran et al., 2017). Many cancer therapies and screening methods concentrate on reducing the rate at which cancer cells proliferate. Early and precise cancer detection is essential for successful therapy, and Positron Emission Tomography (PET), Computed Tomography (CT), and Magnetic Resonance Imaging (MRI) are often utilized in recent examinations and tests (Kim et al., 2010). It has become extremely challenging for researchers to transform findings from studies into effective treatments and favorable results for patients since current approaches are too limited in scope to give comprehensive clinical information on all the different types and stages of tumors (Akhter et al., 2013; Wang et al., 2010). Most current anticancer medications do not distinguish between cancerous and healthy cells, they are physiologically toxic and have undesirable side effects. In addition, a key issue with cancer is that it is often not detected until it has spread throughout the body (a stage known as "metastatic") (Chaturvedi et al., 2019). Various research findings are under consideration for the diagnosis of cancer with the help of nano-technology which includes building a detector that is capable of screening for cancer while wearing it on your wrist (Nie et al., 2007), lanthanide-derived up conversion nanoparticles have the capability to convert low-energy photons into high-energy photons, thereby enabling the detection of tissue depth through auto-fluorescence (Gu et al., 2013). Fluorescence nano-probes that exhibit pH responsiveness have the capability to identify the presence of fibroblast activation protein-a on the membranes of tumor-associated fibroblasts (Ji et al., 2013). To conduct tissue biopsies and detect instances of pancreatic cancer as well as cervical cancer, the gold nanoparticles may also be used in conjunction with antibodies. In addition to this, photoacoustic tomography is another application for gold nanoparticles. Because of this, gold nanoparticles are considered to be a viable option-based diagnostic agent for various malignancies (Purohit and Singh, 2018). Cancer cells can be specifically targeted by liposomes, which degrade quickly, are compatible with human cells, and are far more robust in colloidal solutions (Akbarzadeh et al., 2013; Zhang et al., 2008). To evaluate the efficacy and safety of drug delivery, using anti-cancer drugs and free drug administration, researchers found that liposomes have minimal toxicity at the site of malignancies versus placebo or free medications (Allen and Cullis, 2013; Sutradhar and Amin, 2014). Doxil (Liposomal Doxorubicin), the first FDA-approved nano-drug, treats AIDS patients' Kaposi's sarcoma with PEGylated liposomal formulation (Singh, 2018). Merrimack, a pharmaceutical corporation, has developed a liposomal sphere that encapsulates the drug irinotecan. The pharmaceutical agent under consideration is MM-398. This demonstrates promise in the management of chemotherapy-resistant cancers, such as lung, pancreatic, colorectal, and glioma malignancies (Ko et al., 2013; Roy et al., 2013; Saif, 2014).

8.5.4 ROLE OF NANOTECHNOLOGY IN GENETIC AND METABOLIC DISORDERS

Genetic disorders are sought to be cured through Gene therapy and there have been several research that use gene therapy to cure diseases caused by genetic aberrations, Nucleic Acid Drugs are one of the treatment options promising in hereditary retinopathy and other genetic defects. The method of treating hereditary retinopathy with Nucleic Acid Drugs (NAD) for gene interference is among the most promising in the field of ophthalmology. However, there have been a number of problems with their administration, bioavailability, and usefulness. The problems mentioned earlier can be addressed in one of two ways: either by enhancing the long-term stability of NADs or by choosing more appropriate carriers for NAD distribution. The development of nanotechnology has helped accelerate the adoption of both strategies. Liposomes, nano-capsules, Nano-micelles, dendrimers, nanoparticles, nanoparticles hydrogels, and nano-emulsions are just some of the conventional nano-carriers synthesized by nanotechnology that are widely reported to improve delivery efficiency (Sánchez-López et al., 2017). Second, new technologies like DNA origami have opened up more potential domains of use for gene therapy (Zhan et al., 2023). The role of mi RNAs in disease diagnosis has been well researched and seems to be promising but there is still no diagnostic test available to detect mi RNAs, nanotechnology allows the development of commercial diagnostic tools for mi RNA (Devaraj and Loganathan, 2023).

Synthetic and natural nanoparticles regulate osteoclast differentiation, activity, apoptosis, and osteoblast crosstalk. Synthetic nanoparticles pack antiresorptive medicines or osteoclast gene regulatory molecules. Targeting osteoclast membranes or bone extracellular matrix improves control. Extracellular vesicles, or natural nanoparticles, also help bone homeostasis. Nanoparticle-controlled osteoclast activity may contribute to osteoporosis treatment (Rouco et al., 2023). Nano-biosensors, nanoparticle-based treatments have been successfully used in diabetes and other metabolic and genetic disorders, new research has shown a promising future for nano-medicine, according to research a riboswitch is designed for the treatment of metabolic disorders, Virus-like particles as Nano-carriers were designed and it was demonstrated that VLPs can serve as an exceptional barrier against nuclease-mediated degradation. For inherited metabolic abnormalities, it may be employed as a possible disease-modifying medication. The functioning riboswitch can bind metabolites individually and selectively, stopping the process of their self-assembly and its consequences (Zilberzwige-Tal et al., 2022). Nanotechnology has transformed healthcare since its debut, improving results, and has created the next generation of nanostructures. Nanotechnology has become ubiquitous in the previous two decades, aided by considerable healthcare research. Nanomedicine uses nanotechnology and nano-carriers/

nano-systems to prevent, diagnose, and cure illness. Theranostic nano-systems outperform conventional ones. Gene therapy, targeted medicine delivery, and cancer and genetic disease treatment use nanotechnology. Nano-technology's potential is untapped. To change healthcare, further efforts must be made to overcome these restrictions and maximize its potential (Anjum et al., 2021).

8.6 CONCLUSION

In the modern era, bio-nanotechnologies are revolutionizing day by day and their novel approaches are being discovered at an amazingly fast speed. Nanotechnology plays an important role in reducing environmental pollutants such as removing air pollutants, soil remediation, and wastewater treatment, by monitoring environmental pollutants with biosensors and they also play a major role in agriculture and health management systems. Nano-technology techniques are used as the nanomaterials have a large surface area in comparison to size, so they require less activation energy and have high reactivity. Further, these nano-technology techniques are available easily, cost-effective, increased efficiency, have less environmental impact, and for sustainable in environment. However, new research is still being conducted on nano-technology methods for the benefit of living organisms and human humankind.

KEYWORDS

- **agriculture**
- **bio-fertilizers**
- **bio-nanotechnologies**
- **biosensors**
- **environmental pollutants**
- **nanoparticles**
- **nanotechnology**
- **theranostic nano-systems**

REFERENCES

Abdolahi, M., Sarraf, P., Javanbakht, M. H., Honarvar, N. M., Hatami, M., Soveyd, N., Tafakhori, A., Sedighiyan, M., Djalali, M., Jafarieh, A., & Masoudian, Y. (2018). A novel combination of ω-3 fatty acids and nano-curcumin modulates interleukin-6 gene expression and high sensitivity C-reactive protein serum levels in patients with migraine: A randomized clinical trial study. *CNS Neuroscience & Therapeutics, 17*(6), 430–438.

Abhilash, P., Powell, J. R., Singh, H. B., & Singh, B. K. (2012). Plant-microbe interactions: Novel applications for exploitation in multipurpose remediation technologies. *Trends in Biotechnology, 30*(8), 416–420.

Ahmad, A., Majid, M., & Ooi, B. J. D. (2011). Functionalized PSf/SiO2 nanocomposite membrane for oil-in-water emulsion separation. *Desalination, 268*(1–3), 266–269.

Akbarzadeh, A., Rezaei-Sadabady, R., Davaran, S., Joo, S. W., Zarghami, N., Hanifehpour, Y., Samiei, M., Kouhi, M., & Nejati-Koshki, K. (2013). Liposome: Classification, preparation, and applications. *Nanoscale Research Letters, 8*, 1–9.

Akhter, S., Ahmad, I., Ahmad, M. Z., Ramazani, F., Singh, A., Rahman, Z., Ahmad, F. J., Storm, G., & Kok, R. J. (2013). Nanomedicines as cancer therapeutics: Current status. *Current Cancer Drug Targets, 13*(4), 362–378.

Alghuthaymi, M. A., Almoammar, H., Rai, M., Said-Galiev, E., & Abd-Elsalam, K. A. (2015). Myconanoparticles: Synthesis and their role in phytopathogens management. *Biotechnology & Biotechnological Equipment, 29*(2), 221–236.

Allen, T. M., & Cullis, P. R. (2013). Liposomal drug delivery systems: From concept to clinical applications. *Advanced Drug Delivery Reviews, 65*(1), 36–48.

Altay, V., & Ozturk, M. (2012). Land degradation and halophytic plant diversity of Milleyha wetland ecosystem (Samandağ-Hatay), Turkey. *Pakistan Journal of Botany, 44*, 37–50.

Anjum, S., Ishaque, S., Fatima, H., Farooq, W., Hano, C., Abbasi, B. H., & Anjum, I. (2021). Emerging applications of nanotechnology in healthcare systems: Grand challenges and perspectives. *Pharmaceuticals, 14*(8), 707.

Antunes, J., Justino, C., da Costa, J. P., Cardoso, S., Duarte, A. C., & Rocha-Santos, T. (2018). Graphene immunosensors for okadaic acid detection in seawater. *Microchemical Journal, 138*, 465–471.

Araújo, I. P., Costa, D. B., & De Moraes, R. (2014). Identification and characterization of particulate matter concentrations at construction job sites. *Sustainability, 6*(11), 7666–7688.

Ashraf, M., Ozturk, M., & Ahmad, M. S. A. (2010). *Plant Adaptation and Phytoremediation*. Springer.

Ayub, M., Ashraf, M. Y., Kausar, A., Saleem, S., Anwar, S., Altay, V., & Ozturk, M. (2021). Growth and physio-biochemical responses of maize (*Zea mays* L.) to drought and heat stresses. *Plant Biosystems, 155*(3), 535–542.

Babu, S., Singh, R., Yadav, D., Rathore, S. S., Raj, R., Avasthe, R., Yadav, S. K., Das, A., Yadav, B., & Shekhawat, K. (2022). Nanofertilizers for agricultural and environmental sustainability. *Chemosphere, 292*, 133451.

Bakshi, M., & Abhilash, P. (2020). Nanotechnology for soil remediation: Revitalizing the tarnished resource. In P. Singh et al. (Eds.), *Nano-Materials as Photocatalysts for Degradation of Environmental Pollutants* (pp. 345–370). Elsevier.

Baldissera, M. D., Grando, T. H., Souza, C. F., Cossetin, L. F., Sagrillo, M. R., Nascimento, K., & Monteiro, S. G. (2016). Nerolidol nanospheres increase its trypanocidal efficacy against Trypanosoma evansi: New approach against diminazene aceturate resistance and toxicity. *Experimental Parasitology, 166*, 144–149.

Baruah, A., Chaudhary, V., Malik, R., & Tomer, V. K. (2019). Nanotechnology-based solutions for wastewater treatment. In *Nanotechnology in Water and Wastewater Treatment* (pp. 337–368). Elsevier.

Bauomy, A. A. (2020). Zinc oxide nanoparticles and L-carnitine effects on neuro-schistosomiasis mansoni induced in mice. *Environmental Science and Pollution Research, 27*(15), 18699–18707.

Bhawana, P., & Fulekar, M. (2012). Nanotechnology: Remediation technologies to clean up environmental pollutants. *Research Journal of Chemistry and Sciences, 2231*, 606X.

Bocan, T. M., Stafford, R. G., Brown, J. L., Akuoku Frimpong, J., Basuli, F., Hollidge, B. S., & Smith, D. R. (2019). Characterization of brain inflammation, apoptosis, hypoxia, blood-brain barrier integrity, and metabolism in Venezuelan equine encephalitis virus (VEEV TC-83) exposed mice by *in vivo* positron emission tomography imaging. *Viruses, 11*, 1052.

Cai, C., Zhao, M., Yu, Z., Rong, H., & Zhang, C. (2019). Utilization of nanomaterials for *in-situ* remediation of heavy metal(loid) contaminated sediments: A review. *Science of the Total Environment, 662*, 205–217.

Chaturvedi, V. K., Singh, A., Singh, V. K., & Singh, M. P. (2019). Cancer nanotechnology: A new revolution for cancer diagnosis and therapy. *Current Drug Metabolism, 20*(6), 416–429.

Chinnamuthu, C., & Boopathi, P. M. (2009). Nanotechnology and agroecosystem. *Madras Agricultural Journal, 96*, 17–31.

Connor, E. E., Mwamuka, J., Gole, A., Murphy, C. J., & Wyatt, M. D. (2005). Gold nanoparticles are taken up by human cells but do not cause acute cytotoxicity. *Small, 1*(3), 325–327.

Dangi, K., & Verma, A. K. (2021). Efficient and eco-friendly smart nano-pesticides: Emerging prospects for agriculture. *Materials Today, 45*, 3819–3824.

Darwash, M., Yousuf, A., Ozturk, M., & Bhat, R. A. (2023). *Phytosequestration-Strategies for Mitigation of Aerial Carbon Dioxide and Aquatic Nutrient Pollution*. Springer Briefs.

Das, A., & Das, B. (2019). Nanotechnology is a potential tool to mitigate abiotic stress in crop plants. In A. B. de Oliveira (Ed.), *Abiotic and Biotic Stress in Plants* (p. 85). IntechOpen.

Das, K., Singh, A. A., & Eram, R. (2022). Clearing the air: Technologies for monitoring and control of air pollution. In P. Singh et al. (Eds.), *Environmental Sustainability and Industries* (pp. 85–116). Elsevier.

Devaraj, R., & Loganathan, A. K. (2023). Circulating microRNA as a potential biomarker and its nanotechnology-based detection methods: A literature review. *Nanomedicine Journal*. https://doi.org/10.22038/NMJ.2023.68157.1729.

El-Saadony, M. T., ALmoshadak, A. S., Shafi, M. E., Albaqami, N. M., Saad, A. M., El-Tahan, A. M., & Abd El-Mageed, T. A. (2021). Vital roles of sustainable nano-fertilizers in improving plant quality and quantity—An updated review. *Saudi Journal of Biological Sciences, 28*(12), 7349–7359.

Fakruddin, M., Hossain, Z., & Afroz, H. (2012). Prospects and applications of nanobiotechnology: A medical perspective. *Journal of Nanobiotechnology, 10*(1), 1–8.

Fang, Y., Zhou, Y., & Ramasamy, R. P. (2018). Communication—Direct detection of methyl salicylate using a tri-enzyme-based electrochemical sensor. *Journal of the Electrochemical Society, 165*(9), B358.

Farshchi, F., Saadati, A., & Hasanzadeh, M. (2020). A novel immunosensor for the monitoring of PSA using the binding of the biotinylated antibody to the prostate-specific antigen based on nano-ink modified flexible paper substrate: Efficient method for diagnosis of cancer using biosensing technology. *Heliyon, 6*(7), e04327.

Fernández-Cabada, T., & Ramos-Gómez, M. (2019). A novel contrast agent based on magnetic nanoparticles for cholesterol detection as Alzheimer's disease biomarker. *Nanoscale Research Letters, 14*(1), 1–6.

García-Caparrós, P., De Filippis, L., Gul, A., Hasanuzzaman, M., Ozturk, M., Altay, V., & Lao, M. T. (2020). Oxidative stress and antioxidant metabolism under adverse environmental conditions: A review. *The Botanical Review, 87*, 421–466. https://doi.org/10.1007/s12229-020-09231-1.

Gardner, G. T., & Stern, P. C. (1996). *Environmental Problems and Human Behavior*. Allyn and Bacon.

Gavrilaş, S., Ursachi, C. Ş., Perţa-Crişan, S., & Munteanu, F. D. (2022). Recent trends in biosensors for environmental quality monitoring. *Sensors, 22*(4), 1513.

Ghori, N. H., Ghori, T., Hayat, M. Q., Imadi, S. R., Gul, A., Altay, V., & Ozturk, M. (2019). Heavy metal stress and responses in plants. *International Journal of Environmental Science and Technology, 16*, 1807–1828.

Giraldo, J. P., Wu, H., Newkirk, G. M., & Kruss, S. (2019). Nanobiotechnology approaches for engineering smart plant sensors. *Nature Nanotechnology, 14*(6), 541–553.

Glenn, J., Florescu, E., & Socialomics, M. P. T. (2016). Future work/technology 2050 real-time Delphi study: Excerpt from the 2015-16 State of the Future report. *Journal of Socialomics, 5*(3), 1000171.

González, E., Montenegro, I., González, E., & Forero, E. (2016). Bio-nanotechnology: Challenges and opportunities. In E. González & E. Forero (Eds.), *Bio-Nanotechnology for Sustainable Environmental Remediation and Energy Generation* (p. 19). ACCEFN-NanoCiTec.

Graverini, G., Piazzini, V., Landucci, E., Pantano, D., Nardiello, P., Casamenti, F., & Bergonzi, M. C. (2018). Solid lipid nanoparticles for delivery of andrographolide across the blood-brain barrier: *In vitro* and *in vivo* evaluation. *Colloids and Surfaces B: Bio Interfaces, 161*, 302–313.

Gu, Z., Yan, L., Tian, G., Li, S., Chai, Z., & Zhao, Y. (2013). Recent advances in the design and fabrication of upconversion nanoparticles and their safe theranostic applications. *Advanced Materials, 25*(28), 3758–3779.

Guo, H., White, J. C., Wang, Z., & Xing, B. (2018). Nano-enabled fertilizers control the release and use efficiency of nutrients. *Current Opinion in Environmental Science & Health, 6*, 77–83.

Hameed, A., Qadri, T. N., Mahmooduzzafar, T. O. S., Ozturk, M., Altay, V., & Ahmad, P. (2017). Physio-biochemical and nutritional responses of *Abelmoschus esculentus* (L.) Moench (okra) under mercury contamination. *Fresenius Environmental Bulletin, 26*, 5814–5823.

Haq, S., Bharose, R., Ozturk, M., Altay, V., Bhatti, A. A., Dervash, M. A., & Hakeem, K. R. (2021). Impact of treated sewage water on nutrient status of Alfisols and vegetable crops. *Notulae Botanicae Horti Agrobotanici Cluj-Napoca, 49*(2), 12255–12255.

Imanberdieva, N., Imankul, B., Severoğlu, Z., Altay, V., & Ozturk, M. (2018). Potential impacts of climate change on plant diversity of Sary-Chelek Biosphere Reserve in Kyrgyzstan.

In D. Egamberdieva & M. Ozturk (Eds.), *Vegetation of Central Asia and Environs* (pp. 349–364). Springer Switzerland AG.
Islam, M. S., Ang, B. C., Andriyana, A., & Afifi, A. M. (2019). A review on fabrication of nanofibers via electrospinning and their applications. *SN Applied Sciences, 1*, 1–16.
Jafernik, K., Malinowska, M. A., Bialik-Wąs, K., Sharafan, M., Ekiert, H., Sikora, E., & Szopa, A. (2023). Chitosan nanoparticles: A potential biomedical device. In S. K. Swain & A. Biswal (Eds.), *Chitosan Nanocomposites: Bio Nanomechanical Applications* (pp. 135–159). Springer.
Ji, T., Zhao, Y., Wang, J., Zheng, X., Tian, Y., Zhao, Y., & Nie, G. (2013). Tumor fibroblast specific activation of a hybrid ferritin nanocage-based optical probe for tumor microenvironment imaging. *Small, 9*(14), 2427–2431.
Jiang, D., Zeng, G., Huang, D., Chen, M., Zhang, C., Huang, C., & Wan, J. (2018). Remediation of contaminated soils by enhanced nanoscale zero-valent iron. *Environmental Research, 163*, 217–227.
Kim, D., Jeong, Y. Y., & Jon, S. (2010). A drug-loaded aptamer-gold nanoparticle bioconjugate for combined CT imaging and therapy of prostate cancer. *ACS Nano, 4*(7), 3689–3696.
Ko, A. H., Tempero, M. A., Shan, Y. S., Su, W. C., Lin, Y. L., Dito, E., Ong, A., Wang, Y. W., Yeh, G. C., & Chen, L. T. (2013). A multinational phase 2 study of nano liposomal irinotecan sucrosofate (PEP02, MM-398) for patients with gemcitabine-refractory metastatic pancreatic cancer. *British Journal of Cancer, 109*(4), 920–925.
Kudhier, M., Sabry, R., Al-Haidarie, Y., & Al-Marjani, M. (2018). Significantly enhanced antibacterial activity of Ag-doped TiO_2 nanofibers synthesized by electrospinning. *Materials Technology, 33*(3), 220–226.
Kulkarni, M. B., Ayachit, N. H., & Aminabhavi, T. M. (2022). Recent advancements in nano biosensors: Current trends, challenges, applications, and future scope. *Biosensors, 12*(10), 892.
Kumar, S., Kumar, B., Baruah, A., & Shanker, V. (2013). Synthesis of magnetically separable and recyclable g-C3N4–Fe3O4 hybrid nanocomposites with enhanced photocatalytic performance under visible-light irradiation. *The Journal of Physical Chemistry C, 117*(49), 26135–26143.
Kurbanoglu, S., Erkmen, C., & Uslu, B. (2020). Frontiers in electrochemical enzyme-based biosensors for food and drug analysis. *Trends in Analytical Chemistry, 124*, 115809.
Lal, R. (2015). Restoring soil quality to mitigate soil degradation. *Sustainability, 7*(5), 5875–5895.
Li, M., Duanmu, K., Wan, C., Cheng, T., Zhang, L., Dai, S., & Fei, H. (2019). Single-atom tailoring of platinum nanocatalysts for high-performance multifunctional electrocatalysis. *Nature Catalysis, 2*(6), 495–503.
Lofrano, G., Carotenuto, M., Libralato, G., Domingos, R. F., Markus, A., Dini, L., & Sharma, S. K. (2016). Polymer functionalized nanocomposites for metals removal from water and wastewater: An overview. *Water Research, 92*, 22–37.
Lowry, G. V., Avellan, A., & Gilbertson, L. M. (2019). Opportunities and challenges for nanotechnology in the agri-tech revolution. *Nature Nanotechnology, 14*(6), 517–522.
Lu, H., Wang, J., Stoller, M., Wang, T., Bao, Y., & Hao, H. (2016). An overview of nanomaterials for water and wastewater treatment. *Advances in Materials Science and Engineering, 2016*, 4964828.
Malandrakis, A. A., Kavroulakis, N., & Chrysikopoulos, C. V. (2022). Metal nanoparticles against fungicide resistance: Alternatives or partners? *Pest Management Science, 78*(10), 3953–3956.

Mäler, K. G. (1990). International environmental problems. *Oxford Review of Economic Policy, 6*(1), 80–108.

Manjunatha, S., Biradar, D., & Aladakatti, Y. R. (2016). Nanotechnology and its applications in agriculture: A review. *Journal of Farm Sciences, 29*(1), 1–13.

Marchiol, L. (2018). Nanotechnology in agriculture: New opportunities and perspectives. *New Vision in Plant Science, 9*(4), 161.

Maryam, S., & Gul, A. (2023). Heavy metal remediation by nanotechnology. In L. Newmann et al. (Eds.), *Phytoremediation: Management of Environmental Contaminants* (pp. 571–597). Springer International Publishing.

Masood, N., Ahmed, R., Tariq, M., Ahmed, Z., Masoud, M. S., Ali, I., & Hasan, A. (2019). Silver nanoparticle-impregnated chitosan-PEG hydrogel enhances wound healing in diabetes-induced rabbits. *International Journal of Pharmaceutics, 559*, 23–36.

Mattigod, S. V., Fryxell, G. E., Alford, K., Gilmore, T., Parker, K., Serne, J., & Engelhard, M. (2005). Functionalized TiO2 nanoparticles for use in *in situ* anion immobilization. *Environmental Science & Technology, 39*(18), 7306–7310.

McConnell, E. M., Nguyen, J., & Li, Y. (2020). Aptamer-based biosensors for environmental monitoring. *Frontiers in Chemistry, 8*, 434.

Mittal, D., Kaur, G., Singh, P., Yadav, K., & Ali, S. A. (2020). Nanoparticle-based sustainable agriculture and food science: Recent advances and future outlook. *Frontiers in Nanotechnology, 2*, 579954.

Mohamadian, N., Ghorbani, H., Wood, D. A., & Khoshmardan, M. (2019). A hybrid nanocomposite of poly (styrene-methyl methacrylate-acrylic acid)/clay as a novel rheology-improvement additive for drilling fluids. *Journal of Polymer Research, 26*, 1–14.

Mohamed, E. F. (2017). Nanotechnology: Future of environmental air pollution control. *Environmental Management and Sustainable Development, 6*(2), 429.

Mohamed, E. F., & Awad, G. (2022). Development of nano-sensors and biosensors as an air pollution detection technique for the foreseeable future. *Comprehensive Analytical Chemistry, 99*, 163–188.

Montanarella, F., & Kovalenko, M. V. (2022). Three millennia of nanocrystals. *ACS Nano, 16*(4), 5085–5102.

Montes-Cobos, E., Ring, S., Fischer, H. J., Heck, J., Strauß, J., Schwaninger, M., & Reichardt, H. M. (2017). Targeted delivery of glucocorticoids to macrophages in a mouse model of multiple sclerosis using inorganic-organic hybrid nanoparticles. *Journal of Controlled Release, 245*, 157–169.

Mushtaq, W., Mehdizade, M., Siddiqui, M. B., Ozturk, M., Jabran, K., & Altay, V. (2020). Phytotoxicity of above-ground weed residue against some crops and weeds. *Pakistan Journal of Botany, 52*(3), 851–860.

Naresh, V., & Lee, N. J. S. (2021). A review on biosensors and recent development of nanostructured materials-enabled biosensors. *Sensors, 21*(4), 1109.

Nayak, R. R., Pradhan, N., Behera, D., Pradhan, K. M., Mishra, S., Sukla, L. B., & Mishra, B. K. (2011). Green synthesis of silver nanoparticle by *Penicillium purpurogenum* NPMF: The process and optimization. *Journal of Nanoparticle Research, 13*(8), 3129–3137.

Ng, C. K., Mohanty, A., & Cao, B. (2015). Biofilms in bio-nanotechnology: Opportunities and challenges. In O. M. Singh (Ed.), *Bio-Nanoparticles: Biosynthesis and Sustainable Biotechnological Implications* (pp. 83–100). John Wiley & Sons, Inc.

Ngô, C., & Van de Voorde, M. (2014). *Nanotechnology in a Nutshell*. Springer.

Nie, S., Xing, Y., Kim, G. J., & Simons, J. W. (2007). Nanotechnology applications in cancer. *Annual Review of Biomedical Engineering, 9*, 257–288.

Nigam, V. K., & Shukla, P. (2015). Enzyme based biosensors for detection of environmental pollutants—A review. *Journal of Microbiology and Biotechnology, 25*, 1773–1781.

Omanović-Mikličanina, E., & Maksimović, M. (2016). Nanosensors applications in agriculture and food industry. *Bulletin of Chemical Technologists in Bosnia and Herzegovina, 47*, 59–70.

Ownby, M., Desrosiers, D. A., & Vaneeckhaute, C. (2021). Phosphorus removal and recovery from wastewater via hybrid ion exchange nanotechnology: A study on sustainable regeneration chemistries. *NPJ Clean Water, 4*(1), 6.

Ozturk, M., & Altay, V. (2012). Aegean grasslands as endangered ecosystems in Turkey. *Pakistan Journal of Botany, 44*, 7–18.

Ozturk, M., & Altay, V. (2021). Role of *Quercus coccifera* (= Q. calliprinos) in the light of climate change scenarios in the Mediterranean Basin. *Plant Fungal Research, 4*(2), 8–20.

Ozturk, M., Altay, V., & Dimeyeva, L. A. (2022a). Climate change, biodiversity, and ecology of Caspian Sea: An overview. In M. A. Dervash & A. A. Wani (Eds.), *Climate Change Alleviation for Sustainable Progression* (pp. 78–94). CRC Press.

Ozturk, M., Altay, V., & Karahan, F. (2017a). Studies on trace elements distributed in *Glycyrrhiza* taxa in Hatay-Turkey. *International Journal of Plant Environmental, 3*(2), 1–7.

Ozturk, M., Altay, V., & Kucuk, M. (2019a). Trace elements in the soil-plant systems of copper mine areas—A case study from Murgul Copper Mine from the Black Sea region of Turkey. *Phyton, 88*(3), 223–238.

Ozturk, M., Gucel, S., Altay, V., Ahmad, M. S. A., Ashraf, M. Y., & Ashraf, M. (2019b). Clustering of halophytic species from Cyprus based on ionic contents, biochemical attributes, and micronutrients. *Phyton, 88*(1), 63–68.

Ozturk, M., Hakeem, K. R., Faridah-Hanum, I., & Efe, R. (2015). Climate change impacts on high-altitude ecosystems. In M. Hakeem, I. Faridah-Hanum, & R. Efe (Eds.), *Climate Change and Plant Science* (pp. 259–289). Springer.

Ozturk, M., Metin, M., Altay, V., Bhat, R., Toderich, K., Ejaz, M., Gul, A., Unal, B. T., Hasanuzzaman, M., Nibir, L., Nahar, K., Kawano, T., & Caparrós, P. G. (2022b). Arsenic and human health: Genotoxicity, epigenomic effects and cancer signaling. *Biological Trace Element Research, 200*, 988–1001.

Ozturk, M., Metin, M., Altay, V., Prasad, M. N. V., Gul, A., Bhat, R. A., Darvash, M. A., Hasanuzzaman, M., Nahar, K., Unal, D., Unal, B. T., Caparrós, P. G., Kawano, T., Toderich, K., & Bukhari, A. (2023a). Role of rare earth elements in plants. *Plant Molecular Biology Reporter*. https://doi.org/10.1007/s11105-023-01369-7.

Ozturk, M., Roy, A., Bhat, R. A., Vardar-Sukan, F., & Tonelli, F. M. P. (2023b). *Synthesis of Bionanomaterials for Biomedical Applications*. Elsevier.

Ozturk, M., Uysal, I., Yucel, E., Altay, V., & Karabacak, E. (2017b). Soil-plant interactions in the monumental plane trees (*Platanus orientalis*) grove-Canakkale-Turkey. *Journal of Environmental Biology, 38*(6), 1129–1137.

Primožič, M., Knez, Ž., & Leitgeb, M. (2021). (Bio)Nanotechnology in food science—Food packaging. *Nanomaterials, 11*(2), 292.

Purohit, R., & Singh, S. (2018). Fluorescent gold nanoclusters for efficient cancer cell targeting. *International Journal of Nanomedicine, 13*(S1), 15–17.

Puyol, D., Batstone, D. J., Hülsen, T., Astals, S., Peces, M., & Krömer, J. (2017). Resource recovery from wastewater by biological technologies: Opportunities, challenges, and prospects. *Frontiers in Microbiology, 7*, 2106.

Rafique, M., Tahir, M. B., & Sadaf, I. (2019). Nanotechnology: An innovative way for wastewater treatment and purification. In R. Prasad & T. Karchiyappan (Eds.), *Advanced Research in Nano Sciences for Water Technology* (pp. 95–131). Springer.

Ravikumar, A., Panneerselvam, P., Radhakrishnan, K., Morad, N., Anuradha, C., & Sivanesan, S. (2017). DNAzyme based amplified biosensor on ultrasensitive fluorescence detection of Pb(II) ions from aqueous system. *Journal of Fluorescence, 27*, 2101–2109.

Raza, M. M., Ullah, S., Tariq, A., Abbas, T., Yousaf, M. M., Altay, V., & Ozturk, M. (2019). Alleviation of salinity stress in maize using silicon nutrition. *Notulae Botanicae Horti Agrobotanici Cluj-Napoca, 47*(4), 1340–1347.

Rehman, T., Khan, M. A., Khan, H., Shah, W. A., Altay, V., & Ozturk, M. (2020). Tillage, nutrients, and weed control affects yield of sugarbeet. *Fresenius Environmental Bulletin, 29*(5), 3380–3387.

Rezaei, S., Dabirmanesh, B., Zare, L., Golestani, A., Javan, M., & Khajeh, K. (2020). Enhancing myelin repair in experimental model of multiple sclerosis using immobilized chondroitinase ABC I on porous silicon nanoparticles. *International Journal of Biological Macromolecules, 146*, 162–170.

Rohela, G. K., Srinivasulu, Y., & Rathore, M. S. (2019). A review paper on recent trends in bio-nanotechnology: Implications and potentials. *Nanosci. Nanotechnol. -Asia, 9*(1), 12–20.

Rouco, H., García-García, P., Briffault, E., & Diaz-Rodriguez, P. (2023). Modulating osteoclasts with nanoparticles: A path for osteoporosis management. *Wiley Interdisciplinary Reviews: Nanomedicine and Nanobiotechnology, 15*(4), e1885.

Roy, A. C., Park, S. R., Cunningham, D., Kang, Y. K., Chao, Y., Chen, L. T., & Ramos, F. J. (2013). A randomized phase II study of PEP02 (MM-398), irinotecan or docetaxel as a second-line therapy in patients with locally advanced or metastatic gastric or gastro-oesophageal junction adenocarcinoma. *Annals of Oncology, 24*(6), 1567–1573.

Saif, M. W. (2014). MM-398 achieves primary endpoint of overall survival in phase III study in patients with gemcitabine refractory metastatic pancreatic cancer. *JOP Journal of the Pancreas, 15*(3), 278–279.

Sánchez-López, E., Espina, M., Doktorovova, S., Souto, E. B., & García, M. L. (2017). Lipid nanoparticles (SLN, NLC): Overcoming the anatomical and physiological barriers of the eye—Part II-Ocular drug-loaded lipid nanoparticles. *European Journal of Pharmaceutical Sciences, 110*, 58–69.

Sardar, M. F., Ahmad, H. R., Zia-Ur-Rehman, M., Ozturk, M., & Altay, V. (2018). Absorption of foliar-applied lead (Pb) in rice (*Oryza sativa* L.): A hydroponic experiment. *Fresenius Environmental Bulletin, 27*(10), 6634–6639.

Savage, N., Thomas, T. A., & Duncan, J. S. (2007). Nanotechnology applications and implications research supported by the US Environmental Protection Agency STAR grants program. *Journal of Environmental Monitoring, 9*(10), 1046–1054.

Shahrasbi, S., Pirasteh-Anosheh, H., Emam, Y., Ozturk, M., & Altay, V. (2021). Elucidating some physiological mechanisms of salt tolerance in *Brassica napus* L. seedlings induced by seed priming with plant growth regulators. *Pakistan Journal of Botany, 53*(2), 367–377.

Shehada, N., Brönstrup, G., Funka, K., Christiansen, S., Leja, M., & Haick, H. (2015). Ultrasensitive silicon nanowire for real-world gas sensing: Noninvasive diagnosis of cancer from breath volatolome. *Nano Letters, 15*(2), 1288–1295.

Singh, S. (2018). Liposome encapsulation of doxorubicin and celecoxib in combination inhibits progression of human skin cancer cells. *International Journal of Nanomedicine, 13*(S1), 11–13.

Singh, S. K., & Mishra, D. N. (2019). Nose to brain delivery of galantamine loaded nanoparticles: *In-vivo* pharmacodynamic and biochemical study in mice. *Current Drug Delivery, 16*(1), 51–58.

Singh, S., Kulkarni, P., & Dash, D. (2013). Biomedical applications of carbon-based nanomaterials. In D. Bagchi, M. Bagch, H. Moriyama, & F. Shahidi (Eds.), *Bio-Nanotechnology: A Revolution in Food, Biomedical and Health Sciences* (pp. 443–463). John Wiley & Sons, Ltd.

Sonkaria, S., Ahn, S. H., & Khare, V. (2012). Nanotechnology and its impact on food and nutrition: A review. *Recent Patents on Food, Nutrition & Agriculture, 4*(1), 8–18.

Sun, L., Liu, D., Fu, D., Yue, T., Scharre, D., & Zhang, L. (2021). Fluorescent peptide nanoparticles to detect amyloid-beta aggregation in cerebrospinal fluid and serum for Alzheimer's disease diagnosis and progression monitoring. *Chemical Engineering Journal, 405*, 126733.

Sutradhar, K. B., & Amin, M. (2014). Nanotechnology in cancer drug delivery and selective targeting. *International Scholarly Research Notices, 2014*, 939378.

Tang, T., Valenzuela, A., Petit, F., Chow, S., Leung, K., Gorin, F., & Dhenain, M. (2018). *In vivo* MRI of functionalized iron oxide nanoparticles for brain inflammation. *Contemporary Medical Molecular Imaging, 2018*, 3476476.

Tariq, M., Choudhary, S., Singh, H., Siddiqui, M. A., Kumar, H., Amir, A., & Kapoor, N. (2021). Role of nanoparticles in abiotic stress. In F. Ahmad & M. Sultan (Eds.), *Technology in Agriculture*. IntechOpen. http://dx.doi.org/10.5772/intechopen.99928.

Tawfik, S. M., Elmasry, M. R., Sharipov, M., Azizov, S., Lee, C. H., & Lee, Y. I. (2020). Dual emission nonionic molecular imprinting conjugated polythiophenes-based paper devices and their nanofibers for point-of-care biomarkers detection. *Biosensors and Bioelectronics, 160*, 112211.

Thakur, S., Thakur, S., & Kumar, R. (2018). Bio-nanotechnology and its role in agriculture and food industry. *Journal of Molecular Genetic Medicine, 12*(324), 1747–0862.1000324.

Tran, S., DeGiovanni, P. J., Piel, B., & Rai, P. (2017). Cancer nanomedicine: A review of recent success in drug delivery. *Clinical and Translational Medicine, 6*, 1–21.

Tyagi, I., Gupta, V., Sadegh, H., Ghoshekandi, R. S., & Makhlouf, A. H. (2017). Nanoparticles as adsorbent: A positive approach for removal of noxious metal ions: A review. *Science and Technology Development, 34*(3), 195–214.

Vadlapudi, V., & Kaladhar, D. S. V. G. K. (2014). Green synthesis of silver and gold nanoparticles. *Middle East Journal of Scientific Research, 19*(6), 834–842.

Vaseashta, A., Vaclavikova, M., Vaseashta, S., Gallios, G., Roy, P., & Pummakarnchana, O. J. S. (2007). Nanostructures in environmental pollution detection, monitoring, and remediation. *Scientific Reports: Advances in Materials, 8*(1–2), 47.

Vopálenská, I., Váchová, L., & Palková, Z. J. B. (2015). New biosensor for detection of copper ions in water based on immobilized genetically modified yeast cells. *Biosensors and Bioelectronics, 72*, 160–167.

Wang, D. K., Rahimi, M., & Filgueira, C. S. (2021). Nanotechnology applications for cardiovascular disease treatment: Current and future perspectives. *Nanomedicine: Nanotechnology, Biology, and Medicine, 34*, 102387.

Wang, J., Sui, M., & Fan, W. (2010). Nanoparticles for tumor targeted therapies and their pharmacokinetics. *Current Drug Metabolism, 11*(2), 129–141.

Wu, J., Dong, M., Rigatto, C., Liu, Y., & Lin, F. (2018). Lab-on-chip technology for chronic disease diagnosis. *NPJ Digital Medicine, 1*(1), 7.

Xing, Y., Li, W., Wang, Q., Li, X., Xu, Q., Guo, X., & Lin, H. J. M. (2019). Antimicrobial nanoparticles incorporated in edible coatings and films for the preservation of fruits and vegetables. *Molecules, 24*(9), 1695.

Yadav, A. S., & Srivastava, D. S. (2015). Application of nanotechnology in weed management: A review. *Research Reviews: Journal of Crop Science & Technology, 4*(2), 21–23.

Yadav, K., Singh, J., Gupta, N., & Kumar, V. (2017). A review of nano bioremediation technologies for environmental cleanup: A novel biological approach. *Journal of Materials and Environmental Science, 8*(2), 740–757.

Yalcin, I. E., & Altay, V. (2023). Investigation of water-soil-plant relationships based on hazardous and macro-micro element concentrations on Orontes River, Türkiye. *International Journal of Phytoremediation.* https://doi.org/10.1080/15226514.2023.2202241.

Yan, W., Lien, H. L., Koel, B. E., & Zhang, W. x. (2013). Iron nanoparticles for environmental clean-up: Recent developments and future outlook. *Environmental Science: Processes & Impacts, 15*(1), 63–77.

Yao, J., Yang, M., & Duan, Y. (2014). Chemistry, biology, and medicine of fluorescent nanomaterials and related systems: New insights into biosensing, bioimaging, genomics, diagnostics, and therapy. *Chemical Reviews, 114*(12), 6130–6178.

Yarcı, C., & Altay, V. (2016). Kocaeli ve çevresindeki tarım alanlarının yabani ot florası. *Erzin University Journal of Science and Technology, 9*(2), 148–171.

Yarcı, C., Serin, M., & Altay, V. (2007). The segetal vegetation of Kocaeli Province (Turkey). *Ekoloji, 16*(63), 23–33.

Zeng, J., Wu, J., Li, M., & Wang, P. (2018). A novel magnetic nanoparticle for early detection of amyloid plaques in Alzheimer's disease. *Archives of Medical Research, 49*(4), 282–285.

Zhan, P., Peil, A., Jiang, Q., Wang, D., Mousavi, S., Xiong, Q., & Lin, C. (2023). Recent advances in DNA origami-engineered nanomaterials and applications. *Chemical Reviews, 123*(7), 3976–4050.

Zhang, L., Gu, F. X., Chan, J. M., Wang, A. Z., Langer, R. S., & Farokhzad, O. C. (2008). Nanoparticles in medicine: Therapeutic applications and developments. *Clinical Pharmacology & Therapeutics, 83*(5), 761–769.

Zhang, Y., Zhang, L., Yang, L., Vong, C. I., Chan, K. F., Wu, W. K. K., & Wong, S. H. (2019). Real-time tracking of fluorescent magnetic spore-based microrobots for remote detection of *C. diff* toxins. *Science Advances, 5*(1), eaau9650.

Zheng, Z., Zhu, S., Lv, M., Gu, Z., & Hu, H. (2022). Harnessing nanotechnology for cardiovascular disease applications—A comprehensive review based on bibliometric analysis. *Nano Today, 44*, 101453.

Zilberzwige-Tal, S., Gazit, D., Adsi, H., Gartner, M., Behl, R., Bar-Yosef Laor, D., & Gazit, E. (2022). Engineered riboswitch nanocarriers as a possible disease-modifying treatment for metabolic disorders. *ACS Nano, 16*(8), 11733–11741.

CHAPTER 9

Endophytes-Mediated Secondary Metabolites Synthesis in Plants: An Omic Approach

SURYANARAYAN BHASKAR MEGHANA,[1]
SAMBAMURTHY KUMARASWAMY SUKRUTHA,[2,*]
MALLAPPA KUMARA SWAMY,[1,*] BOREGOWDA PURUSHOTHAM,[1] and MOHD. SAYEED AKHTAR[3]

[1]*Department of Biotechnology, East West First Grade College of Science, Bangalore, Karnataka, India*

[2]*Department of Microbiology, East West First Grade College of Science, Bangalore, Karnataka, India*

[3]*Department of Botany, Gandhi Faiz-e-Aam College, Shahjahanpur, Uttar Pradesh, India*

*Corresponding author.

ABSTRACT

Endophytes play a pivotal role in influencing the fitness of the host plant. This interaction, scaffolds for the synthesis of novel and commercially valuable secondary metabolites having a plethora of applications. An increase in the synthesis of therapeutically important metabolites mediated by plant-endophyte interaction could be an economical and sustainable approach. Nevertheless, the inclusive plant-endophyte interaction remains unclear. Of late, Omics studies encompassing transcriptional/genomics, translational/proteomics, and metabolomics studies have emerged as promising tools in unraveling the role of endophytes in secondary metabolite synthesis in medicinal plants. The present chapter provides comprehensive insights

Agricultural Biotechnology: Issues, Challenges, and Recent Developments. Mohd. Sayeed Akhtar and Khalid Rehman Hakeem (Eds.)
© 2025 Apple Academic Press, Inc. Co-published with CRC Press (Taylor & Francis)

into the application of various Omics tools in ascertaining the influence of endophytic microorganisms on secondary metabolite synthesis in various medicinal plants, the scientific gaps, and future prospects within this field to the scientific community.

9.1 INTRODUCTION

Endophytes are a diverse group of microorganisms such as bacteria, fungi, algae, and actinomycetes that reside asymptomatically in healthy plant tissues (Yadav and Meena, 2021). The word "Endophyte" is derived from the two Greek words-*endon* = within and *phyte* = plant). In the year 1866, it was first coined by De Bary. Endophytes are in a mutualistic relationship without causing any apparent damage to their host plants. According to their colonizing behaviors, endophytes are classified into facultative and obligate categories. Endophytes that colonize the plants at certain stages of their life cycles are known as facultative endophytes. Whereas, obligate types live inside the plants throughout their life. Endophyte communities differ in number, species, composition, host, and tissue. In plant-endophyte relationships, the former protects, and provides shelter and nutrients, whereas the latter is considered the powerhouse of an array of novel natural/bioactive compounds (Schouten, 2019). These bioactive compounds provide plethora of benefits to plant communities such as tolerance to biotic and abiotic stress, increase in biomass (Wu et al., 2021), promotion of plant fitness (Caruso et al., 2020), biological control agents (Exzra et al., 2009), disease resistance (Rao et al., 2020), nutrient uptake (Yadav and Meena, 2021), promote hormonal secretion (Soni et al., 2021), and seed germination (Yang et al., 2018) and carves the way to their life. They are packed with a myriad of secondary metabolites that include terpenoids (Jayaram et al., 2021), alkaloids (Venieraki et al., 2017), flavonoids (Rai et al., 2021), steroids (Adeleke et al., 2021), phenyl propanoids (Caruso et al., 2020), quinines (Singh et al., 2021), indole derivatives (Rai et al., 2021), amides (Srinivas et al., 2015), isocoumarin derivatives (Rao et al., 2020). Synthesis of extracellular enzymes such as pectinases, asparaginases, amylases, and cellulases has also been explored for their widespread agricultural, industrial, and biotechnological applications. The present chapter provides comprehensive insights into the application of various Omics tools in ascertaining the influence of endophytic microorganisms on secondary metabolite synthesis in various medicinal plants, the scientific gaps, and future prospects within this field to the scientific community.

9.2 PLANT-ENDOPHYTES INTERACTION

The endophytic consortium has the ability to synthesize the same or similar kinds of natural/bioactive compounds as their host. The crosstalk between the endophytes and the medicinal plant's results in the synthesis of value-added secondary metabolites has attracted researchers globally (Kusari et al., 2012). Interactions between the endophytic microfloras with the host plants vary from microbe to microbe and host to host. Nevertheless, our understanding of this relationship remains unclear. Research works on plant-endophyte mutualism have been focused on conventional approaches. Therefore, uncovering the relationships of the plant and endophytes for the production of value-added products can be achieved through the solicitation of modern biological tools (Firakova et al., 2007). Furthermore, the identification and characterization of genes intricated in such beneficial interactions through the amalgamation of omics tools opens up a new avenue for research. Solicitation of omics tools such as genomics, transcriptomics, and proteomics aids in uncovering the influence of endophytic consortium on the plant genes and proteins for value-added secondary metabolite synthesis (Figure 9.1).

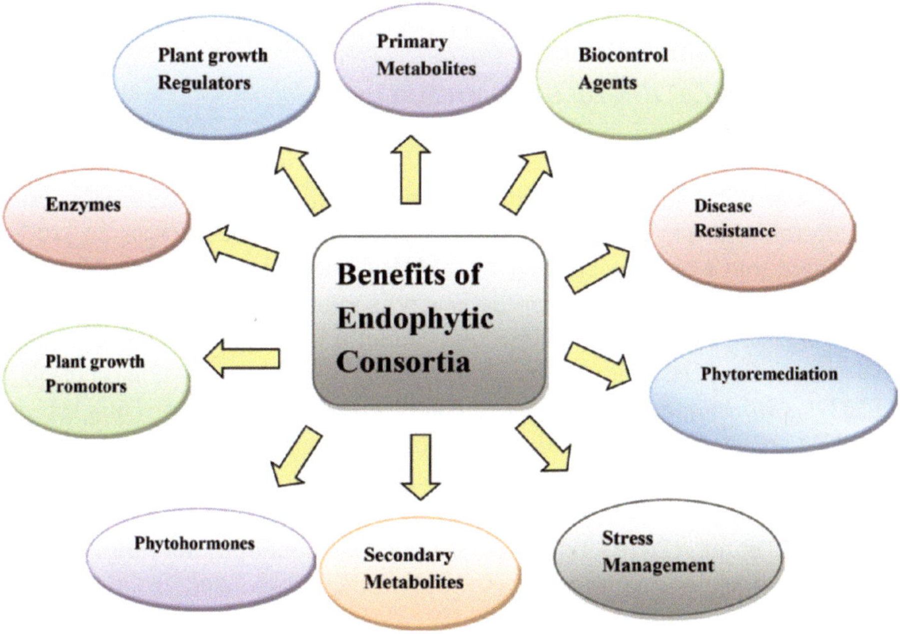

FIGURE 9.1 Benefits of endophytic consortia on medicinal plants.

9.3 APPLICATION OF OMICS TOOLS

Advancements in analytical technology have opened up a new avenue in the field of molecular biology and genomics (Yadav and Meena, 2021). Genomic studies provide insight into the nature of genes. However, functional expression of modulation of genes and abundance of ribonucleic acid (RNA) transcripts under various environmental conditions can be ascertained through transcriptome analysis (Joyce et al., 2006). One of the earliest transcriptomic studies was the upregulation of genes involved in the pyrimidine metabolism of *Epulorhiza* sp., isolated from the roots of an orchid, *Anoectochilus roxburghii* (Li et al., 2012).

Proteomics study deals with the study of functional gene expression products. The proteomic study can be annotated well with genomic information. Further, complementary data generated from the metagenomic study could help us to understand the distribution of endophytic microflora. All the above techniques are inter-connected, and thus the combinatorial approach pays the path to unravel the complexities involved in the plant-endophyte relationship. The application of "Omics" tools helps us to discover the genes, transcripts, and proteins from an array of endophytes and provides a scope to understand their influence on the modulation of genes, responsible for secondary metabolite production in various medicinal plants (Figure 9.2).

9.3.1 APPLICATION OF GENOMICS STUDY

Pandey et al. (2016) reported that endophytes isolated from the leaves, roots, capsules, and seeds of the *Papaver somniferum* were inoculated to the *P somniferum* seeds, free from endophytes. Gene expression studies indicated that a capsule endophyte, SM1B (*Acinetobacter*), upregulated the expression of the key genes for the Benzylisoquinoline alkaloid biosynthesis and had no effect on the baine 6-O-demethylase (T6ODM) and codeine O-demethylase (CODM) gene expression. Whereas, capsule endophyte SM3B (*Marmoricola* sp.), upregulated the expression of T6ODM and CODM, respectively. Overall, the results demonstrate the specific role of endophytes in the modulation of host plant productivity and BIA production (Pandey et al., 2016).

Singh et al. (2019) reported that selected endophytic fungi such as *Curvularia* sp. and *Choanephora infundibulifera* isolated from *C. roseus* cultivar Dhawal increased serpentine content by 211.7–337.6%, while *Aspergillus japonicus* and *Pseudomonas* sp. enhanced zajmalicine content by

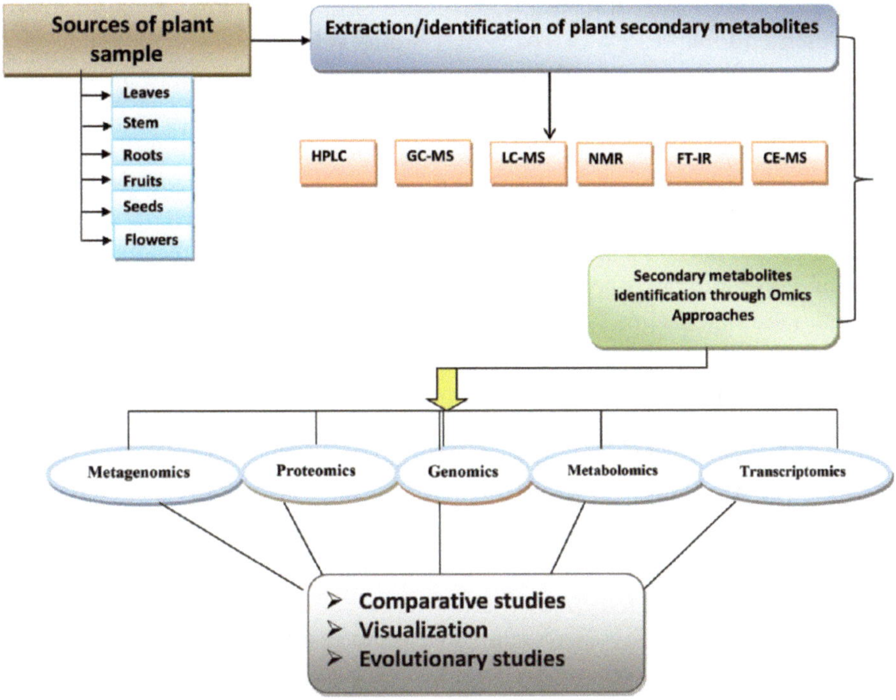

FIGURE 9.2 Application of "omics" tools.

123.4–203.8% in cultivar Prabal by modulating the expression of Geraniol 10-hydroxylase, tryptophan decarboxylase (TDC) and strictosidine synthase and regulatory genes involved in terpenoid indole alkaloid biosynthetic pathway in the root. Endophytic *Actinobacterium* and *Pseudonocardia* strain YIM63111 upregulated cytochrome P450 monooxygenase and cytochrome P450 oxidoreductase genes, thus resulting in an increase in Artemisinin content in *Artemisia annua* (Li et al., 2012). Polysaccharide fraction, a potent elicitor of root endophyte *Trichoderma atroviride* D16 upregulated transcriptional activity of gene involved in tanshinones biosynthetic pathway for synthesis of tanshinones in the hairy root of *Salvia miltiorrhiza* (Ming et al., 2013). In another study, fungal endophytes such as *Curvularia* sp. CATDLF5 and *Choanephora infundibulifera* CATDLF6 induced vindoline production in *C. roseus* by upregulating the expression of transcriptional activator and terpenoid indole alkaloid pathway genes such as geranoil-10 hydroxylase, TDC, strictosidine synthase, 16-hydroxytabersonine-O-methyltransferase, desacetoxyvindoline-4-hydroxylase, deacetylvindoline-4-O-acetyltransferase. Further, it downregulated the expression of Cys2/His2-type zinc finger protein family transcriptional repressors genes (Pandey et al., 2016). Selected endophytes (WPL16, WPL17, and WPS23) were inoculated in *W. somnifera* plants to study the gene expression for withanolide biosynthesis. Results indicated that gene expression of *FPPS*, *SMT*, *ODM*, *SQS*, *SE*, *DXS*, *DXR*, *CPR1 CAS*, *HMGR*, *SGT*, and *CPR2* was upregulated in the leaves as well as in the roots of *in-vitro*-grown normal and composite plants. However, inoculated plants expressed the genes for *CAS*, *SMT*, *HMGR*, *SGT*, *DXS*, *DXR*, *CPR1*, and *CPR2* in the leaves compared to the control plants. Nevertheless, upregulation of *SQS*, *SE*, *FPPS*, and *ODM* genes was observed in the plants inoculated with WPL17 and WPS23 endophytic consortium (Pandey et al., 2013).

Inoculation of SM3B (*Marmoricola* sp.) has the ability to upregulate the T6ODM and CODM genes. On the other hand, SM1B (*Acinetobacter*) endophyte upregulates the expressions of most of the genes of the BIAs pathway except T6ODM and CODM. Conversely, co-inoculation (SM3B and SM1B) resulted in the upregulation of genes for the synthesis of morphine (TYDC, NCS, 6OMT, CNMT, NMCH, SalSyn, SalR, SalAT, COR) than that of single-inoculations. Co-inoculation increased the synthesis of morphine (22.50%), papaverine (36%), and noscapine (54%) content than the single inoculation and endophyte-free control plants, respectively. Further, the workers observed that co-inoculation improved transpiration rate, stomatal conductance, photosynthetic pigments, and photosynthetic rate. It also increased biomass (73.3%), capsule weight (66%), and seed yield (79%) more than single-inoculation (Ray et al., 2019). Inoculation of endophytic fungi, *Gilmaniella* sp. AL12 in *Atractylodes lancea*, induced high HMGR gene expression. This resulted in an increase in jasmonic acid (JA) and volatile oil accumulation (Ren et al., 2012). When compared to non-endophyte treated control plants, inoculation of endophytic bacteria such as *Bacillus amyloliquefaciens* and *Pseudomonas fluorescens* either in single or in combination to *Withania somnifera* improved the tolerance/resistance against, a leaf spot disease fungus *Alternaria alternata*. High expression levels of plant defense-related genes such as lignin-forming anionic peroxidase (39.9 fold), PR-3 a class IV chitinase (18.70 fold), PR-3, a class II chitinase (16.22 fold), PR-12 defensin (14.64 fold), hevein-like protein (13.1 fold), lipoxygenase (10.9 fold), β-1,3-glucanase (10.4 fold) and PR-1 (7.23 fold) was observed in combination-treated plants. Contrastingly, in pathogen-infected control plants, as the disease progressed, a gradual decrease in the expression of plant defense-related genes was noted (Mishra et al., 2018a). In a similar study, the application of endophytic bacteria namely *B amyloliquefaciens*

(MPE20) and *P. fluorescens* (MPE115) individually or in combination with *W. somnifera*, challenged under biotic stress caused by a leaf spot disease fungus, *Alternaria alternata* significantly modulated withanolide biosynthetic pathway. The endophytic bacterial consortia upregulated the gene expression of 3-hydroxy-3-methylglutaryl co-enzyme A reductase, 1-deoxy-D-xylulose-5-phosphate reductase, farnesyl diphosphate synthase, squalene synthase, cytochrome p450, sterol desaturase, sterol Δ-7 reductase and sterol glycosyl transferases genes. Further, it also induced withaferin A (1.52–1.96 fold), withanolide A (3.32–5.96 fold), and withanolide B (12.49–21.47 fold) synthesis (Mishra et al., 2018b).

Under field conditions, inoculation of native endophytic fungi such as *Aspergillus terreus* strain 2aWF, *Penicillium oxalicum* strain 5aWF, and *Sarocladium kiliense* strain 10aWF to four weeks old *W somnifera* plant, induced the accumulation of withanolide A content (97 to 100%) in leaves. However, an increase in withanolide A content (52%–65%) was observed in the roots treated with 5aWF and 10aWF, respectively. Compared to control plants, a significant increase in the expression levels of withanolide biosynthetic pathway genes such as *HMGR, DXR, FPPS, SQS, SQE, CAS, SMT1, STE1, CYP710A1,* and host-resistant related gene, NPR1was observed (Kushwaha et al., 2019). In another study, treatment for 24 hours with mycelial extract (1% w/v) and culture filtrate (5% v/v) obtained from native endophytic *A. terreus* 2aWF in the root cell suspension cultures of *W. somnifera* significantly enhanced the withanolide A (10.29 μg/g FCB and 12.20 μg/g FCB). Withanolide biosynthetic pathway genes such as *HMGR, DXS, DXR, FPPS, SQS, SQE,* and *CAS* genes were upregulated on treatment with 3% cell homogenate and culture filtrate in *W. somnifera* suspension culture or callus treatment compared to control cell suspension culture (Kushwaha et al., 2019). When compared to the control plants, maximum production of withaferin A was achieved in callus and cell suspension cultures of *W. somnifera* with cell homogenate (3%), culture filtrate (3%), and fungal disc of endophytic fungus *Piriformospora indica*. The expression of genes for withanolides synthesis such as *hmgr, fpps, ss, se, cas, dxs,* and *dxr* was increased (Ahlawat et al., 2016).

Kushwaha et al. (2019) studied the effect of co-inoculation of native fungal endophytes such as *A. terreus* (2aWF), *P. oxalicum* (5aWF), and *S. kiliense* (10aWF) along with the biocontrol agent, *Trichoderma viride* (TV) on the synthesis of Withaferin A in *W. somnifera*. Results indicated that, compared to the control plants, shoot (65–150%), root weight (35–75%), and plant height (15–35%) was increased. In the leaves of TV-treated plants, withanolide A content was significantly increased by 260%. Whereas, in the co-inoculation treatments, withanolide A production was enhanced up to 109–242% in leaves and 19–73% in the roots, respectively. Furthermore, a significant increase in withaferin A content (27%) was noticed in the leaves of co-inoculation. However, no effect on withanolide A and withaferin A production was observed in the roots treated with *Trichoderma viride*. Compared to the control plants, co-inoculation treatments in roots and leaves upregulated the gene expression of *HMGR, DXR, FPPS, SQS, SQE, CAS, SMT1, STE1*, and *CYP710A1*. In addition, upregulation of the expression of host-resistant-related gene NPR1 by 9 and 3–7 times in TV and co-inoculation treatment, respectively. Overall, these results suggest that co-inoculation of TV along with native endophytes of *W. somnifera* promoted withanolide and withaferin A production in the leaves and roots of *W. somnifera* (Kushwaha et al., 2019).

In another study, Mastan et al. (2019) reported the influence of innate bacterial endophyte, *Alcaligenes faecalis* (CFRB1) on forskolin synthesis in *Coleus forskohlii*. Inoculation of *Alcaligenes faecalis* CFRB1 to *C. forskohlii*, under field conditions, markedly upregulated the forskolin biosynthesis genes such as CfTPS1, CfTPS2, CfTPS3, CfTPS4, acyltransferase (CfACT1-8), diterpene synthases (CfTPSs) and cytochrome P450 (CfCYP76AH15) compared to the control plants. In addition, the inoculated plants also displayed/exhibited/showed reduced severity towards root-rot and nematode infection (78 and 81%, respectively). In a similar study, compared to the control plants, treatment of root endophyte *Fusarium redolens* (RF1), shoot endophytes namely, *Phialemoniopsis cornearis* (SF1) and *Macrophomina pseudophaseolina* (SF2) induced forskolin content in *C. forskohlii* by 52 to 88% in pot and 60 to 84% under field conditions. Endophyte-treated plants led to high expression of diterpene synthases (CfTPSs) like CfTPS1 (2.5 fold with RF1 and SF2 treated plants), CfTPS2 (three- to seven-fold with RF1, SF1, and SF2 treated plants), CfTPS3 (two- to seven-fold with RF1, SF1 and SF2 treated plants) and CfTPS4 (2.0 to 11.8-fold with RF1, SF1 and SF2 treated plants). Treatment with SF2 resulted in a significant increase in the expression of key diterpene synthases (CfTPS2 gene) in the root cork of *C. forskohlii*. Furthermore, endophyte treatments conferred antagonistic activity against nematode galls (80%) and plant pathogens like *Fusarium oxysporum*, *Colletotricum gloeosporioides,* and *Sclerotium rolfsii* (Mastan et al., 2019). An oleaginous fungal endophyte, *Mortierella alpina* CS10E4, isolated from *Crocus sativus* L. stimulated apocarotenoid biosynthesis genes such as phytoene synthase (4.5 fold), beta carotene hydroxylase (5 fold), two isoforms of carotenoid cleavage dioxygenase genes *CsCCD2* and *CsCCD4*

(4 and 2.4 fold, respectively). Stimulation of these genes enhanced apocarotenoids production such as crocin and safranal by two- and four-fold compared to the control plants, respectively. In addition, the treated plants also reduced corm rot disease, caused by *Fusarium oxysporum* by 53% (Wani et al., 2017).

Treatment with native endophytic fungi such as CATDLF5 (Curvularia sp.) and CATDLF6 (Choanephora infundibulifera) favored serpentine production by 212–338%, while CATDRF2 (Aspergillus japonicus) and CATDS5 (Pseudomonas sp.) enhanced ajmalicine content by 123–204% in alkaloid rich-genotype of *C. roseus* cultivar Dhawal, respectively. Further, it also induced the expression of terpenoid indole alkaloid structural and regulatory genes in the root. High expression of structural and regulatory genes for terpenoid indole alkaloid synthesis such as geraniol 10-hydroxylase, TDC, and strictosidine synthase was observed in the roots. Further, upregulation of Octadecanoid-derivative responsive Catharanthus AP2/ERF domain transcription activators like ORCA3 and downregulation of transcriptional repressor, ZCTs (Cys2/His2-type zinc finger protein family) upregulated the expression of genes for secondary metabolite production in endophyte inoculated plants (Singh et al., 2020). In a similar study, under field conditions, treatment with innate endophytes Fusarium redolens (RF1), Phialemoniopsis cornearis (SF1), Macrophomina pseudophaseolina (SF2), and Trichoderma viride (TV1) enhanced the forskolin content by 37, 50, 44, and 27%, respectively, over control plants. However, co-inoculation of endophytes RF1 + TV1, SF1 + TV1, and SF2 + TV1 significantly increased the forskolin content by 94, 82, and 75% in *C. forskohlii* roots compared to uninoculated control plants. Overall, these results suggest that, co-inoculation of endophytes induced forskolin content (Mastan et al., 2021).

9.3.2 APPLICATION OF PROTEOMICS AND METABOLOMICS STUDY

Treatment with polysaccharide fraction of endophytic fungus *T. atroviride* enhanced tanshinones accumulation in *S. miltiorrhiza* hairy roots by peroxide reaction, protein phosphorylation, Ca^{2+} triggering, and JA signal transduction, leading to an increase in leucine-rich repeat (LRR) protein synthesis (Maggini et al., 2019). Interaction of a consortium of native endophytic bacteria augmented higher alkamide levels in *Echinacea purpurea.* The proteomic study indicated that, the enzymatic source for the amine moiety synthesis of the alkamides that is, branched-chain amino acids (BCCA) decarboxylase gene (GenBank Accession #LT593930) was upregulated in the *E. purpurea* tissues than in the control ones (Tshikhudo et al., 2019).

9.4 CONCLUSION

Synthesis of tailor-made secondary metabolites in medicinal plants using endophytic consortia incites the development of new scientific approaches. This could prime the propagation of medicinal plants for their value-added metabolites to a new era for natural drug discovery. Of late, studies on the effect of endophytic microflora eliciting secondary metabolite production are well established. However, the influence of endophytic microorganisms on the modulation of gene and protein expression under different environmental conditions in medicinal plants is in still under infancy stage. Therefore, profound knowledge of crosstalk between the plant-endophyte interactions is inevitable to comprehend the pivotal role of endophytes for plant welfare through the means of employing "Omics study." The complementary information obtained through the "Omics" studies is critical in framing the models for a better understanding of the influence of endophyte microflora on the modulation of gene and protein expression, thus leading to enhanced secondary metabolite synthesis in medicinal plants. The scope for exploitation of these tools, especially in the field of natural product discovery is highly warranted.

Data generated via "Omics tools" provides us a trivial insight into information regarding the plant-endophyte interaction. Growing lines of evidence suggest that the interaction is well studied using genomics studies only. Therefore, there lies a huge scientific gap in understanding and elucidating this relationship using other "Omics tools" such as transcriptomics, proteomics metagenomics, and so forth. Genomic studies introduce the hidden information in a microenvironment in the form of molecular machinery. Nevertheless, the gene expression and its function remain unclear, whereas, genome expression under the influence of different environmental conditions can be understood through transcriptome studies. On the other hand, proteomic studies provide insight into functional gene products expressed in plant-endophyte interactions. Further, supplementing the metagenomics study would provide complete information on the influence of endophytic consortia on medicinal plants. Henceforth, all these tools must be integrated with other systemic approaches such as system biology, to unravel the several myriads of myths related to signaling mechanisms.

Thus, combined data analysis using various "Omics" tools could uncover the riddle of crosstalk between the plant endophyte relationships.

KEYWORDS

- **endophytes**
- **functional gene products**
- **genomics**
- **medicinal plant**
- **microenvironment**
- **omics**
- **plant-endophyte interactions**
- **secondary metabolites**

REFERENCES

Adeleke, B. S., & Babalola, O. O. (2021). Pharmacological potential of fungal endophytes associated with medicinal plants: A review. *Journal of Fungi, 7*, 147. https://doi.org/10.3390/jof7020147.

Ahlawat, S., Saxena, P., Ali, A., & Abdin, M. Z. (2016). Piriformospora indica elicitation of withaferin A biosynthesis and biomass accumulation in cell suspension cultures of *Withania somnifera. Symbiosis, 69*, 37–46. https://doi.org/10.1007/s13199-015-0364-9.

Alam, B., Lǐ, J., Gě, Q., Khan, M. A., Gōng, J., Mehmood, S., Yuán, Y., & Gǒng, W. (2021). Endophytic fungi: From symbiosis to secondary metabolite communications or vice versa? *Frontiers in Plant Science, 12*, 791033. https://doi.org/10.3389/fpls.2021.791033.

Carroll, G. C. (1988). Fungal endophytes in stems and leaves: From latent pathogen to mutualistic symbionts. *Ecology, 69*, 2–9.

Caruso, G., Abdelhamid, M. T., Kalisz, A., & Sekara, A. (2020). Linking endophytic fungi to medicinal plants therapeutic activity: A case study on Asteraceae. *Agriculture, 10*(7), 286.

Ezra, D., Lousky, T., & Elad, Y. (2009). Endophytes as biological control agents for plant pathogens. In *Biological Control of Fungal and Bacterial Plant Pathogens* (pp. 11–14).

Fontana, D. C., de Paula, S., Torres, A. G., de Souza, V. H. M., Pascholati, S. F., Schmidt, D., & Dourado Neto, D. (2021). Endophytic fungi: Biological control and induced resistance to phytopathogens and abiotic stresses. *Pathogens, 10*, 570. https://doi.org/10.3390/pathogens10050570.

Jayaram, H., Marigowda, V., & Thara Saraswathi, K. J. (2021). Secondary metabolite production and terpenoid biosynthesis in endophytic fungi *Cladosporium cladosporioides* isolated from wild *Cymbopogon martinii* (Roxb.) Wats. *Microbiology Research, 12*, 812–828.

Jia, M., Chen, L., Xin, H. L., Zheng, C. J., Rahman, K., Han, T., & Qin, L. P. (2016). A friendly relationship between endophytic fungi and medicinal plants: A systematic review. *Frontiers in Microbiology, 7*, 906.

Joyce, A. R., & Palsson, B. O. (2006). The model organism as a system: Integrating 'omics' data sets. *Nature Reviews Molecular Cell Biology, 7*, 198–210.

Kushwaha, R. K., Singh, S., Pandey, S. S., Kalra, A., & Babu, C. S. V. (2019). Fungal endophytes attune withanolide biosynthesis in *Withania somnifera*, prime to enhanced withanolide A content in leaves and roots. *World Journal of Microbiology and Biotechnology, 35*, 20. https://doi.org/10.1007/s11274-019-2593-1.

Kushwaha, R. K., Singh, S., Pandey, S. S., Kalra, A., & Babu, C. S. V. (2019). Innate endophytic fungus, *Aspergillus terreus*, as a biotic elicitor of withanolide A in root cell suspension cultures of *Withania somnifera*. *Molecular Biology Reports, 46*(2), 1895–1908. https://doi.org/10.1007/s11033-019-04641-w.

Kushwaha, R. K., Singh, S., Pandey, S. S., Rao, D. K. V., Nagegowda, D. A., Kalra, A., & Babu, C. S. V. (2019). Compatibility of inherent fungal endophytes of *Withania somnifera* with *Trichoderma viride* and its impact on plant growth and withanolide content. *Journal of Plant Growth Regulation, 38*, 1228–1242. https://doi.org/10.1007/s00344-019-09928-7.

Li, J., Zhao, G. Z., Varma, A., Qin, S., Xiong, Z., Huang, H. Y., Zhu, W. Y., Zhao, L. X., Zhang, S. X., & Li, W. (2012). An endophytic *Pseudonocardia* species induces the production of artemisinin in *Artemisia annua*. *PLoS ONE, 7*, e51410. https://doi.org/10.1371/journal.pone.0051410.

Maggini, V., De Leo, M., Mengoni, A., Gallo, E. R., Miceli, E., Reidel, R. V. B., Biffi, S., Pistelli, L., Fani, R., Firenzuoli, F., & Bogani, P. (2019). Plant-endophytes interaction influences the secondary metabolism in *Echinacea purpurea* (L.) Moench: An *in vitro* model. *Scientific Reports, 7*, 16924. https://doi.org/10.1038/s41598-017-17110-16924.

Mastan, A., Bharadwaj, R. K., Kushwaha, R. K., & Babu, C. S. V. (2019). Functional fungal endophytes in *Coleus forskohlii* regulate labdane diterpene biosynthesis for elevated forskolin accumulation in roots. *Microbial Ecology, 78*(4), 914–926. https://doi.org/10.1007/s00248-019-01376-w.

Mastan, A., Rane, D., Dastager, S. G., & Babu, C. S. V. (2019). Plant probiotic bacterial endophyte, *Alcaligenes faecalis*, modulates plant growth and forskolin biosynthesis in *Coleus forskohlii*. *Probiotics and Antimicrobial Proteins, 12*, 481–493. https://doi.org/10.1007/s12602-019-09582-1.

Mastan, A., Rane, D., Dastager, S. G., & Babu, C. S. V. (2021). Molecular insights of fungal endophyte co-inoculation with *Trichoderma viride* for the augmentation of forskolin biosynthesis in *Coleus forskohlii*. *Phytochemistry, 184*, 112654. https://doi.org/10.1016/j.phytochem.2021.112654.

Ming, Q., Su, C., Zheng, C., Jia, M., Zhang, Q., Zhang, H., Rahman, K., Han, T., & Qin, L. (2013). Elicitors from the endophytic fungus *Trichoderma atroviride* promote *Salvia miltiorrhiza* hairy root growth and tanshinone biosynthesis. *Journal of Experimental Botany, 64*(18), 5687–5694.

Mishra, A., Singh, S. P., Mahfooz, S., Bhattacharya, A., Mishra, N., Shirke, P. A., & Nautiyal, C. S. (2018b). Bacterial endophytes modulate the withanolide biosynthetic pathway and physiological performance in *Withania somnifera* under biotic stress. *Microbiological Research, 17–28.*

Mishra, A., Singh, S. P., Mahfooz, S., Singh, S. P., Bhattacharya, A., Mishra, N., & Nautiyal, C. S. (2018a). Endophyte-mediated modulation of defense-related genes and systemic resistance in *Withania somnifera* (L.) Dunal under *Alternaria alternata* stress. *Applied and Environmental Microbiology, 84*, e02845–17. https://doi.org/10.1128/AEM.02845-17.

Pádua, A. P. S. L. D., Freire, K. T. L. D. S., Oliveira, T. G. L. D., Silva, L. F. D., Araújo-Magalhães, G. R., Agamez-Montalvo, G. S., Silva, I. R. D., Bezerra, J. D. P., Souza-Motta, C. M. D. (2018). Fungal endophyte diversity in the leaves of the medicinal plant *Myracrodruon urundeuva* in a Brazilian dry tropical forest and their capacity to produce L-asparaginase. *Acta Botanica Brasilica, 33*, 39–49.

Pandey, S. S., Singh, S., Babu, C. S. V., Shanker, K., Srivastava, N. K., Shukla, A. K., & Kalra, A. (2016). Fungal endophytes of *Catharanthus roseus* enhance vindoline content by modulating structural and regulatory genes related to terpenoid indole alkaloid biosynthesis. *Scientific Reports, 6*, 26583. https://doi.org/10.1038/srep26583.

Pandey, S. S., Singh, S., Babu, C. S. V., Shanker, K., Srivastava, N. K., & Kalra, A. (2016). Endophytes of *Opium poppy* differentially modulate host plant productivity and genes for the biosynthetic pathway of benzylisoquinoline alkaloids. *Planta, 243*(5), 1097–114. https://doi.org/10.1007/s00425-016-2467-9.

Pandey, S. S., Singh, S., Pandey, H., Srivastava, M., Ray, T., Soni, S., Pandey, A., Shanker, K., Babu, C. S. V., Banerjee, S., & Gupta, M. M. (2018). Endophytes of *Withania somnifera* modulate plant content and the site of withanolide biosynthesis. *Scientific Reports, 8*, 5450. https://doi.org/10.1038/s41598-018-23716-5.

Rai, N., Keshri, P. K., Verma, A., Kamble, S. C., Mishra, P., Barik, S., Singh, S. K., & Gautam, V. (2021). Plant-associated fungal endophytes as a source of natural bioactive compounds. *Mycology, 12*, 139–159. https://doi.org/10.1080/21501203.2020.1870579.

Rao, H. Y., Kamalraj, S., & Jayabaskaran, C. (2020). Fascinating fungal endophytes associated with medicinal plants: Recent advances and beneficial applications. In A. Kumar & V. K. Singh (Eds.), *Microbial Endophytes: Prospects for Sustainable Agriculture* (pp. 263–289).

Ray, T., Pandey, S. S., Pandey, A., Srivastava, M., Shanker, K., & Kalra, A. (2019). Endophytic consortium with diverse gene-regulating capabilities of benzylisoquinoline alkaloids biosynthetic pathway can enhance endogenous morphine biosynthesis in *Papaver somniferum*. *Frontiers in Microbiology, 10*, 925. https://doi.org/10.3389/fmicb.2019.00925.

Ren, C. G., & Dai, C. C. (2012). Jasmonic acid is involved in the signaling pathway for fungal endophyte-induced volatile oil accumulation of *Atractylodes lancea* plantlets. *BMC Plant Biology, 12*, 128. https://doi.org/10.1186/1471-2229-12-128.

Schouten, A. (2019). Saving resources: The exploitation of endophytes by plants for the biosynthesis of multifunctional defense compounds. In A. Schouten (Ed.), *Endophyte Biotechnology: Potential for Agriculture and Pharmacology* (pp. 122–144). CABI International.

Sharma, H., Rai, A. K., Dahiya, D., Chettri, R., & Nigam, P. S. (2021). Exploring endophytes for *in vitro* synthesis of bioactive compounds similar to metabolites produced *in vivo* by host plants. *AIMS Microbiology, 7*(2), 175–199.

Singh, A., Singh, D. K., Kharwar, R. N., White, J. F., & Gond, S. K. (2021). Fungal endophytes as efficient sources of plant-derived bioactive compounds and their prospective applications in natural product drug discovery: Insights, avenues, and challenges. *Microorganisms, 9*(1), 197.

Singh, S., Pandey, S. S., Shanker, K., & Kalra, A. (2020). Endophytes enhance the production of root alkaloids ajmalicine and serpentine by modulating the terpenoid indole alkaloid pathway in *Catharanthus roseus* roots. *Journal of Applied Microbiology, 128*(4), 1128–1142.

Soni, S. K., Singh, R., Ngpoore, N. K., Niranjan, A., Singh, P., Mishra, A., & Tiwari, S. (2021). Isolation and characterization of endophytic fungi having plant growth promotion traits that biosynthesize bacosides and withanolides under *in vitro* conditions. *Brazilian Journal of Microbiology, 52*(4), 1791–1805.

Srinivas, R. P., Nigam, A., Aruna, J., Alam, M. A., Ishara, L., Chamith, Y. H., & Chikkaswamy, B. K. (2015). Antimicrobial activity in cultures of endophytic fungi isolated in some medicinal plant species. *International Journal of Advanced Research in IT and Engineering, 4*(2), 1–24.

Sunitha, V. H., Ramesha, A., Savitha, J., & Srinivas, C. (2012). Amylase production by endophytic fungi *Cylindrocephalum* sp. isolated from medicinal plant *Alpinia calcarata* (Haw.) Roscoe. *Brazilian Journal of Microbiology, 43*, 1213–1221.

Toghueo, R. M. K., Ejiya, I. E., Sahal, D., Yazdani, S. S., & Boyom, F. F. (2017). Production of cellulolytic enzymes by endophytic fungi isolated from Cameroonian medicinal plants. *International Journal of Current Microbiology and Applied Sciences, 6*(2), 1264–1271.

Tshikhudo, P. P., Ntushelo, K., Mudau, F. N., Salehi-Rad, M., Sharifi-Rad, J., Martins, N., Martorell, M., & Sharifi-Rad, J. (2019). Understanding *Camellia sinensis* using omics technologies along with endophytic bacteria and environmental roles on metabolism: A review. *Applied Sciences, 9*, 281. https://doi.org/10.3390/app9020281.

Venieraki, A., Dimou, M., & Katinakis, P. (2017). Endophytic fungi residing in medicinal plants have the ability to produce the same or similar pharmacologically active secondary metabolites as their hosts. *Hellenic Plant Protection Journal, 10*(2), 51–66.

Wani, Z. A., Kumar, A., Sultan, P., Bindu, K., Hassan, S. R., & Ashraf, N. (2017). *Mortierella alpina* CS10E4, an oleaginous fungal endophyte of *Crocus sativus* L., enhances apocarotenoid biosynthesis and stress tolerance in the host plant. *Scientific Reports, 7*, 8598. https://doi.org/10.1038/s41598-017-08866-0.

Wu, W., Chen, W., Liu, S., Wu, J., Zhu, Y., Qin, L., & Zhu, B. (2021). Beneficial relationships between endophytic bacteria and medicinal plants. *Frontiers in Plant Science, 12*, 758.

Yadav, G., & Meena, M. (2021). Bioprospecting of endophytes in medicinal plants of Thar Desert: An attractive resource for biopharmaceuticals. *Biotechnology Reports*, e00629.

Yang, J. W., Ling, H., Zhang, Y., Zeng, X., & Guo, S. X. (2018). Effects of endophytic fungi on seed germination of medicinal plants of *Orchidaceae*: A review. *Mycosystema.*

CHAPTER 10

Nanobiosensors to Detect Environmental Pollution

ALVINA GUL,[1,*] NOOR UL HUDA,[1] TUBA SHARF BATOOL,[2] WARDA JABBAR,[2] VOLKAN ALTAY,[3] FAZILET VARDAR SUKAN,[4] and MUNIR OZTURK[5,*]

[1]*Atta-Ur-Rahman School of Applied Biosciences, National University of Sciences and Technology, Islamabad, Pakistan*

[2]*Department of Plant Biotechnology, Atta-Ur-Rahman School of Applied Biosciences, National University of Sciences and Technology, Islamabad, Pakistan*

[3]*Hatay Mustafa Kemal University, Faculty of Arts and Sciences, Department of Biology, Hatay, Türkiye*

[4]*Faculty of Engineering and Natural Sciences; Director of SUNUM Nanotechnology Research and Application Center, Sabanci University, Tuzla, Istanbul, Türkiye*

[5]*Ege University, Department of Botany and Center for Environmental Studies, Izmir, Türkiye*

*Corresponding author.

ABSTRACT

The presence of nanoparticles in nature is not recent since they naturally exist in great diversity from those produced by human activity to those of natural origin. This chapter discusses the use of nano-technology-based sensors for the detection of environmental pollution. The chapter focuses on the use of nano-biosensors, which combine nanotechnology with biological components, as promising tools for the detection and monitoring

Agricultural Biotechnology: Issues, Challenges, and Recent Developments. Mohd. Sayeed Akhtar and Khalid Rehman Hakeem (Eds.)
© 2025 Apple Academic Press, Inc. Co-published with CRC Press (Taylor & Francis)

of environmental pollutants. The chapter highlights the dependence between nanoparticle synthesis and application design, as well as the importance of considering environmental considerations in the generation of nanoparticles. The use of nanotechnology is seen as a possible solution to problems that have influenced humanity for centuries, such as pollution, and the rise of the internet is intended to contribute to a model of analysis and follow-up of contaminant particles in an effective and reliable way.

10.1 INTRODUCTION

Nanoparticles, the particles that make up matter can be classified by their size. Coarse particles cover a range between 10,000 and 2500 nm and, fine between 2500 and 100 nm. Until the 70s, they were called ultrafine particles those between 1 and 100 nm. Later they designated the prefix nano- (from the Greek "*nanos*" which means "dwarf") and expressed the billionth part of the unit. Therefore, a nanomaterial is defined as one that possesses certain structural characteristics, of which at least one of its dimensions is in the range of 1–100 nm. Thus, nanotechnology is the research area that studies, designs and manufactures materials at the nanoscale with various applications (Kajitvichyanukul et al., 2022; Ozturk et al., 2023a).

Nanotechnology is an area of extensive development in recent years, due to its applications and scope. In addition, the unique properties of nanomaterials (nanoparticles, nanocrystals, metal colloids, nano-rods, dendrimers, and nano-capsules) receive particular attention for their possible use in pharmaceuticals, bioengineering, remediation of water and contaminated soil. The presence of nanoparticles in nature is not recent since they naturally exist in great diversity from those produced by human activity to those of natural origin. They first arise under synthesis or accidentally, that is, as a product of industrial activities, burning of fuels or daily activities, and secondary products of mining and metallurgical activity (Ozturk et al., 2023a, b). Those natural nanoparticles result from environmental factors, such as emissions of gases, aerosols, particles from volcanoes, and weathering of rocks such as allophane, ferrihydrite, goethite, magnetite, and anthracite. There are two types of materials synthesis methods including that of synthesis under the control of the "bottom-up" approach and the miniaturization of the "top-down" approach. The first, goal is to synthesize nanoparticles by the deposition and growth of crystals from solutions in the liquid or vapor phase, such as the methods of sol-gel and aerosol pyrolysis, or through precipitation processes. With the top-down method, the nanoparticles are obtained through miniaturization processes, applying different techniques, such as mechanical grinding or thermal decomposition (Yadav, 2021).

Various studies emphasize the dependence between nanoparticle synthesis and application design. This is because the interaction and stability of the materials that make up the nanoparticle can be decisive in the beneficial or detrimental effect of the treatment with nanoparticles. In recent years, the application of innovative technology for eco-friendly approaches has been aimed at satisfying needs including contaminant attenuation, organic molecule sensors, and heavy metal detection, among others. Likewise, nano-technology can provide strategies for saving in the consumption of fertilizers in the soil, reduction of losses due to physical and chemical decomposition of these inputs, attenuation of the effect of environmental contaminants in water and soil, and even alternatives to increase carbon sequestration in the floor. Numerous publications propose nano-technological systems for the control of contaminants and recalcitrant molecules in aqueous matrices and soils (Acharya and Pal, 2020).

The detection of environmental pollutants is of great importance, as it allows for the effective management of potential hazards that could negatively impact human health and the environment. Nano-biosensors, which combine nanotechnology with biological components, have been extensively studied and developed as promising tools for the detection and monitoring of environmental pollutants. In recent years there have begun to hear about nano-technoscience (nano-technology and nano-science), as a possible solution to problems that have influenced for decades and centuries to humanity originated even from the great revolution from the 18th century to our days; affecting our health, environment, and habitat as the so-called global warming, caused largely by pollution that humans generate as garbage, waste toxic, smoke caused by means of transportation and factories, among others (Svendsen et al., 2020). With the use of nano-technology, technologies of information and communications (ICT), and the rise of the internet, it is intended to contribute to a model of analysis and follow-up of contaminant particles in an effective way and reliable to able to create mechanisms that help reduce environmental pollution, considering that there are current systems that do it but oriented to microsensors.

Nanotechnology is a generic and evolving term that encompasses the development of many products and processes, with the common feature of the nano-meter scale (1 to 100 nm). For a particle to belong to this scale, it must have nano-metric dimensions in more than one of its dimensions.

Usually, nanoparticles have nano-metric dimensions in their three dimensions, while the terms "nano-structured material" and "nanomaterial" are more general and apply to materials whose manufacture, or whose dimensions are controlled at the nano-metric level (Usman et al., 2020). There are three types of nanoparticles: natural ones, such as those produced in volcanic eruptions; incidental ones, such as engine combustion emissions; and manufactured ones, generated on purpose with a purpose. There are two reasons why environmental considerations are critical in the generation of nanoparticles. One is the fact that it is a nascent technology that has not done any damage yet. In recent decades, ecological awareness has developed in terms of the effects that different materials can have on ecosystems, which can be seen in the regulations that have emerged regarding their use. Likewise, the techniques associated with the generation and use of nanoparticles have been recently developed, which provides a unique opportunity to control the development of this technology from its inception (Qian et al., 2020).

Secondly, there is a lack of data regarding the possible effects that nanoparticles have on organisms and the environment, should they be released. For example, it has been estimated that in 2010 between 260,000 and 390,000 tons of nano-structured materials were released, which ended up in landfills (63–91%), soils (8–28%), bodies of water (0.4–7%) and in the atmosphere (0.1–1.5%). Little is known about how they behave in different environments and the effects they have on living things (Vázquez-Núñez et al., 2020). Since nanoparticles are produced with relatively new techniques, there is no specific regulation on them. In this regard, nano-ethics is an area of ethics that studies nanotechnology and its products and whose purpose is to serve as a guide to develop training, prohibition, and restriction regulations regarding the use of nanomaterials. In this chapter, the focus will be on the role of enzymes in coping with pollution via nanotechnology, particularly in the development of nano-biosensors. The advantages of using enzymes, the limitations and difficulties faced in this field, as well as the prospects of these sensors are also discussed.

10.2 USE OF NANO-TECHNOLOGY FOR SOIL REMEDIATION

Dutta et al. (2022) added nanoparticles of gold and silver on the soil surface and stabilized phenanthrene which is a carcinogenic polycyclic aromatic hydrocarbon. They suggested that the binding of phenanthrene to metallic nanoparticles is the result of the absorption phenomenon. This mechanism originates due to the properties of the nucleus and the concentration of reagents in the nanoparticles, evidencing the influence of nanoparticles on transport, bioavailability, and toxic properties of contaminants in the soil. Kumar and Sundramoorthy (2019) propose that single-walled carbon nanotubes (single-walled carbon nanotubes (SWCNTs)) have the potential to reduce the mobility and bioavailability of herbicides in soil for agricultural use. It is due to the capacity of SWCNTs as adsorbents of organic compounds, particularly diuron, and norflurazon, contaminating herbicides in agricultural soils. Based on the correlation results of the surface area of the SWCNTs and the concentration of hydrogen/carbon/oxygen present in its structure, the authors concluded that SWCNTs are capable of adsorbing diuron and norflurazon under the influence of carboxylic functional groups present on the surface of SWCNTs. Despite the promising results of this study, it is necessary to evaluate the risk and possible adverse effects of the application of a large amount of SWCNT in soils with these contaminants.

An alternative to control soil lead contamination, coming from urban areas, is the application of stable apatite nanoparticles Ca_5 (PO_4). After 30 days of application, the concentration of Pb (II) leachates in soil with apatite nanoparticle treatment decreased in contrast to the concentration in untreated soil. This was demonstrated by the decrease of the tclp (toxicity characteristic leaching procedure) index in soils contaminated with Pb (II) post-apatite nanoparticle treatment (Kumar et al., 2019). Laboratory tests showed that the nanoparticles could effectively decrease up to 66% of the tclp fraction in soil contaminated with Pb. A new class of stable iron-manganese binary oxide nanoparticles prepared with water-soluble starch and carboxy methyl cellulose (CMC) were presented as an alternative to immobilize arsenic. An experiment in this respect was based on the sorption capacity and immobilization of arsenite (As^{3+}) in the soil (Pramanik et al., 2020). Although the adsorption capacity of As^{3+} was similar, both with naked or stabilized Fe-Mn nanoparticles, the latter were added to the soil and *in situ* immobilization of As^{3+}. The presence of stabilized Fe-Mn nanoparticles in columns of soil with As^{3+} reduced between 91% and 96% of As^{3+} leachates and 78% of the tclp index of As^{3+} present in the soil bed, however, the few investigations on the use of nano-technology in the soil and the complexity of this system, limit the comprehensive understanding of the behavior of nano-structures (Shukla et al., 2019). Consequently, the evaluation of environmental risks is difficult (imbalance of sensitive microbial communities, alterations in nutrient cycles, trophic magnification), and toxicology result from the introduction of this technology on the ground.

10.2.1 USE OF NANOMATERIALS AS SORBENTS IN WATER

Titanium dioxide (TiO_2) is a commonly applied semiconductor that recently gains importance with notable results in environmental aspects, such as photocatalyst in reduction reactions of inorganic contaminants, in the photodegradation of various organic contaminants, and in activation of microorganisms (Saleh, 2021). Titanate nanotubes (TNTs) are proposed as an alternative to contaminated water treatment with potentially toxic elements (Pte). TNTs are generally synthesized by the TiO_2 and NaOH hydrothermal reaction. The advantages offered by this nanomaterial are based on its special physicochemical properties, such as uniform microstructure, small diameter of the tube, large specific surface area, and so forth. Besides, TNTs are abundant in hydroxyl groups (-OH) that are in the interlaminations and the surface, which is related to its intense ionic exchange activity: between H^+/Na^+ and ethylpropyltryptamine (EPT) ions, and with their cation adsorbent capacity in the aqueous medium. Consequently, TNTs are applied to remove various contaminants from water residuals, such as cations of EPT (e.g., Pb^{2+} and Cd^{2+}, Ag^+, Cu^{2+}, and Ni^{2+}), anions (e.g., As^{3-} and As^{5-}), stabilize radioactive ions (e.g., Sr^{2+}, Ba^{2+}, and Cs^+), and even amino aromatic dyes. Awad et al. (2020) showed that the adsorption of our EPTs on the catalytic surface of the TNTs follows the sequence: $Pb^2 > Cd^2 > Cu^2 > Cr^3$. However, the presence of inorganic ions (e.g., Fe^{3+}, Al^{3+}, Na^+, K^+, Mg^{2+}, Ca^{2+}) in an aqueous solution has inhibitory effects on absorption, because they compete for adsorption sites with EPT ions. This poses a limitation in the application. Of these nanostructures, wastewater often contains these ions.

Wadhawan et al. (2020) reported that titanium nanoparticles require high concentration and ionic affinity to achieve aggregation. These factors must be considered to interpret the behavior of nanoparticles, as well as their removal from the wastewater. In another study by Pal et al. (2022), carbon oxide nanoparticles used titanium oxide (TiO_2) to remove fluoroquinolone antibiotics, particularly ofloxacin, from residual water of nosocomial and veterinary origin. The specific analysis of the concentration of ofloxacin in solution after removing the solid nanoparticles showed the presence of dimethylated groups and ionized carboxylates of fluoroquinolone. These compounds are products of the degradation of ofloxacin on the surface of the TiO_2 nanoparticles. However, studies on the adsorption of fluoroquinolones on titanium oxide showed the influence of organic matter (humic acid) present in the aqueous medium and the chemical carrier of the drug, which could alter the transport and fate of norfloxacin in the atmosphere (Peng et al., 2012).

Similarly, Xu et al. (2012) proposed the use of magnetic nanoparticles as an alternative for the elimination of contaminants in wastewater. Among some ideal characteristics concerning the rest of the nanostructures, the size, greater surface area with respect to its total volume, magnetic properties, low toxicity, chemical stability, and compatibility with biomolecules are also mentioned (Alfansuri et al., 2022). At present, nano-metric scale filters are being developed to have high chemical selectivity and controllable flow rate and direct the separation of chemical reagents, prolonged-release drugs, and wastewater treatment. In this area, it is essential to study the interaction of the molecules with the pore surface and the affinity of the filter materials. In addition, while manufacturing nano-filters, the range of porosity (1–10 nm), the affinity of the polymer-like membranes, the macromolecular architecture of mesopores, and the cavity of carbon nanotubes (CNTs) should be considered. The functionality of CNT filters is based on pumping ions across the membrane by applying pressure or an electric field. However, in the large-scale application and expectations of nanotechnology, there are still drawbacks due to the few studies on the formation mechanisms, stabilization, release, transport, deposition as well as toxicological aspects of nanoparticles in the aquatic and terrestrial environment (Ong et al., 2018).

10.3 USE OF BIOSENSORS AT NANO-METER SCALE

Biosensors are described as analytical detection devices comprising a chemical origin recognition biological entity coupled to a transducer. This system allows the quantitative development of a complex biochemical parameter, that is, it transduces the recognition event (Banica, 2012). These devices are useful in the detection of organic and inorganic molecules, in the control of biological species, monitoring of organic contaminants, EPT, detection of pathogens in the food (Rico-Sánchez et al., 2021), and even, in the identification of the potential risk of bioterrorism, among other applications. However, the main challenges faced by many biosensors include their low sensitivity, as well as their specificity, and complex cleaning methods.

The development of nanotechnology presents some solutions promising to alleviate these problems (Zhang et al., 2011). This can be applied in biotechnology, medicine, the food industry, agriculture, materials engineering, and basic sciences, among other areas of knowledge and application. Recently based biosensors are developed in nanotechnology comprising nanoparticles with a core magnetic, covered by silica, more immobile layers of different metals on the outer surface, and, finally, a coating of organic or inorganic molecules of synthetic nature capable of joining to biomolecules. These can be physically immobilized through hydrophobic, ionic, and or van der Waals forces to a solid matrix. They also bind to chemically activated surfaces by covalent immobilization. However, in some applications, there are failures due to weak interaction of the biomolecules with the surface of the biosensor (Gokaltun et al., 2017). Sensors can be fabricated under ambient conditions of pressure and temperature with high precision and stability. Xu et al. (2014) developed a disposable nano-biosensor for the rapid detection of toxic compounds in water, such as 2-anthramine, acridine orange, and 2-naphthylamine. This device consists of immobile DNA on modified gold nanoparticles with cysteamine. Nanoparticles improve DNA immobilization, leading to an increase in the guanine signal. The interaction of the analyte with the immobilized DNA was quantified by variations in the electrochemical signal of guanine translated to voltage. The results of the genotoxicity tests in water samples from lakes integrate the set of promising technologies for the detection of contaminants in aquatic environments. Enzyme sensors offer versatility and high specificity in the detection of organic compounds, such as penicillin (Wu et al., 2014). However, the use of nano-graphene in the biosensor showed advantages in terms of speed of response, repeatability, and detection limit.

10.3.1 USE OF NANO-BIOSENSORS TO DETECT POLLUTION AND OFFER REMEDIATION

The nano-biosensor (biological nano-sensor), is a device capable of detecting in real-time and with high sensitivity and selectivity, chemical, and biological agents. The term "nano-biosensor" designates those biosensors whose properties come modulated by the nano-technological scale with which they are made. It is expected that nano-biosensors have a sensitivity much higher than that of devices conventional. Furthermore, they could easily be introduced into the human body so they could provide much data more dependable on the state of health of a patient. The next step to follow is for the nano-sensors can be fabricated and/or manipulated to respond to different stimuli. Although the theory says that nanoparticles can be developed so that the polluting particles adhere to these, and then destroy them; in our case, it does not go to that point. Our case goes to how these particles can respond to contaminants in nano-sensors. Among the "polluting particles" that affect the environment, are the material particulate matter (solids such as ash, shoot, organic, dust, and salt and liquids such as aerosol and mist), chemical material (exhaust gases, primary types such as chlorofluorocarbon, methane, carbon dioxide (CO_2), carbon monoxide, nitrogen, nitrous oxide, oxide of nitrogen and sulfur oxide and type gases secondary as aldehyde, ozone, and nitrateperoxyacetyl) and steam (conformed by the water and volatile organic compound (VOC); Zamora-Ledezma et al., 2021).

10.3.1.1 CONTAMINANT DETECTION AND PREVENTION

One of the great challenges of environmental sciences is to be able to detect the presence of pollutants in different ecosystems and even in the human body (Altay and Ozturk, 2012; Darwash et al., 2023; Haq et al., 2021; Yalcin and Altay, 2023). The new detection systems for toxic substances must increasingly be done with lower detection levels to anticipate damage or environmental impact. Detection is not an easy task, especially in systems as complex and with so many interactions between the different phases such as ecosystems. It is of great interest to have real-time pollutant monitoring systems that can transmit data from a remote area to the decision-making center. Single-walled carbon nanotubes have been used as chemical sensors thanks to the change in electrical resistance that these semiconductors undergo in the presence of gaseous substances such as nitrogen dioxide or ammonia. These sensors have a faster response at room temperature to these gases than conventional sensors. On the other hand, boron-doped silicon nanowires have been used in ultra-sensitive and real-time sensors for the detection of chemical and biological compounds (Daoudi et al., 2021). These nanowires can be amine-functionalized and used to detect pH changes, while biotin-functionalized nanowires can detect concentrations in the picomolar range of streptavidin, a protein with a high affinity for biotin. Like almost all nanomaterials, these nanowires can be functionalized with antibodies or antigens and recognized by their very high-affinity counterparts. Metabolic indicators such as calcium can also be detected in real time. These properties can also be used to detect pathogens or other chemical or biological agents in water, air, or food (Fallah et al., 2021). Both in medicine and in the diagnosis of the state of health of ecosystems, nanomaterials can be used as indicators of exposure to pollutants, in time and concentration. This would make it possible to detect the sources of contamination and to assess early the importance of the environmental or health impact.

The pollution reduction strategies include and use of renewable and less toxic raw materials, and the design of new products with environmentally friendly production processes. Industrial ecology is a new branch of engineering that aims to redesign and replace all the production processes that are currently in use, to produce a great diversity of satisfiers, with clean, sustainable processes with low or no environmental impact.

Nano-structured catalysts can make synthesis processes more efficient, reducing polluting by-products and energy consumption. An example is the use for separation and catalysis of zeolites (nano- and microporous aluminosilicates). These zeolites have been used to oxidize aromatic hydrocarbons, such as toluene, to produce benzaldehyde (Panov et al., 2000). The oxidation reaction starts with the presence of visible light, reducing energy consumption for catalysis, and undesirable side reactions are eliminated, increasing the yield of the desired product. The use of the nano-structured zeolites presented a conversion of 85% to benzaldehyde, compared to only 35% when conventional zeolites were used. On the other hand, the emergence of bionanotechnology, which combines the properties of biological materials with the properties of nanostructured materials, has opened new horizons and applications of nanotechnology. The production of nanomaterials with biopolymers or materials inspired by biological systems is an example of the manufacture of environmentally friendly microelectronic devices. It is estimated that manufacturing a 32-megabyte microchip weighing only 2 g requires 1.7 kg of oil and chemicals and 32 liters of water (Williams, 2003). Using biomolecular nanolithography can replace the way to make semiconductor chips. Using this technique, well-defined chip architectures were produced on a biopolymer (polylysine) quench where metal nanoparticles were scattered in lines and meshed at room temperature. These semiconductor chips are fully biodegradable.

Nanotechnology can also be used to replace toxic materials in widespread use today. For example, previously computer monitors were made of cathode ray tubes (CRTs) and contained many toxic materials. The manufacture of new liquid crystal monitors, which are smaller, consume less power, and do not contain lead (Socolof et al., 2005). The use of carbon nanotubes in computer monitors could, in the future, reduce environmental impact by eliminating toxic compounds and reducing the need for raw materials and energy, in addition to improving performance. Field emission monitors (feds), which use carbon nanotubes (Talin et al., 2001), are the latest development in monitor and display technology and are now commercially available. Although the toxicity of carbon nanotubes is still controversial, the number of nanotubes in a fed monitor is very small, around 0.5 g per monitor, compared to the kilograms of lead in a CRT monitor (Socolof et al., 2005).

10.3.1.2 TREATMENT AND REMEDIATION OF CONTAMINATED SITES

Since its inception, nanotechnology has been proposed as a possible solution to wastewater treatment and remediation of contaminated sites. The oxidizing or reducing of the capacity of nanomaterials, or as nutrients, has been suggested as an alternative for the transformation of pollutants and toxic substances, as well as to stimulate microbial growth. Due to their small size and large specific surface area, nanomaterials are more reactive and can be more easily dispersed. For example, conventional methods of oxidation of contaminants using Fenton's reagent or zero valence iron for the remediation of sites contaminated with solvents, such as trichloroethane or other chlorinated solvents, result in undesirable by-products such as dichloroethanes and vinyl chloride. The use of bimetallic nanoparticles almost eliminates the production of these undesirable by-products (Osman et al., 2023).

Nanoparticles that are activated by light such as titanium dioxide (TiO_2) and zinc oxide (ZnO), which are semiconductors with a wide bandgap, continue to be highly studied for the removal of contaminants. These particles are cheap and can be produced in large quantities, as well as being low toxic. For example, ZnO nanoparticles are capable of both sensing and photocatalyzing the destruction of dangerous polychlorinated phenols. There is great interest in manipulating the surface of these particles with organic and inorganic pigments or dyes to have a response not only in the UV but also in the visible spectrum and, in this way, make photocatalysis more efficient for the transformation of pollutants into the environment. It must be borne in mind that only 5% of the solar radiation that reaches the earth's surface is in the UV range.

Nanotechnology also has potential use in the remediation of contaminated sites *in situ*. For example, these nanoparticles can easily be added and spread in reactors for the treatment of soils, sediments, and solid waste. They can also be immobilized in solid matrices such as carbon, zeolites, or membranes for the treatment of gaseous effluents (Sahoo et al., 2021). There are reports that show that the injection of iron nanoparticles into the subsoil stimulates the degradation of polluting organochlorine compounds, such as trichloroethane. With this technology, the toxicity of contaminants in the soil is reduced. On the other hand, it has been shown that bimetallic iron/palladium, iron/silver, or zinc/palladium nanoparticles are powerful reductants and catalyze the transformation of a wide variety of contaminants, such as polychlorinated biphenyls, which are highly recalcitrant to microbial degradation and are considered strong carcinogens. They are also capable of transforming other halogenated compounds such as solvents and pesticides, reducing almost all halogenates tested to hydrocarbons, which are substances that are easy to be degraded by the natural microflora of the site. In addition, these iron-based nanoparticles have been able to reduce a wide variety of polluting compounds such as perchlorates, nitrates, dichromates, nickel, mercury, and uranium dioxide (Pasinszki and Krebsz, 2020).

Nano-structured materials, such as carbon nanotubes, have been proposed as excellent adsorbents for pollutants such as dangerous dioxins. Dioxins are considered the most toxic substances ever synthesized by man. The absorption of dioxins by carbon nanotubes is at least three times higher than that of activated carbon, with the Langmuir constant of nanotubes being orders of magnitude larger than that of activated carbon (Yang et al., 2006). This property of nanomaterials makes them potential candidates for air or water decontamination treatment. Another example of the uses in environmental processes of nano-structured materials is dendrimers. These nano-structured particles are highly branched polymers with controlled composition and architecture. Dendrimers at the nano-meter scale can be used as chelating agents in polymeric membrane ultrafiltration systems. Dendrimers can be designed to encapsulate zero valence metals and thus be able to "dissolve" them in specific media or adsorb them on certain surfaces.

The conservation of soil, water, and air quality is one of the most important challenges of the 21st century. Society has been characterized, especially in the last 200 years, in altering ecosystems with the dispersion of large amounts of pollutants derived from industrial activities, transportation, as well as from each of the aspects of human activity. The four most important challenges of environmental sciences are: (i) prevent pollution with the implementation of new less or non-polluting industrial processes; (ii) find efficient and low-cost methods of remediation and restoration of contaminated sites; (iii) predict in advance the environmental impact of new products launched on the market; and (iv) have highly sensitive pollutant detection methods and be able to detect pollution sources early to avoid damage (Singh, 2006). Paradoxically, the solution to these great challenges could be found in the technology of extremely tiny materials: nanotechnology. Nanotechnology is the revolutionary science of material manipulation at the molecular or atomic level that has already impacted all disciplines, such as chemistry, physics, biology, and engineering. Undoubtedly, nanotechnology will be able to participate in the improvement of environmental processes. However, like any new technology, there is a reasonable risk that it may also create new environmental problems.

10.4 NANO-TECHNOLOGY AND ENVIRONMENT

The term "nanotechnology" was used for the first time in 1974 by Norio Taniguchi, a researcher at the University of Tokyo, who thus pointed out the ability to handle materials at the nanometric level. The initial references to nanotechnology were presented in 1959 by physicist Richard Feynman in a conference entitled "There's Plenty of Room at the Bottom," in which he envisioned the possibility of manipulating materials on an atomic and molecular scale (Toumey, 2008). According to Aguilar (2012), the bulk of investment in nano-technology is focused on the development of a nano-technological tools or instruments (e.g., tunneling microscope or scanning probe microscopy (SPM), matrices or arrays, molecular labels, and microfluidics) which are estimated to concentrate 4% of the total funds; (i) new materials (e.g., textiles, ceramics, etc.) with 12%; (ii) new devices (e.g., sensors) with 32%; and (iii) nano-biotechnological innovations with 52%, which is distributed in the development of new medicines, diagnostic procedures, for the administration of medicines and for the discovery of bio-pharmaceuticals.

Additionally, there are numerous patents and products derived from nanotechnology that are commercialized or that are still in the pre-commercialization phases. The degree of return on investment is still small, especially in the medium and long term, since various data indicate that, in terms of the global market, there is a business trend with nano-applications in around 15% of the companies (Cozzens et al., 2013). The development of nanotechnology has been led by highly industrialized countries, especially the United States, Japan, and the European Union, and emerging countries such as China and Korea. Such countries are considered leaders in this field for several reasons, among which are the allocation of considerable budgets by governments to strengthen research and development. However, this great investment is directed mostly to the creation and development of nano-metric materials and their applications, leaving behind the study of the consequences that this new technology can bring to the environment and living beings.

10.5 APPLICATIONS OF NANO-BIOSENSORS

Medicine, engineering, computing, mechanics, physics, and chemistry are some of the disciplines that are using the possibilities of nanotechnology. Below is a brief review of the most cited applications in the specialized literature:

1. **Nanotechnology Applied in Water:** The ability to recycle water from any source for different uses would save enormous amounts of water and allow the use of hitherto untapped water resources. Porous filters of a nano-meter scale can eliminate 100% of viruses and bacteria. An electrical separation technology that attracts ions to special sheets can remove salts and heavy metals.
2. **Nanotechnology is Applied to the Energy Sector:** It is necessary to develop new technologies that allow sustaining the growing consumption of electrical energy worldwide and, at the same time, which are friendly to the environment. In this sense, solar cells emerge as a promising device for the generation of clean energy using alternative resources.
3. **Nanotechnology Applied to Computing Devices:** Using semiconductor nanotubes, logic-operating computing circuits and transistors have been developed, increasing speed, decreasing power consumption, and increasing performance. The development of nano-transistors and nano-memories is crucial to absorb the growing and immense capacities of data processing and memory capacity demanded by multimedia developments.
4. **Nanotechnology Applied to Agriculture:** Regarding agro-industrial production, on the one hand, the nano-structuring of all kinds of agrochemicals (fertilizers, herbicides, and pesticides, among others) is experienced. On the other hand, progress is also being made in the design of functional materials for specific applications such as irrigation systems, and "smart" plastics for packaging, among others.

5. **Nanotechnology Applied to Medicine:** Among the most promising and dynamic applications of nanotechnology are those in the health sector. Research on diseases such as cancer, HIV aids, diabetes, osteoarthritis, degenerative diseases, and cardiovascular and nervous disorders, among many others, stands out (Nasrollahzadeh et al., 2019).

Some more applications of nano-biosensors in the environment are shown in Figure 10.1.

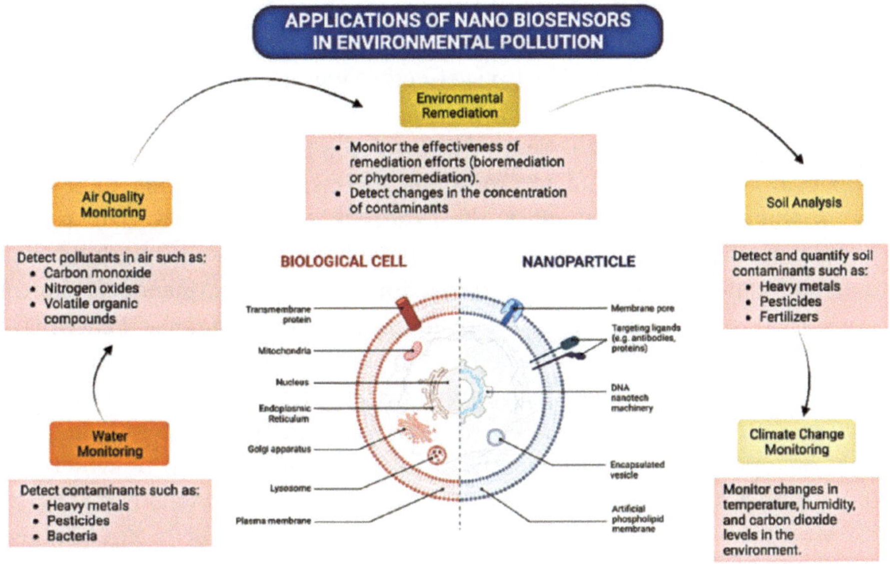

FIGURE 10.1 Representation of applications of nanotechnology in the environment, including water and air monitoring, environmental remediation, soil analysis, and climate change monitoring.

10.6 ENVIRONMENTAL PROBLEMS ASSOCIATED WITH NANOPARTICLES

Several authors maintain that the strength of nanotechnologies is based on the possibility of (i) making known products more efficient; (ii) making multifunctional products; and (iii) significantly reducing and substituting the amount of raw material in many industrial branches. But these changes do not occur without linked, collateral, and unpredictable effects. And here lie the weaknesses of nano-technologies: (a) the fact that the matter worked at the nano-scale has unknown toxicological properties; (b) there is the question of who will be affected in terms of health by these technologies; and (c) the employment situation and the social division of labor are considered, both at the regional, national, and international levels. Each type of nanoparticle must be studied regarding its possible risks, but the individual findings should not be generalized to any type of nanoparticle, since, due to its unique properties, the toxicity can be different. In this sense, Helland et al. (2008) points out that there are intrinsic properties of nanoparticles that must be considered in terms of their toxicity. One is the way that it can affect how the nanoparticle interacts with biological membranes.

The term "nano-toxicity" has been coined to refer to the adverse effects that manufactured nanoparticles have on living organisms and ecosystems (Surendhiran et al., 2020). The toxic effects of nanoparticles, in general, on the different systems at the organism level have been recently reviewed, specifically, those of gold nanoparticles, from which it is concluded that the current information is insufficient to determine their status regarding human and environmental health. Likewise, the factors that contribute to their toxicity and the mechanisms by which they act have been proposed, concluding that conventional particle toxicology procedures are a recommendable starting point to study nanoparticles. An example that warns against generalizing the results is the study of the toxicity of spheroidal gold nanoparticles in mice where it was shown that only a certain range of diameters of these nanoparticles was toxic to mice. In this study, mice were inoculated with gold nanoparticles with a diameter between 3 and 100 nm, of which only those with a diameter between 8 and 37 nm were toxic (Irvine et al., 2015).

Most man-made nanoparticles are not found naturally in the environment, so organisms may lack the necessary immune or purification systems to deal with them. It has been reported that 12 nm nanoparticles can cross the blood-brain barrier and that nanoparticles of 30 nm or less can be endocytosed by cells, however, their effects on environmental and human health have not yet been determined (Sokolova et al., 2020). This is partly since most of the studies have been carried out *in vitro*. One aspect to consider for the study of the toxicity of gold nanoparticles is their ability to form complexes with other molecules of biological interest. On the one hand, many of the applications of AuNPs require their coating with a stabilizing or recognition agent; in this case, the type of complex that they will form is known from their design (Medici et al., 2021). On the other hand, gold nanoparticles can incidentally form complexes with the components of the medium. For example, with the supplements of the culture medium in cytotoxicity tests, of the physiological medium in the case of organisms exposed to these nanoparticles or of the environment in ecosystems (Sheikhi et al., 2019). The formation of these complexes can modify the physiological properties of gold nanoparticles, such as charge, surface chemistry, and aggregation state, which affect the interaction with cells. This, in turn, increases the complexity of the system regarding the interpretation of data in toxicity tests and its extrapolation to the impact of gold (González-Vega et al., 2022).

10.7 NANOPARTICLES IN THE ENVIRONMENT

Ettlinger et al. (2022) present different methods to investigate the toxicity of these particles, *in vitro* and *in vivo* and point out the importance of considering accessory reagents that may be present as impurities in the medium where the nanoparticles are found and that may have toxic effects by themselves. They present an extensive compilation of *in vitro* and *in vivo* toxicity studies on cells and organisms, also including parameters on the gold nanoparticles used, such as size and functional groups. New methods are constantly being advanced in the study of toxicity at the cellular level. A recent study indicates that nanoparticles can be bio-accumulated and amplified in a terrestrial food chain. In their study, the researchers grew tobacco plants in a medium containing gold nanoparticle and then fed *Manduca sexta* larvae with these plants; the accumulation of these nanoparticles in the tissues of the plants and in the larvae was demonstrated. In the latter, there was also an amplification of up to an average factor of 11.6 but that varied according to the diameter of the nanoparticle (Mansuri and Kumar, 2022).

The effects of nanoparticles on the environment and selected biological matrices have been estimated and outlined in general terms, however, each ecological system of interest must be studied to know the cycles in which the present nanoparticles enter. For example, practical quantitative models have been proposed to study the behavior of nanoparticles in general in aquatic systems and their impact on aquatic organisms (Al-Thawadi, 2020). In this sense, Schwirn et al. (2020) raises key questions and proposes the sedimentation and dissolution processes as determining factors in the effective concentration of nanoparticles in the aqueous medium. They point out that the presence of biological films (biofilms) increases the removal of nanoparticles from the environment, in addition, they present a compendium of natural aquatic systems where the exposure of organisms to various types of nanoparticles has been studied. Likewise, it has been pointed out that studies should be carried out on nano-structured materials before allowing their widespread use, however, governments have been late in ordering studies on the life cycle of products and, except for those related to medical applications, there has been ignored this call for attention. Emphasizing the soil as an ecological system, the author also reviews the possible advantages of the applications of nano-structured materials in agriculture, such as the use of nano-fertilizers and the improvement of pesticides with the incorporation of metallic nanoparticles.

10.7.1 PROPOSED MEASURES

10.7.1.1 GENERAL ASPECTS

Three general questions have been suggested to answer to know the environmental effects of nanoparticles: (i) How do nanoparticles change over time once they are present in the environment? (ii) What effects could they have on organisms? and (iii) What effects could they have on ecosystems? To answer these questions, studies must be carried out on its life cycle, quantity, mobility, reactivity, bioavailability, stability, solubility, toxicity, ecotoxicity, and persistence (Malhi et al., 2020). Even clearly defining the objectives of the study in the medical, biological, or ecological fields, the panorama faced by the agencies when trying to elaborate a regulation that regulates the use of nanoparticles is intricate, since, in addition, they must consider economic implications and the needs of society.

10.7.1.2 ETHICAL ASPECTS

For its part, nano-ethics contributes with other questions that are not oriented toward the scientific field, such as those mentioned above, but towards the elaboration of public policies. These questions include: what are the potential health and safety risks with nano-technology applications? Who is responsible if something happens? What are the rights of an individual affected by a nano-technology process or product (such as nanoparticles)? How can society be protected from the risks of nanotechnology (Kah et al., 2019). Actions to answer these questions would focus on gathering information, sharing this information, educating the public, and developing laws and regulations.

10.7.1.3 NEED FOR RULES FOR THE REGULATION

The challenges faced by agencies such as the FDA for the regulation of nanomaterials have been specified by several authors. One of the main problems regarding the establishment of legal regulations regarding nanomaterials is that there is still no universally accepted definition for nanomaterial, which includes nanoparticles (Foulkes et al., 2020). Although nano-materials have been developed for a few decades, it was not until April 29, 2009, when the European parliament required a comprehensive and scientific definition of the term nano-material, as well as a correction in the legislation of the European Union due to the potential risks to human health and the environment (Gottardo et al., 2021). Until now, no government has established a regulatory framework for nanotechnology. However, political considerations have been raised that can guide the competent authorities in the development and application of nanotechnology in general and, specifically, in relation to nanoparticles. These measures include:

- Standardize the nomenclature and protocols, to be able to compare the results of the tests.
- Cooperation between the public and private sectors.
- Sensitize regulatory and environmental agencies regarding the benefits and potential risks of nanotechnology.
- Financially support research and development of nanotechnologies that contribute to sustainable development.
- Evaluate the possible environmental and human health impacts of nanoparticles, especially those that are already being mass-produced.
- Identify, evaluate, and publicize good practices, handling methods, and safety procedures related to nanomaterials.
- Apply, where possible, established policies regarding environmental and health considerations related to chemical products.
- Educate the public about the benefits and risks of nanotechnology.
- Promote cooperation between governments and intergovernmental organizations to share information regarding the impacts of nanotechnology on the environment and on human health.

The European Union took nanotechnology into account in its work program for 2013 (Rodríguez et al., 2019). It recognizes that research and innovation are key to competitiveness and social progress. In this context, it provides some guidelines on how to improve safety in relation to the production and use of nanoparticles. In the first place, it refers to the fact that risk management must be part of the culture of all those involved in the provision of the material, which would be achieved through hazard identification, knowledge of potential adverse effects, measurement, and exposure control. The lack of information about the risks associated with nano-structured materials is evident in the proposals of the work program for 2013 of the European Union referring to where research should be focused, which are oriented towards the development of prevention tools. These proposals also provide more specific guidelines to guide these investigations. Proposals include:

- Development of good practices regarding the management of nanomaterials and waste treatment.
- Development of risk mitigation methods in techniques that involve nanomaterials and that may result in exposure in the work environment or in the ecosystem.
- Development of tools to calculate exposure to nanomaterials and their effects on humans and ecosystems.
- Development of methods that allow the long-term prediction of the eco-toxicological effects of exposure to nanomaterials.
- Incorporate the knowledge generated into the design of processes and products at the nanoscale.

10.7.1.4 PRACTICAL PROPOSALS

Practical and specific measures have been proposed for the management of occupational risk in relation to manufactured nanoparticles and, specifically, in relation to laboratory personnel, where guidelines are given for the classification of "nano-" laboratories in three levels. For each level (Organization, 2020) proposes prevention measures regarding laboratory logistics, access restriction, the use of ventilation and vacuum, and personal protective equipment, among others. Although several standards have been proposed, it is not known if they are followed and if validation methods need to be developed to determine their effectiveness. Regarding the protection of the population, it has been pointed out that the Food and Drug Administration (FDA) does not have a specific regulatory framework for nano-structured products. However, this body hopes to be able to regulate them under the name of "combination products," for which a statute has been established. Soares et al. (2018) points out that the existing toxicity tests may not be sufficient to establish the risks, and that the possible regulation of a nanostructured material also depends on a requirement from the manufacturer, who must declare that nanotechnology is involved in the product.

Another advance was obtained in the discussion of "nano-reference values," at the international workshop on nano-reference values in September 2011, where exposure limits were proposed for many types of nanoparticles based on aspects such as composition and bio-persistence (Mihalache et al., 2017). Recently, a report from the European network on the health and environmental impact of nanomaterials was presented on the dangers and fate of nanomaterials in the environment, which summarizes the advances in the field, as well as the different workshops and publications regarding the subject (Linkov et al., 2009).

10.8 MICROALGAE: SUITABLE ORGANISMS FOR THE DESIGN OF NANO-SENSORS

In recent years, both molecules and cellular materials have been used for the preparation of biosensors: plant tissues; animal cells. In addition, the use of whole microorganisms such as bacteria or microalgae has been revealed as a good alternative. The recent use of microalgae whole for the development of biosensors has positioned them as a microorganism with the that more promising results are obtained in relation to both the high sensitivity and the reproducibility of the data they generate.

Microalgae are the main components of phytoplankton and the beginning of the trophic chain in aquatic ecosystems; therefore, their availability and abundance are very high, being found in all aquatic environments and in many terrestrial environments. Most species of microalgae can grow in almost any environmental condition and survive with low concentrations of nutrients and in extreme environmental conditions of pH, temperature, and salinity that could be lethal to other organisms. Therefore, it is not surprising that microalgae have been chosen as ideal candidates for the development of biosensors capable of responding to critical changes in aquatic ecosystems. It is estimated that there are about 90,000 species of microalgae (not including cyanobacteria) of which approximately half have been described, the enormous potential of microalgae for the development of biosensors is yet to be explored (Martins et al., 2022).

The functioning of microalgae biosensors is based on the inhibitory effect that some toxic substances cause in the photosynthetic activity of the whole organism. The photosynthetic activity status of green algae can be monitored by three main methods: (i) measurement of the intensity of fluorescence emitted by their chlorophyll molecules; (ii) the production of oxygen that is generated as a product of photosynthesis; and (iii) measurement of alkalization of means. Basically, the functioning of a photosynthetic organism of plankton is summarized as follows: sunlight (photosynthetically active) is absorbed through an uptake “antenna” formed by proteins associated with light-sensitive pigments. A small part of the absorbed energy is dissipated as heat, while another part is emitted as fluorescence. The most important part of the absorbed energy is used in the assimilation of CO_2 and the photolysis of water, to carry out photosynthesis, with the consequent synthesis of organic matter and the production of O_2.

$$6CO_2 + 6H_2O \rightarrow C_6H_{12}O_6 + 6O_2$$

In addition, the photosynthetic activity gives rise to the alkalinization of the culture medium because of CO_2 consumption and the consequent displacement of the oxygen dissolution equilibrium in water (equations); said alkalization is inhibited in the presence of the toxic agent.

$$CO_2\,(g) + H_2O\,(l) \rightarrow HCO_3^-\,(aq) + H_3O^+\ (2)\ 6CO_2\,(g)$$
$$+ 6H_2O\,(l) \leftarrow C_6H_{12}O_6 + 6O_2\ (3)$$

In summary, in the presence of toxic agents, the activity of microalgae is affected. The decrease in photosynthetic function can be assessed through the decrease in fluorescence emission, the decrease in oxygen production, and/or the alkalinization of the medium. Any of the parameters derived from these methods can be used, in principle, to detect the presence of a toxic agent. The most widely used method to detect photosynthetic inhibition in the presence of a toxic agent is the measurement of chlorophyll fluorescence emission in photosystem 2 (PS2).

This type of biosensor has been described, for example, to detect the presence of certain herbicides that inhibit the transport of electrons to PS2 during photosynthesis. It has been estimated that 50% of the herbicides are used to cause an effect at this level. There are many bibliographic references in relation to the measurement of photosynthetic inhibition (produced by a toxin such as an herbicide, a heavy metal, etc.) That occurs because of the decrease in photosystem 2 (PS2) activity. Microalgae from the *Chlorophyta* division such as *Chlorella vulgaris* and *Dictyosphaerium chlorelloides* have been used for the detection of both organic and inorganic toxic compounds both in water and immobilized on supports (Zhang et al., 2019).

10.8.1 IMMOBILIZATION OF MICROALGAE

One of the limiting steps in the development of microalgae biosensors is to immobilize the biomass in a compatible material, to avoid losses without altering either the stability or the activity of the cells. Most immobilization techniques rely on the use of organic materials (polyvinyl alcohols or polysulfones) which can sometimes be toxic to algae. Although other more biocompatible supports have been tested, such as calcium alginate, they suffer from a lack of instability over time, which limits their use for designs intended for long-term use. To improve the stability of the supports, different types of immobilization techniques: microencapsulation within semipermeable membranes; adsorption to cellulose derivatives; treatment in gel matrices; crosslinking in glutaraldehyde, co-crosslinking in mixed albumin-glutaraldehyde matrices, colonization on porous silicones and semipermeable membranes. However, of all the options, the one that most generally presents optimal properties are the silica-gel matrices; in addition to presenting high mechanical and chemical stability, they have the additional advantage of being a transparent material, a necessary requirement for cells to adequately develop their photosynthetic activity (Mallick, 2006).

10.8.2 A LIMITATION OF MICROALGAE NANO-SENSORS: THE PROBLEM OF SPECIFICITY

Despite the important advantages, previously described, related to the use of microalgae biosensors (speed, repeatability, accuracy, direct analysis of the concentrations without pretreatment of samples or with minimal pretreatment of the same, which makes possible the detection *in situ* of concentrations real environmental), there is an aspect that still compromises its use. Normally, biosensors have limited specificity, because a global signal is obtained for a set of substances that are present in the samples and that could eventually be affecting synergistically. In other words, the safest application is often the one that tries to determine "contamination levels" in general, or water contamination in general terms, without being able to specify the concrete measure of the concentration of a certain toxicant. This data may be sufficient in some cases, but as environmental regulations become more restrictive and demanding, the exact determination of a certain parameter may become an unavoidable requirement (Liu et al., 2020).

The specificity in microalgae biosensors can be achieved through a genetic selection procedure to obtain specific biosensors. The applicability of the biosensors thus obtained is potentially very high, since in principle it would only be restricted by the natural limits of each species. The operational principle on which the operation of highly specific microalgae biosensors is based consists of the simultaneous use of two different genotypes to detect a certain contaminant. The sensitive genotype would make it possible to obtain the necessary sensitivity against a certain toxin and the resistant genotype would be the one that would reveal the specificity. As an example, if a concentration of x mg/l of a specific toxin is present in a water sample, the biosensor would work in this way: on the one hand, the sensitive genotype would alert to the presence of a toxin (it would be affected its photosynthetic activity); for its part, the presence of the specific resistant mutant would reveal what toxin it is and, in most cases, even its concentration (Araújo et al., 2022).

10.8.3 ROLE OF ENZYMES IN NANO-BIOSENSORS

Enzymes are biological catalysts that can accelerate chemical reactions without being consumed. They are widely used in various fields, including biotechnology, medicine, and environmental science, due to their high specificity and efficiency. In the context of nano-biosensors, enzymes can be immobilized onto nanomaterials to enhance the sensitivity and selectivity of the sensor. One of the most common approaches to immobilize enzymes on nanomaterials is through physical adsorption or covalent attachment. In physical adsorption, enzymes are attached to the surface of nanomaterials through weak interactions such as hydrogen bonding, van der Waals forces, or electrostatic interactions. On the other hand, covalent attachment involves the formation of a covalent bond between the enzyme and the nano-material surface.

Enzymes can also be encapsulated within or onto nanomaterials to protect them from harsh environments and enhance their stability. For instance, silica nanoparticles have been used to encapsulate enzymes, which can protect the enzyme from degradation and maintain their activity (Rad et al., 2012). Enzymes can be used to detect a wide range of environmental pollutants, including heavy metals, pesticides, and organic compounds. For example, enzymes such as acetylcholinesterase and butyrylcholinesterase have been used to detect organophosphate pesticides through inhibition assays. The inhibition of these enzymes by organophosphate pesticides can be measured by monitoring the activity of the enzyme, which decreases with increasing pesticide concentration.

10.8.3.1 ADVANTAGES OF ENZYME-BASED NANO-BIOSENSORS

Enzyme-based nano-biosensors offer several advantages over conventional sensors. Firstly, enzymes have high specificity and sensitivity, which allows for the detection of low concentrations of pollutants. Secondly, enzymes can be easily immobilized onto nanomaterials, which provide a large surface area for interaction with the analyte. This results in a higher signal-to-noise ratio and improved sensitivity. Moreover, enzyme-based nano-biosensors can be easily regenerated, which makes them more cost-effective and environmentally friendly than conventional sensors. Additionally, the use of enzymes in nano-biosensors can reduce the need for hazardous chemicals, which makes them safer to use (Cao et al., 2011). Enzyme-based nano-biosensors have several advantages over conventional sensors:

1. **High Specificity and Sensitivity:** Enzymes have high specificity and sensitivity, which allows for the detection of low concentrations of pollutants. This is particularly important in environmental monitoring, where the detection of trace amounts of pollutants can have significant impacts on human health and the environment.
2. **Large Surface Area for Interaction:** Enzymes can be easily immobilized onto nanomaterials, which provide a large surface area for interaction with the analyte. This results in a higher signal-to-noise ratio and improved sensitivity.
3. **Easy Regeneration:** Enzyme-based nano-biosensors can be easily regenerated, which makes them more cost-effective and environmentally friendly than conventional sensors. This is particularly important in environmental monitoring, where sensors are often deployed for extended periods of time.
4. **Safer to Use:** The use of enzymes in nano-biosensors can reduce the need for hazardous chemicals, which makes them safer to use. This is particularly important in environmental monitoring, where the use of hazardous chemicals can have negative impacts on human health and the environment.
5. **Versatility:** Enzymes can be used to detect a wide range of environmental pollutants, including heavy metals, pesticides, and organic compounds. This makes enzyme-based nano-biosensors versatile and applicable to a wide range of environmental monitoring applications.

Overall, the advantages of enzyme-based nano-biosensors make them a promising tool for the detection and monitoring of environmental pollutants. Further research and development in this field will likely lead to improved sensitivity, selectivity, and stability, which will enable more accurate and reliable environmental monitoring.

10.8.3.2 LIMITATIONS AND DIFFICULTIES

Despite the advantages of enzyme-based nano-biosensors, there are several limitations and difficulties that must be addressed. One of the main challenges is the stability of enzymes, particularly in harsh environments. Enzymes can denature or lose their activity when exposed to high temperatures, extreme pH, or organic solvents. This can limit their use in certain applications. Another challenge is the selectivity of enzymes, which can be affected by interfering substances present in the sample matrix. For instance, enzymes used to detect heavy metals can be inhibited by other metal ions, which can lead to false-positive results. Therefore, it is important to carefully select the enzyme and optimize the assay conditions to ensure high selectivity (Bai, 2021). Furthermore, the immobilization of enzymes onto nanomaterials can affect their activity and stability. The choice of immobilization method, nano-material surface chemistry, and enzyme loading density can all influence the performance of the sensor. While enzyme-based nano-biosensors offer many advantages for the detection of environmental pollutants, there are also several limitations and difficulties associated with their use. These include:

1. **Enzyme Stability:** One of the primary challenges associated with the use of enzymes in nano-biosensors is their stability. Enzymes can be denatured by changes in temperature, pH, and exposure to organic solvents. These factors can reduce the sensitivity and specificity of the sensor and can lead to false readings.

2. **Limited Lifespan:** Enzymes have a limited lifespan and can degrade over time due to natural processes or exposure to harsh environmental conditions. This can reduce the accuracy and reliability of the sensor and can lead to increased maintenance costs.

3. **High Cost:** Enzymes can be expensive to produce and purify, which can increase the cost of developing and deploying nano-biosensors. This can limit the scalability of the technology and may make it difficult to deploy in resource-limited settings.

4. **Limited Detection Range:** Enzymes have specific catalytic activities, which can limit their ability to detect a wide range of

environmental pollutants. This can lead to false negatives, where pollutants go undetected by the sensor.

5. **Interference:** Enzyme-based nano-biosensors can be subject to interference from other substances in the environment, which can affect the accuracy and reliability of the sensor. For example, the presence of other enzymes or organic compounds can interfere with the detection of the target pollutant.
6. **Sample Matrix Effects:** Enzyme-based nano-biosensors can be subject to sample matrix effects, which can affect the sensitivity and specificity of the sensor. Sample matrix effects can arise due to differences in the physical and chemical properties of the sample matrix, such as pH, ionic strength, and the presence of organic compounds.
7. **Ethical Concerns:** There are ethical concerns associated with the use of enzymes in nano-biosensors. For example, the use of enzymes derived from animals may raise ethical concerns related to animal welfare, while the use of enzymes derived from genetically modified organisms (GMOs) may raise concerns related to biosafety and environmental impact.

10.9 CONCLUSION

Although the use of nanostructures is increasingly frequent in consumer products, risks that can affect the environment and human health are mostly still unknown. One important reason is the lack of information and detection techniques and identification on the number of nanomaterials in circumstances relevant to the environment. They must also discriminate between nanoparticles of an anthropogenic origin from those that come from nature. This way it will be possible to determine the influence and environmental implications of the use of nanotechnology in various sectors of human endeavor.

From the analysis of existing information up to now, nanotechnology is a great opportunity to improve production processes, making them less polluting, more sustainable, and with lower energy consumption. Nanomaterials can also help us to restore contaminated sites, eliminating toxic substances. It can help to detect through devices, based on nanotechnology, the levels of contamination and the presence of toxic substances in our bodies.

However, all these advantages can also be accompanied by the danger of a massive and indiscriminate dispersal of nanomaterials. International committees and workshops are of great help in terms of proposing lines of action that tend to minimize information gaps regarding environmental effects, however, it is difficult to reach a consensus, and it is necessary to monitor the implementation of the proposals. The proposed standards regarding exposure to nanoparticles focus preferably on the protection of laboratory personnel, workers, and users, relegating the environment to the background. Specific information has been accumulated about the toxicity in cells and animals of nanoparticles, such as gold, together with exposure limits to these and other nanoparticles. With this information, it is possible to start building a database. However, it is necessary to standardize the nomenclature, the toxicity tests, and their interpretation, considering that each type of nanoparticle must be studied without generalizing or extrapolating the results from one type to another. Overall, while enzyme-based nano-biosensors offer many advantages for the detection of environmental pollutants, there are also several limitations and difficulties associated with their use. These challenges will need to be addressed through further research and development to improve the accuracy, reliability, and scalability of the technology.

KEYWORDS

- **contamination**
- **environmental pollution**
- **enzyme biosensors**
- **nano-biosensors**
- **nanoparticles**
- **nanotechnology**
- **toxic substances**

REFERENCES

Acharya, A., & Pal, P. K. (2020). Agriculture nanotechnology: Translating research outcome to field applications by influencing environmental sustainability. *Nano Impact, 19*, 100232. https://doi.org/10.1016/j.impact.2020.100232.

Aguilar, Z. (2012). *Nanomaterials for Medical Applications*. Newnes.

Alfansuri, T., Wibowo, N. A., Istiqomah, N. I., Tumbelaka, R. M., & Suharyadi, E. (2022). Detection of green-synthesized magnetite nanoparticles using spin-valve GMR-based sensor and their potential as magnetic labels. *Journal of Magnetism and Magnetic Materials, 560*, 169645. https://doi.org/10.1016/j.jmmm.2022.169645.

Altay, V., & Ozturk, M. (2012). Land degradation and halophytic plant diversity of Milleyha Wetland Ecosystem (Samandağ-Hatay), Turkey. *Pakistan Journal of Botany, 44*, 37–50.

Al-Thawadi, S. (2020). Microplastics and nano plastics in aquatic environments: Challenges and threats to aquatic organisms. *Arabian Journal of Science and Engineering, 45*, 4419–4440. https://doi.org/10.1007/s13369-020-04523-0.

Araújo, R. G., González-González, R. B., Martinez-Ruiz, M., Coronado-Apodaca, K. G., Reyes-Pardo, H., Morreeuw, Z. P., Oyervides-Muñoz, M. A., Sosa-Hernández, J. E., Barceló, D., & Parra-Saldívar, R. (2022). Expanding the scope of nano biocatalysis and nano sensing: Applications of nanomaterial constructs. *ACS Omega, 7*, 32863–32876. https://doi.org/10.1021/acsomega.2c05283.

Awad, A. M., Jalab, R., Benamor, A., Nasser, M. S., Ba-Abbad, M. M., El-Naas, M., & Mohammad, A. W. (2020). Adsorption of organic pollutants by nanomaterial-based adsorbents: An overview. *Journal of Molecular Liquids, 301*, 112335. https://doi.org/10.1016/j.molliq.2020.112335.

Bai, Y. (2021). *Fabrication of Inkjet-Printed Enzyme-Based Biosensors Towards Point-Of-Care Applications*. The University of Western Ontario, Canada.

Banica, F. G. (2012). *Chemical Sensors and Biosensors: Fundamentals and Applications*. John Wiley & Sons.

Cao, S., Chen, J., Jin, X., Wu, W., & Zhao, Z. (2011). Enzyme-based biosensors: Synthesis and applications. *Biosensors & Nanomaterials, 2011*, 95–115.

Cozzens, S., Cortes, R., Soumonni, O., & Woodson, T. (2013). Nanotechnology and the millennium development goals: Water, energy, and agri-food. *Journal of Nanoparticle Research, 15*, 1–14. https://doi.org/10.1007/s11071-013-0857-8.

Daoudi, K., Ramachandran, K., Alawadhi, H., Boukherroub, R., Dogheche, E., El Khakani, M. A., & Gaidi, M. (2021). Ultra-sensitive and fast optical detection of the spike protein of the SARS-CoV-2 using AgNPs/SiNWs nanohybrid-based sensors. *Surface and Interface Analysis, 27*, 101454. https://doi.org/10.1016/j.surfin.2021.101454.

Darwash, M., Yousuf, A., Ozturk, M., & Bhat, R. A. (2023). *Phytosequestration–Strategies for Mitigation of Aerial Carbon Dioxide and Aquatic Nutrient Pollution*. Springer Brief.

Dutta, V., Devasia, J., Chauhan, A., Jayalakshmi, M., Vasantha, V., Jha, A., Nizam, A., Lin, K. Y. A., & Ghotekar, S. (2022). Photocatalytic nanomaterials: Applications for remediation of toxic polycyclic aromatic hydrocarbons and green management. *Chemical Engineering Journal Advances, 2022*, 100353. https://doi.org/10.1016/j.cej.2022.100353.

Ettlinger, R., Lächelt, U., Gref, R., Horcajada, P., Lammers, T., Serre, C., Couvreur, P., Morris, R. E., & Wuttke, S. (2022). Toxicity of metal–Organic framework nanoparticles: From essential analyses to potential applications. *Chemical Society Reviews, 51*, 464–484. https://doi.org/10.1039/D1CS00696B.

Fallah, Z., Zare, E. N., Khan, M. A., Iftekhar, S., Ghomi, M., Sharifi, E., Tajbakhsh, M., Nikfarjam, N., Makvandi, P., & Lichtfouse, E. (2021). Ionic liquid-based antimicrobial materials for water treatment, air filtration, food packaging, and anti-corrosion coatings. *Advances in Colloid and Interface Science, 294*, 102454. https://doi.org/10.1016/j.cis.2021.102454.

Foulkes, R., Man, E., Thind, J., Yeung, S., Joy, A., & Hoskins, C. (2020). The regulation of nanomaterials and nanomedicines for clinical application: Current and future perspectives. *Biomaterials Science, 8*, 4653–4664. https://doi.org/10.1039/D0BM00760A.

Gokaltun, A., Yarmush, M. L., Asatekin, A., & Usta, O. B. (2017). Recent advances in non-biofouling PDMS surface modification strategies applicable to microfluidic technology. *Technology, 5*, 1–12. https://doi.org/10.1142/S2339547817500013.

González-Vega, J. G., García-Ramos, J. C., Chavez-Santoscoy, R. A., Castillo-Quiñones, J. E., Arellano-Garcia, M. E., & Toledano-Magaña, Y. (2022). Lung models to evaluate silver nanoparticles' toxicity and their impact on human health. *Nanomaterials, 12*, 2316. https://doi.org/10.3390/nano12142316.

Gottardo, S., Mech, A., Drbohlavová, J., Małyska, A., Bøwadt, S., Sintes, J. R., & Rauscher, H. (2021). Towards safe and sustainable innovation in nanotechnology: State-of-play for smart nanomaterials. *Nano Impact, 21*, 100297. https://doi.org/10.1016/j.impact.2021.100297.

Haq, S., Bharose, R., Ozturk, M., Altay, V., Bhatti, A. A., Dervash, M. A., & Hakeem, K. R. (2021). Impact of treated sewage water on nutrient status of alfisols and vegetable crops. *Notulae Botanicae Horti Agrobotanici Cluj-Napoca, 49*(2), 12255. https://doi.org/10.15835/nbha49212255.

Helland, A., Scheringer, M., Siegrist, M., Kastenholz, H. G., Wiek, A., & Scholz, R. W. (2008). Risk assessment of engineered nanomaterials: A survey of industrial approaches. *Environmental Science & Technology, 42*, 640–646. https://doi.org/10.1021/es0704098.

Irvine, D. J., Hanson, M. C., Rakhra, K., & Tokatlian, T. (2015). Synthetic nanoparticles for vaccines and immunotherapy. *Chemical Reviews, 115*, 11109–11146. https://doi.org/10.1021/cr500061e.

Kah, M., Tufenkji, N., & White, J. C. (2019). Nano-enabled strategies to enhance crop nutrition and protection. *Nature Nanotechnology, 14*, 532–540. https://doi.org/10.1038/s41565-019-0437-1.

Kajitvichyanukul, P., Nguyen, V. H., Boonupara, T., Thi, L. A. P., Watcharenwong, A., Sumitsawan, S., & Udomkun, P. (2022). Challenges and effectiveness of nanotechnology-based photocatalysis for pesticides-contaminated water: A review. *Environmental Research, 212*, 113336. https://doi.org/10.1016/j.envres.2022.113336.

Kumar, A., Gupta, K., Dixit, S., Mishra, K., & Srivastava, S. (2019). A review on positive and negative impacts of nanotechnology in agriculture. *International Journal of Environmental Science and Technology, 16*, 2175–2184. https://doi.org/10.1007/s13762-019-02342-0.

Kumar, T. V., & Sundramoorthy, A. K. (2019). Electrochemical biosensor for methyl parathion based on single-walled carbon nanotube/glutaraldehyde crosslinked acetylcholinesterase-wrapped bovine serum albumin nanocomposites. *Analytica Chimica Acta, 1074*, 131–141. https://doi.org/10.1016/j.aca.2019.01.031.

Linkov, I., Steevens, J., Adlakha-Hutcheon, G., Bennett, E., Chappell, M., Colvin, V., Davis, J. M., Davis, T., Elder, A., & Foss Hansen, S. (2009). Emerging methods and tools for environmental risk assessment, decision-making, and policy for nanomaterials: Summary of NATO Advanced Research Workshop. *Journal of Nanoparticle Research, 11*, 513–527. https://doi.org/10.1007/s11071-009-9533-8.

Liu, F., Zhu, S., Li, D., Chen, G., & Ho, S. H. (2020). Detecting ferric iron by microalgal residue-derived fluorescent nanosensor with an advanced kinetic model. *Science, 23*, 101174. https://doi.org/10.1016/j.scitotenv.2020.101174.

Malhi, Y., Franklin, J., Seddon, N., Solan, M., Turner, M. G., Field, C. B., & Knowlton, N. (2020). *Climate Change and Ecosystems: Threats, Opportunities and Solutions*. The Royal Society.

Mallick, N. (2006). Immobilization of microalgae. In *Immobilization of Enzymes and Cells* (pp. 373–391).

Mansuri, A., & Kumar, A. (2022). *Natural Organic Matter: A Ubiquitous Adsorbent in Aquatic Systems to Probe Nanoparticles Behavior and Effects in Ecosystem–Environmental Toxicology and Ecosystem*. CRC Press.

Martins, R., Figueiredo, J., Sushkova, A., Wilhelm, M., Tedim, J., & Loureiro, S. (2022). "Smart" nanosensors for early detection of corrosion: Environmental behavior and effects on marine organisms. *Environmental Pollution, 302*, 118973. https://doi.org/10.1016/j.envpol.2022.118973.

Medici, S., Peana, M., Coradduzza, D., & Zoroddu, M. A. (2021). Gold nanoparticles and cancer: Detection, diagnosis, and therapy. *Seminars in Cancer Biology, 76*, 27–37.

Mihalache, R., Verbeek, J., Graczyk, H., Murashov, V., & Van Broekhuizen, P. (2017). Occupational exposure limits for manufactured nanomaterials, a systematic review. *Nanotoxicology, 11*, 7–19. https://doi.org/10.3109/17435390.2016.1188685.

Nasrollahzadeh, M., Sajadi, S. M., Sajjadi, M., & Issaabadi, Z. (2019). Applications of nanotechnology in daily life. *Interface Science and Technology, 28*, 113–143.

Ong, C. B., Ng, L. Y., & Mohammad, A. W. (2018). A review of ZnO nanoparticles as solar photocatalysts: Synthesis, mechanisms, and applications. *Renewable and Sustainable Energy Reviews, 81*, 536–551. https://doi.org/10.1016/j.rser.2017.08.073.

Osman, A. I., Elgarahy, A. M., Eltaweil, A. S., El-Monaem, A., Eman, M., El-Aqapa, H. G., Park, Y., Hwang, Y., Ayati, A., & Farghali, M. (2023). Biofuel production, hydrogen production, and water remediation by photocatalysis, biocatalysis, and electrocatalysis. *Environmental Chemistry Letters, 2023*, 1–65. https://doi.org/10.1007/s10311-023-01570-7.

Ozturk, M., Metin, M., Altay, V., Prasad, M. N. V., Gul, A., Bhat, R. A., Darvash, M. A., Hasanuzzaman, M., Nahar, K., Unal, D., Unal, B. T., Caparrós, P. G., Kawano, T., Toderich, K., & Bukhari, A. (2023a). Role of rare earth elements in plants. *Plant Molecular Biology Reporter*. https://doi.org/10.1007/s11105-023-01369-7.

Ozturk, M., Roy, A., Bhat, R. A., Vardar-Sukan, F., & Tonelli, F. M. P. (2023b). *Synthesis of Bionanomaterials for Biomedical Applications*. Elsevier.

Pal, S., Ahamed, Z., & Pal, P. (2022). Removal of antibiotics and pharmaceutically active compounds from water environment: Experiments towards industrial scale up. *Separation and Purification Technology, 2022*, 121249. https://doi.org/10.1016/j.seppur.2022.121249.

Panov, A., Larsen, R., Totah, N., Larsen, S., & Grassian, V. (2000). Photooxidation of toluene and p-xylene in cation-exchanged zeolites X, Y, ZSM-5, and Beta: The role of zeolite physicochemical properties in product yield and selectivity. *The Journal of Physical Chemistry B, 104*, 5706–5714. https://doi.org/10.1021/jp000219a.

Pasinszki, T., & Krebsz, M. (2020). Synthesis and application of zero-valent iron nanoparticles in water treatment, environmental remediation, catalysis, and their biological effects. *Nanomaterials, 10*, 917. https://doi.org/10.3390/nano10050917.

Peng, H., Feng, S., Zhang, X., Li, Y., & Zhang, X. (2012). Adsorption of norfloxacin onto titanium oxide: Effect of drug carrier and dissolved humic acid. *Science of The Total Environment, 438*, 66–71. https://doi.org/10.1016/j.scitotenv.2012.08.014.

Pramanik, P., Krishnan, P., Maity, A., Mridha, N., Mukherjee, A., & Rai, V. (2020). Application of nanotechnology in agriculture. *Environmental Nanotechnology, Monitoring & Management, 4*, 317–348. https://doi.org/10.1016/j.enmm.2020.100056.

Qian, Y., Qin, C., Chen, M., & Lin, S. (2020). Nanotechnology in soil remediation—Applications vs. implications. *Ecotoxicology and Environmental Safety, 201*, 110815. https://doi.org/10.1016/j.ecoenv.2020.110815.

Rad, A. S., Jahanshahi, M., Ardjmand, M., & Safekordi, A. A. (2012). Hydrogen peroxide biosensor based on an enzymatic modification of electrode using deposited silver nanolayer. *International Journal of Electrochemical Science, 7*, 2623–2632.

Rico-Sánchez, A. E., Rodríguez-Romero, A. J., Sedeño-Díaz, J. E., López-López, E., & Sundermann, A. (2021). Aquatic macroinvertebrate assemblages in rivers under the influence of mining activities. *Research Square*. https://doi.org/10.21203/rs.3.rs-568053/v1.

Rodríguez, H., & Eizagirre, A. (2019). Dynamics of responsible innovation constitution in European Union research policy: Tensions, possibilities and constraints. In R. von Schomberg & J. Hankins (Eds.), *International Handbook on Responsible Innovation* (pp. 283–304). Edward Elgar Publishing.

Sahoo, T., Sahu, J. R., Panda, J., Hembram, M., Sahoo, S. K., & Sahu, R. (2021). Nanotechnology: An efficient technique of contaminated water treatment. In M. Kumar, D. D. Snow, R. Honda, & S. Mukherjee (Eds.), *Contaminants in Drinking and Wastewater Sources: Challenges and Reigning Technologies* (pp. 251–270). Springer Singapore.

Saleh, T. A. (2021). Protocols for the synthesis of nanomaterials, polymers, and green materials as adsorbents for water treatment technologies. *Environmental Technology & Innovation, 24*, 101821. https://doi.org/10.1016/j.eti.2021.101821.

Schwirn, K., Voelker, D., Galert, W., Quik, J., & Tietjen, L. (2020). Environmental risk assessment of nanomaterials in the light of new obligations under the REACH regulation: Which challenges remain and how to approach them? *Integrated Environmental Assessment and Management, 16*, 706–717. https://doi.org/10.1002/ieam.4388.

Sheikhi, A., Hayashi, J., Eichenbaum, J., Gutin, M., Kuntjoro, N., Khorsandi, D., & Khademhosseini, A. (2019). Recent advances in nanoengineering cellulose for cargo delivery. *Journal of Controlled Release, 294*, 53–76. https://doi.org/10.1016/j.jconrel.2018.12.013.

Shukla, P., Chaurasia, P., Younis, K., Qadri, O. S., Faridi, S. A., & Srivastava, G. (2019). Nanotechnology in sustainable agriculture: Studies from seed priming to post-harvest management. *Nanotechnology and Environmental Engineering, 4*, 1–15. https://doi.org/10.1007/s40740-019-0118-5.

Singh, Y. (2006). *Environmental Science*. New Age International.

Soares, S., Sousa, J., Pais, A., & Vitorino, C. (2018). Nanomedicine: Principles, properties, and regulatory issues. *Frontiers in Chemistry, 6*, 360. https://doi.org/10.3389/fchem.2018.00360.

Socolof, M. L., Overly, J. G., & Geibig, J. R. (2005). Environmental life-cycle impacts of CRT and LCD desktop computer displays. *Journal of Cleaner Production, 13*, 1281–1294. https://doi.org/10.1016/j.jclepro.2004.08.015.

Sokolova, V., Mekky, G., Van Der Meer, S. B., Seeds, M. C., Atala, A. J., & Epple, M. (2020). Transport of ultrasmall gold nanoparticles (2 nm) across the blood-brain barrier in a six-cell brain spheroid model. *Scientific Reports, 10*, 18033. https://doi.org/10.1038/s41598-020-74996-2.

Surendhiran, D., Cui, H., & Lin, L. (2020). Mode of transfer, toxicity, and negative impacts of engineered nanoparticles on environment, human, and animal health. In *The ELSI Handbook of Nanotechnology: Risk, Safety, ELSI and Commercialization* (pp. 165–204).

Svendsen, C., Walker, L. A., Matzke, M., Lahive, E., Harrison, S., Crossley, A., Park, B., Lofts, S., Lynch, I., & Vázquez-Campos, S. (2020). Key principles and operational practices for improved nanotechnology environmental exposure assessment. *Nature Nanotechnology, 15*, 731–742. https://doi.org/10.1038/s41565-020-0686-2.

Talin, A., Dean, K., & Jaskie, J. (2001). Field emission displays: A critical review. *Solid-State Electronics, 45*, 963–976. https://doi.org/10.1016/S0038-1101(01)00159-2.

Toumey, C. P. (2008). Reading Feynman into nanotechnology: A text for a new science. *Techné: Research in Philosophy and Technology, 12*, 133–168. https://doi.org/10.5840/techne20081213.

Usman, M., Farooq, M., Wakeel, A., Nawaz, A., Cheema, S. A., Rehman, H., Ashraf, I., & Sanaullah, M. (2020). Nanotechnology in agriculture: Current status, challenges, and future opportunities. *Science of The Total Environment, 721*, 137778. https://doi.org/10.1016/j.scitotenv.2020.137778.

Vázquez-Núñez, E., Molina-Guerrero, C. E., Peña-Castro, J. M., Fernández-Luqueño, F., & De La Rosa-Álvarez, M. G. (2020). Use of nanotechnology for the bioremediation of contaminants: A review. *Processes, 8*, 826. https://doi.org/10.3390/pr8060826.

Wadhawan, S., Jain, A., Nayyar, J., & Mehta, S. K. (2020). Role of nanomaterials as adsorbents in heavy metal ion removal from wastewater: A review. *Journal of Water Process Engineering, 33*, 101038. https://doi.org/10.1016/j.jwpe.2020.101038.

Williams, E. (2003). Environmental impacts in the production of personal computers. In R. Kuehr & E. Williams (Eds.), *Computers and the Environment: Understanding and Managing Their Impacts* (pp. 41–72). Springer Netherlands.

World Health Organization. (2020). *Rational Use of Personal Protective Equipment for Coronavirus Disease (COVID-19) and Considerations During Severe Shortages: Interim Guidance, 6 April 2020*. World Health Organization.

Wu, S., Weng, Z., Liu, X., Yeung, K., & Chu, P. K. (2014). Functionalized TiO_2-based nanomaterials for biomedical applications. *Advanced Functional Materials, 24*, 5464–5481. https://doi.org/10.1002/adfm.201302679.

Xu, H. B., Ye, R. F., Yang, S. Y., Li, R., & Yang, X. (2014). Electrochemical DNA nanobiosensor for the detection of genotoxins in water samples. *Chinese Chemical Letters, 25*, 29–34. https://doi.org/10.1016/j.cclet.2013.10.013.

Xu, P., Zeng, G. M., Huang, D. L., Feng, C. L., Hu, S., Zhao, M. H., Lai, C., Wei, Z., Huang, C., & Xie, G. X. (2012). Use of iron oxide nanomaterials in wastewater treatment: A review. *Science of The Total Environment, 424*, 1–10. https://doi.org/10.1016/j.scitotenv.2012.02.009.

Yadav, A. N. (2021). Nanotechnology for agro-environmental sustainability. *Journal of Applied Biology & Biotechnology, 9*, 1–6. https://doi.org/10.7324/JABB.2021.90901.

Yalcin, I. E., & Altay, V. (2023). Investigation of water-soil-plant relationships based on hazardous and macro-micro element concentrations on Orontes River, Türkiye. *International Journal of Phytoremediation*. https://doi.org/10.1080/15226514.2023.2202241.

Yang, K., Zhu, L., & Xin, G. B. (2006). Adsorption of polycyclic aromatic hydrocarbons by carbon nanomaterials. *Environmental Science & Technology, 40*, 1855–1861. https://doi.org/10.1021/es052384b.

Zamora-Ledezma, C., Negrete-Bolagay, D., Figueroa, F., Zamora-Ledezma, E., Ni, M., Alexis, F., & Guerrero, V. H. (2021). Heavy metal water pollution: A fresh look about hazards, novel and conventional remediation methods. *Environmental Technology & Innovation, 22*, 101504. https://doi.org/10.1016/j.eti.2021.101504.

Zhang, W., Zhu, Z., & Cheng, C. Y. (2011). A literature review of titanium metallurgical processes. *Hydrometallurgy, 108*, 177–188. https://doi.org/10.1016/j.hydromet.2011.07.003.

Zhang, Z., Yan, K., Zhang, L., Wang, Q., Guo, R., Yan, Z., & Chen, J. (2019). A novel cadmium-containing wastewater treatment method: Bio-immobilization by microalgae cell and their mechanism. *Journal of Hazardous Materials, 374*, 420–427. https://doi.org/10.1016/j.jhazmat.2018.05.019.

CHAPTER 11

Role of Medicinal Plants in Drug Delivery Systems in Sustainable System

GHLOMREZA ABDI,[1,*] MARJAN ASSEFI,[2] SOHILA NANKALI,[3] NADEEM KIZILBASH,[4] JAVID SHAGHAGHI,[3] MARYAM ABBASI TARIGHAT,[5] and AB WAHEEDWANI[6]

[1]*Department of Biotechnology, Persian Gulf Research Institute, Persian Gulf University, Bushehr, Iran*

[2]*University of North Carolina, Greensboro, USA*

[3]*North Central University, Minneapolis, USA*

[4]*Department of Chemistry, Faculty of Physical Science, Quaid-i-Azam University, Islamabad, Pakistan*

[5]*Faculty of Nano- and Bio-Science and Technology, Persian Gulf University, Bushehr, Iran*

[6]*Department of Horticulture, School of Agriculture, Lovely Professional University, Jalandhar, Punjab, India*

*Corresponding author.

ABSTRACT

There are many difficulties with natural medications that should be defeated like the trouble of leading clinical examination in homegrown drugs, improvement of basic bioassays for organic normalization, pharmacological and toxicological assessment techniques' turn of events, examination of their locales of retention, poisonous home-grown drugs being used, finding different creature models for harmfulness and wellbeing assessment, lawful and administrative parts of natural medications, and so forth. We researched the role of medicinal plants in drug delivery.

Agricultural Biotechnology: Issues, Challenges, and Recent Developments. Mohd. Sayeed Akhtar and Khalid Rehman Hakeem (Eds.)
© 2025 Apple Academic Press, Inc. Co-published with CRC Press (Taylor & Francis)

11.1 INTRODUCTION

From the backdrop of development home grown medicines were utilized to fix human ailments in each conceivable condition. In present-day times we have the choice to utilize them over manufactured particles since natural medications have fewer side effects (Alexis et al., 2008; Atmakuri and Dathi, 2010). From the writing, it is exceptionally certain that home grown drugs show their pharmacological activity either due to explicit constituents or because of mix of constituents (Alexis et al., 2008). However, how much constituents' changes clump to the group due to biological elements, book of an assortment of plant. The pharmacological impact of the medication can be achieved just when its fixation ranges within the helpful reach. Any change above or beneath the restorative focus led to either poisonous impacts or no reaction. So, the titration of the portion as well as the assurance of the portion is necessary. To conquer such cases, and to upgrade the viability of natural medication novel drug delivery framework (NDDS) assumes a significant part, which is a remarkable mix of different parts of science like polymer innovation, pharmaceutics, immunology, sub-atomic biology, and so forth. The professionals of the field to comprehend the issues related to the utilization of homegrown items (e.g., poor solidness in gastric climate, high degree of first-past digestion, and so forth) which makes an obstruction in their successive utilization of engineered particles by the utilization of nano-innovation, where vesicular frameworks help in the designated conveyance of the ideal constituents (Terreno et al., 2008). Rate and degree are the two most significant boundaries among the medication conveyance in addition to assuming that they are enhanced (Chanchal and Swarnlata, 2008) with situated drug conveyance, the viability of the treatment expanded a great deal. Among different possibilities of homegrown, NDDS the vital ones are:

- They can upgrade the solvency of the constituents;
- They can limit the related poisonous impacts;
- Improvement in pharmacological activities.

11.2 LIPOSOMES

Liposomes are circular, colloidal bilayer vesicular frameworks comprised of biodegradable material known as phospholipids (Terreno et al., 2008). The phospholipids are the super primary units of these vesicles and are amphipathic particles shaving bipolarity in their construction. Due to this, upon hydration with fluid media, they embrace a trademark circular shape and build a fluid center within them. The polar head of the phospholipid is situated towards the watery media while the hydrophobic tail is located inside the layer. Therefore, liposomal vesicles have both hydrophilic as well as hydrophobic locale. This property created a flexible transporter framework as they can ensnare both hydrophilic as well as hydrophobic medication within it (Saraf, 2010; Wen et al., 2010). Lipophilic medication possesses the lipoidal space within the bilayered layer while hydrophilic medication involves the watery area for example watery center. The normal size of liposomes differs from 0.05–5.0 μ. Among the different benefits related to liposomes are as follows (He et al., 2008).

- Solubility improvement;
- Enhancement of bioavailability;
- Programmed focusing;
- Prolongation of the span of activity;
- Improve the tissue macrophage take-up of the captured constituents.

11.3 NANOPARTICLES

They are the nanosized particulates (10–100 nm) which can be nanospheres, nano-cases, or Solid Lipid Nanoparticles (SLNs). Nanospheres are lattice-based frameworks in which the medication is consistently scattered in the transporter framework (Hong et al., 2008). The framework can be either manufactured or regular. Among engineered commonly biodegradable polymers are utilized which is generally regarded as safe (GRAS) by FDA for human utilization. The models incorporate polylactic corrosive, poly-α-cyanoacrylate alkyl esters, polyvinyl alcohols, glycolic corrosive polymers, and so forth (Li et al., 2009). Among normal polymers that are used as lattices in nanoparticulates are two significant classes; (i) polysaccharides (chitosan, cellulose, and its subordinates, dextran, alginate, and so on); and (ii) Proteins (gelatin, egg whites, what is more, a few sorts of proteins from the vegetative beginning; Chen, 2008).

In the event of nano-circles, the medication discharge is constrained by two boundaries to be specific, disintegration and dissemination. Their delivery can show a burst discharge instrument as well as a surface disintegration system (Gortzi et al., 2008). Nano-cases in opposition to nano-circles are supply-type frameworks in which the medication center is encircled by a polymeric layer, while SLNs are explicitly intended to give modified conveyance of lipoidal drugs within the body (Chen et al., 2010). They are bilayered frameworks intended for the conveyance of those medications from which appropriate helpful results cannot be filtered in view of both their high sub-atomic weight and their polar nature (Rane and Prabhakar, 2009).

Phytosome helps in defeating these impediments in drug conveyance in order to improve the bioavailability of such synthetic substances. They include restricting medication particles to phospholipids (phosphatidyl choline) in a stoichiometric proportion (1:1 or 1:2) so as to structure lipid complexes. As they include the bond arrangement, they are more stable as for liposomes where there is just an entanglement of these constituents in the watery space. They can entangle higher measures of medication, overcome strength-related issues, and improve cutaneous assimilation (Rong and Juqun, 2007).

11.4 MICROSPHERES

These are circular framework-based frameworks changing 1–300 μ in size, in which medication is consistently scattered in the polymeric network (Priprem et al., 2008). Different strategies that can be embraced in the union of microspheres are single emulsion procedure, two-fold emulsion method, polymerization strategies (typical as well as interfacial), shower drying also, shower coagulating, stage detachment coacervation strategy, and dissolvable extraction technique. First arranged discharge energy is for the most part continued in such frameworks where the delivery rate restricting advances are dissemination and dissolution (Juqun and Rong, 2007). The external disintegration media, first and foremost, will diffuse the grid and make the captured medication solubilize in it and then the medication is set free from the framework this is one kind of component in another sort of the framework comprising polymer shows surface disintegration conduct where the surface dissolve layer by layer and the arrival of the drug occurs. There fundamental elements are there which impact the delivered sum as well as its rate (Youfang et al., 2009). They are as follows:

1. **Size:** The smaller the size more the surface region, the lesser the way length to diffuse or the fewer layers expected to dissolve for drug discharge.
2. **Type of Lattice:** It relies upon the manner by which the grid shows its delivery.

3. **Polymer Focus:** It is conversely relative to how much medication is delivered.

11.5 TRANSFEROSOMES AND ETHOSOMES

Transferosomes and Ectosomes are phospholipid vesicles planned to oversee the medication by means of a transdermal course (Chang et al., 2011). Both have a typical reasoning of improving the infiltration through layer corneum boundary however the method of activity is different. Transferosomes do as such by using the hydration and osmotic tension of the skin while in the case of ectosomes, they have a high happy of ethanol (20–45%). Ethanol being a synthetic saturation enhancer upsets the film obstruction and thereby improves the dissolvability. Besides, it makes the vesicle adaptable without modifying the security part. Transferosomes are utilized to convey the natural constituents in the upper layers of skin while for more profound layers and fundamental conveyance, ectosomes are the better other option (Liu et al., 2008). Typically, they are given as cream or gel, being harmless they have better tolerant compliance. There are huge models where the adequacy of homegrown constituent intervened treatment is expanded by taking on these conveyance frameworks, for example, colchicine, curcumin, and matrine. They incorporate transdermal medication conveyance with the guide of patches embed-based drug conveyance and miniature pellets. Reports demonstrate the way that they can likewise be a decent method for drug conveyance of home-grown constituents to work on the viability of the treatment. Transdermal medication conveyance framework is a painless method for drug conveyance which can be of either solid sort or repository type. They should deliver drugs at a foreordained rate over the site of use (Min et al., 2008). Among their benefits critical are their capacity to control the delivery, plausibility of withdrawal of treatment whenever required, convenience, and capacity to drag out the term of action. They can even consolidate the vesicular framework within them. Numerous other novel medication conveyance frameworks can be used to improve the viability of home-grown medicines.

Sublingual dissolving tablets can be utilized for the organization of phytoconstituents for the speedy beginning of activity, since sublingual mucosa is wealthy in blood supply, the drug straightforwardly sidesteps the primary past digestion which is the fundamental issue related to the home-grown drugs. Mucoadhesive medication conveyance framework can likewise be used to improve the viability of the treatment, the reason is that whether the medication conveyance is a unit dose structure or multi-particulate framework it makes the measurements structure find itself around the assimilation window of the medication particle which might prompt the improvement of bioavailability. They do not have issues connected with oxidation, and so on as related to liposomes since liposomes contain lipids which contain twofold limits in their design (Bisht et al., 2007). They are inclined to free extreme chain-intervened oxidation responses. Previously mentioned models are a couple of different other kinds of NDDS that are as yet forthcoming to be used for the conveyance of home-grown atoms. Focusing on this is another space that can be used to expand the restorative viability of the conveyance framework.

The eventual fate of medication was established previously, before physicists embraced combining engineered silver slugs for all those diseases, and before drug organizations hitched our aggregate well-being to what has become a multibillion-dollar cart (Leonard et al., 2010). Previously, practically every one of the prescriptions was from the plants; the plant being man's just physicist for a very long time (Yen et al., 2008). Spices are organizing a rebound, homegrown "renaissance" is going on all around the globe and that is just the beginning, and more individuals are observing natural treatments to treat different sorts of illnesses instead of standard medication. There are three primary explanations behind the prominence of home-grown medications (Wang et al., 2010).

- There is a developing worry over the dependence and well-being of medications and medical procedures.
- Present-day medication is neglecting to treat a considerable lot of the most widely recognized ailments.
- Numerous regular measures are being displayed to deliver improved results than medications or medical procedures without incidental effects.

The methodology by which a medication is delivered can significantly impact its efficacy. Certain medications possess an optimal concentration range within which maximum advantage is derived and focus above or underneath this reach can be harmful or produce no remedial advantage by any stretch of the imagination (Trickler et al., 2008). Once more, the remarkably slow progress in the effectiveness of treating severe diseases has indicated an increasing necessity for a multidisciplinary approach in delivering therapeutics to specific areas within tissues. From this, revolutionary concepts regarding the management of pharmacokinetics, pharmacodynamics, elusive toxicity, immunogenicity, bio-recognition, and efficacy of pharmaceuticals were formulated. These novel methodologies, commonly referred to as drug delivery systems (DDS), are founded upon interdisciplinary approaches that integrate the fields of polymer science, pharmaceutics, bioconjugate science, and molecular science (Hou and Zhou, 2008).

A novel pharmaceutical delivery system is an innovative approach to drug delivery that addresses the limitations of traditional pharmaceutical delivery systems. Current medication addresses a specific ailment by targeting the affected area within a patient's body and transporting the medication to that particular site. Drug delivery system refers to the process through which an optimal dosage of the concerned medication is administered to the patient, ensuring it reaches the precise "site of action" and initiates its therapeutic effect without any delay. Emerging medication delivery systems aim to overcome all the obstacles associated with conventional DDS. There exist various approaches through which novel drug delivery can be achieved (Dong et al., 2009; Tzu et al., 2008). Various drug delivery and drug targeting systems are currently being developed to minimize drug degradation and loss, prevent adverse side effects, and enhance drug bioavailability and the concentration of the drug accumulated in the desired region. Among the various drug transporters that can be identified are soluble polymers, micro-particles composed of insoluble or biodegradable regular and synthetic polymers, microcapsules, cells, cell phantoms, lipoproteins, liposomes, and micelles. These transporters can be designed to degrade gradually, respond to stimuli (such as pH or temperature), and can even be specifically targeted (e.g., by forming them with explicit antibodies against specific trademark parts of the area of interest; Fu et al., 2008; Hu and Jiang, 2012; Shen et al., 2008). Two significant approaches can be identified for addressing the optimal sites for drug releases (i) uninvolved; and (ii) dynamic focusing. An example of latent focusing is the particular collection of chemotherapeutic specialists in strong growths because of the enhanced permeability of cancerous tissues in comparison to healthy tissues, the contrast between the two becomes apparent. A methodology that could potentially enable dynamic targeting involves the surface functionalization of drug carriers with ligands that are specifically recognized by receptors on the outer membrane of the desired cells. Given the highly specific nature of ligand-receptor interactions, this approach could facilitate a more precise localization of the desired site (Borodina et al., 2008; Ma et al., 2000; Tanwar et al., 2006).

Controlled drug discharge and ensuing biodegradation are significant for creating effective details. Potential delivery systems include: (i) desorption of surface-bound/adsorbed drugs; (ii) dispersion through the transporter lattice; (iii) dissemination (on account of nano-capsules) through the transporter wall; (iv) transporter network disintegration, and (v) a joined disintegration/dispersion process. The method of conveyance can be the contrast between a medication's prosperity and disappointment, as the decision of a medication is in many cases impacted by how the medication is administered (Wen et al., 2010). Sustained (or nonstop) arrival of a medication includes polymers that discharge the medication in a controlled manner. This controlled release occurs either through diffusion from the polymer or through degradation of the polymer over time. Pulsatile discharge is much of the time the favored technique for drug conveyance, as it intently emulates the way by which the body normally delivers chemicals like insulin (Shen et al., 2008).

Natural medicines have been generally utilized all around the world since old times and have been acknowledged by medical practitioners and individuals for their better restorative worth as they have less antagonistic impacts as contrasted and present-day prescriptions. The medications of Ayurvedic beginning can be used in a superior structure with improved viability by consolidating in current dose structures. Notwithstanding, phytotherapeutics need a logical way to deal with conveying the parts in a clever way to increment patient consistency and keep away from the rehashed organization. This can be accomplished by planning novel medication conveyance frameworks for natural constituents. Novel medication conveyance frameworks not only lessen the rehashed organization to defeat resistance but additionally help to expand the restorative worth by diminishing harmfulness and expanding the bioavailability, and so forth. In recent times, there has been a shift in focus among pharmaceutical researchers toward the development of a drug delivery system for herbal medicines, utilizing a scientific approach (Zhao et al., 2002).

Our nation has a tremendous information base of Ayurveda whose potential is just being acknowledged as of late. Be that as it may, the medication conveyance framework utilized for controlling the medication to the patient is conventional and obsolete, bringing about diminished viability of the medication. On account of natural concentrates, there is an extraordinary chance that many mixtures will be obliterated in the exceptionally acidic pH of the stomach. Different parts may be utilized by the liver before arriving at the blood. Subsequently, the necessary measure of the medication may not arrive in the blood. In the event that the medication does not arrive at the blood level, which is known as the "base compelling level" then there will be no remedial impact.

Phytopharmaceuticals are drugs utilizing customary mixtures obtained from botanicals rather than synthetic substances. Regular fixings are all the more effectively and all the more promptly processed by the body. Subsequently, they produce less, if any, aftereffects and give expanded retention in the circulatory system bringing about additional exhaustive and powerful medicines. Drugs produced using synthetic mixtures are inclined to unfavorable incidental effects. The human body will tend to dismiss specific synthetic mixtures which does not happen normally. These dismissals happen as incidental effects; some as gentle as minor cerebral pains, and others as serious as to be possibly deadly. It means a lot to note while phytopharmaceuticals produce less to no side outcomes, compound cooperations with other professionally prescribed medications can happen. Moreover, as they are single and decontaminated compounds, they can be effectively normalized making it simpler to consolidate them in current medication conveyance frameworks contrasted with spices.

Lipid-based drug conveyance frameworks have been explored in different examinations and have shown their true capacity in controlled and designated drug conveyance. Pharmacosomes are amphophilic phospholipid edifices of medications bearing dynamic hydrogen that tight spot to phospholipids. They confer better biopharmaceutical properties to the medication, bringing about superior bioavailability. Phytosomes are novel mixtures containing lipophilic buildings of parts of plants beginning with Silybum Marianum, Ginkgo Biloba, ginseng, and so forth, with phospholipid (Chen, 2008). They are additionally called Phyto lipids conveyance frameworks. They have high lipophilicity and further developed bioavailability and restorative properties. These are progressed types of natural concentrate that have improved pharmacokinetic and pharmacological boundaries, whose results can profitably be utilized in the treatment of intense liver illnesses, either metabolic or infective beginning. Phytosomes are created by a patent cycle in which individual parts of homegrown removes like flavonolignans and terpenoids are bound on a sub-atomic level to the phospholipids like phosphatidylcholine through a polar end. Phytosomes are utilized as a medicament and have a wide extension in cosmetology. Numerous areas of pyrosomes are to be uncovered later on in the possibility of drug application. Phytosomes structure an extension between the traditional conveyance framework and the clever conveyance system (Gortzi et al., 2008). If the actual spices or the cleansed phytopharmaceuticals or pyrosomes are consolidated in clever medication conveyance frameworks, we can get the advantages of both. In this way, it is essential to consolidate the clever medication conveyance framework in Indian Ayurvedic drugs to battle serious sicknesses.

As home-grown novel medication conveyance frameworks have a parcel of potential, a few specialists are pursuing creating novel medication conveyance frameworks like mouth dissolving tablets, supported and broadened discharge definitions, mucoadhesive frameworks, transdermal dose structures, microparticles, microcapsules, nanoparticles, inserts and so forth of spices. Some of them are at the research facility stage and some have reached the stamp. A portion of the examination work done in this space is summed up below. Asoka Life Science Limited sent off Res-Q, the world's most memorable poly-natural mouth-dissolving tablet, a quick mouth-dissolving drug. It has a clever medication conveyance framework that bestows expanded viability. In the Ayurvedic medication section, this is the principal endeavor to make meds more viable in overseeing constant afflictions. Res-Q is a poly-natural medication exceptionally compelling for lung issues and other respiratory illnesses like asthma. This extraordinary mouth-dissolving drug conveyance framework guarantees that the medication arrives at the blood straightforwardly and that first-pass digestion is avoided. It breaks up in the mouth by blending in with the spit and gets assimilated. This Res-Q gives alleviation from respiratory misery in 15 minutes or less. Along these lines, this medication looks like the viability of Sorbit rate, a progressive mouth-dissolving drug utilized in heart trouble.

A patent portrays an orally administrable definition for the controlled delivery or stable stockpiling of a granulated spice, containing a granulated spice and a transporter, the planned arrival of 75% of the dynamic fixings somewhere in the range of 4 and 18 hr after organization. The active components are selected from the assemblage consisting of hypericin, hyperforin, and echinacosides. The innovation looks to give work on natural arrangements, whose arrangements offer a helpful oral measurement type of spices for providing ideal plasma centralizations of the organically dynamic mixtures that work with client consistency. The oral-controlled and stable delivery measurements type of granulated spice is in either framework details like lattice tablets or in multi-particulate definitions like microcapsules put into two-piece cases that are finished to get a medication conveyance framework, which will guarantee a consistent stock of the dynamic parts for a supported period. Herbal drugs have been broadly utilized all around the world since old times and have been perceived by doctors and patients for their better remedial worth as they have less unfavorable impacts as contrasted and current meds. The medications of Ayurvedic beginning can be used in a superior structure with improved viability by consolidating in present-day dose structures. In any case, phytotherapeutics needs a logical way to deal with conveying the parts in an original way to increment patient consistency and keep away from the rehashed organization. This can be accomplished by planning novel medication conveyance frameworks for homegrown constituents.

11.6 CONCLUSION

Novel pharmaceutical delivery systems serve to diminish the repetitive administration required to surmount noncompliance, while simultaneously contributing to the enhancement of therapeutic value through the alleviation of toxicity and augmentation of bioavailability, among numerous other advantages. In recent times, pharmaceutical scientists have redirected their attention towards devising a drug delivery system for herbal medicines by means of a scientific methodology. The astute examination can augment not only the capture but also the sustenance in the market. However, there exist numerous challenges associated with herbal medicines that need to be overcome, such as the difficulty in conducting clinical research on herbal drugs, the development of standardized bioassays for biological evaluation, the advancement of pharmacological and toxicological assessment methods, the examination of their sites of absorption, the utilization of toxic herbal drugs, the exploration of substitute animal models for the purpose of evaluating toxicity and ensuring safety, along with the examination of the legal and regulatory components about herbal medicines.

KEYWORDS

- **bioavailability**
- **drug delivery**
- **herbal medicines**
- **liposomes**
- **microspheres**
- **nanoparticles**
- **pharmacology**

REFERENCES

Alexis, F., Basto, P., Levy-Nissembaum, E., Radovic-Moreno, A. F., Zhang, L., Pridgen, E., Wang, A. Z., Marein, S. L., Westerhof, K., Morner, L. K., & Farokhzad, O. C. (2008). HER-2-targeted nanoparticle affibody bioconjugates for cancer therapy. *ChemMedChem, 3*(12), 1839–1843. https://doi.org/10.1002/cmdc.200800253.

Atmakuri, L. R., & Dathi, S. (2010). Current trends in herbal medicines. *Journal of Pharmacy Research, 3*, 109–113. https://doi.org/10.1016/j.jpr.2010.02.005.

Bisht, S., Feldmann, G., Soni, S., Ravi, R., Karikar, C., & Maitra, A. (2007). Polymeric nanoparticle encapsulated curcumin: A novel strategy for human cancer therapy. *Journal of Nanobiotechnology, 5*, 3. https://doi.org/10.1186/1477-3155-5-3.

Borodina, T. N., Rumsh, L. D., Kunizhev, S. M., Sukhorukov, G. B., Vorozhtsov, G. N., Feldman, B. M., Rusanova, A. V., Vasil'eva, T. V., Strukova, S. M., & Markvicheva, E. A. (2008). Entrapment of herbal extracts into biodegradable microcapsules. *Biochemistry (Moscow) Supplement Series B: Biomedical Chemistry, 2*, 176–182. https://doi.org/10.1134/S1990750808010089.

Chanchal, D., & Swarnlata, S. (2008). Novel approaches in herbal cosmetics. *Journal of Cosmetic Dermatology, 7*, 89–95. https://doi.org/10.1111/j.1473-2165.2008.00372.x.

Chang, C. H., Huang, W. Y., Lai, C. H., Hsu, Y. M., Yao, Y. H., Chen, T. Y., Wu, J. Y., Peng, S. F., & Lin, Y. S. (2011). Development of novel nanoparticles shelled with heparin for berberine delivery to treat *Helicobacter pylori*. *Acta Biomaterialia, 7*(2), 593–603. https://doi.org/10.1016/j.actbio.2010.08.017.

Chen, C. (2008). Inhibiting the vascular smooth muscle cell proliferation by EPC and DPPC liposome encapsulated magnalol. *Journal of Chinese Institute of Chemical Engineers, 39*, 407–411. https://doi.org/10.1016/j.jcich.2008.07.002.

Chen, J., Chen, Z., & Wang, W. (2010). Ammonium sulfate gradient loading of brucine into liposome: Effect of phospholipid concentration on entrapment efficiency and physicochemical properties *in vitro*. *Drug Development and Industrial Pharmacy, 36*, 245–253. https://doi.org/10.3109/03639040903291525.

Dong, X., Mattingly, C. A., Tseng, M. T., Cho, M. J., Liu, Y., Adams, V. R., & Mumper, R. J. (2009). Doxorubicin and paclitaxel-loaded lipid-based nanoparticles overcome multidrug resistance by inhibiting P-glycoprotein and depleting ATP. *Cancer Research, 69*, 3918–3926. https://doi.org/10.1158/0008-5472.CAN-08-4367.

Fu, Z. Y., Zhang, J. Y., Wang, W. M., & Wang, H. (2008). Microencapsulation of Radix Salvia Miltiorrhiza nanoparticles by spray drying. *Powder Technology, 184*, 114–121. https://doi.org/10.1016/j.powtec.2007.07.010.

Gortzi, O., Lalas, S., & Chinou, I. (2008). Reevaluation of bioactivity and antioxidant activity of *Myrtus communis* extract before and after encapsulation in liposome. *European Food Research and Technology, 226*, 583–590. https://doi.org/10.1007/s00217-007-0726-2.

He, Z. F., Liu, D. Y., Zeng, S., & Ye, J. T. (2008). Study on preparation of ampelopsin liposomes. *Journal of Chinese Medicinal Materials, 33*, 27–30. https://doi.org/10.3321/j.issn:1001-4454.2008.01.010.

Hong, W., Chen, D. W., Zhao, X. L., Qiao, M. X., & Hu, H. Y. (2008). Preparation and study *in vitro* of long-circulating nanoliposomes of curcumin. *Zhongguo Zhong Yao Za Zhi, 33*, 889–892. https://doi.org/10.13860/j.issn.1001-5302.2008.11.013.

Hou, J., & Zhou, S. W. (2008). Formulation and preparation of glycyrrhizic acid solid lipid nanoparticles. *ACTA Academiae Medicinae Militaris, 30*, 1043–1045. https://doi.org/10.3969/j.issn.0379-2861.2008.11.008.

Hu, T., & Jiang, J. G. (2012). Application of nanotechnology in traditional Chinese medicine. *Current Nanoscience, 8*, 474–484. https://doi.org/10.2174/157341312800008469.

Juqun, X., & Rong, G. (2007). Interactions between flavonoids and hemoglobin in lecithin liposomes. *International Journal of Biological Macromolecules, 40*, 305–311. https://doi.org/10.1016/j.ijbiomac.2007.06.005.

Leonard, K., Ahmad, B., Okamura, H., & Kurawaki, J. (2010). *In situ* green synthesis of biocompatible ginseng-capped gold nanoparticles with remarkable stability. *Colloids and Surfaces B: Biointerfaces, 80*, 213–218. https://doi.org/10.1016/j.colsurfb.2010.05.007.

Li, D. C., Zhong, X. K., Zeng, Z. P., Jiang, J. G., Li, L., Zhao, M. M., et al. (2009). Application of targeted drug delivery system in Chinese medicine. *Journal of Controlled Release, 138*, 103–112. https://doi.org/10.1016/j.jconrel.2009.05.008.

Liu, M., Li, H., Luo, G., Liu, Q., & Wang, Y. (2008). Pharmacokinetics and biodistribution of surface modification polymeric nanoparticles. *Archives of Pharmacal Research, 31*(4), 547–554. https://doi.org/10.1007/s12272-008-1044-2.

Ma, B., Duan, X., & Wang, Z. (2000). Clinical and experimental study on Shuanghua aerosol in treating infantile upper respiratory tract infection. *Chinese Journal of Integrated Traditional and Western Medicine, 20*, 653–655. https://doi.org/10.1007/s11605-009-1042-1.

Min, K. H., Park, K., Kim, Y. S., Bae, S. M., Lee, S., Jo, H. G., Park, R. W., Kim, I. S., Jeong, S. Y., Kim, K., & Kwon, I. C. (2008). Hydrophobically modified glycol chitosan nanoparticles-encapsulated camptothecin enhance the drug stability and tumor targeting in cancer therapy. *Journal of Controlled Release, 127*(3), 208–218. https://doi.org/10.1016/j.jconrel.2008.09.007.

Priprem, A., Watanatorn, J., Sutthiparinyanont, S., Phachonpai, W., & Muchimapura, S. (2008). Anxiety and cognitive effects of quercetin liposomes in rats. *Nanomedicine: Nanotechnology, Biology, and Medicine, 4*, 70–78. https://doi.org/10.1016/j.nano.2007.12.002.

Rane, S., & Prabhakar, B. (2009). Formulation and evaluation of pH-sensitive, long-circulating liposomes for paclitaxel delivery. *International Journal of PharmTech Research, 1*, 914–917. https://doi.org/10.1016/j.ijpharm.2008.12.019.

Rong, G., & Juqun, X. (2007). Studies on molecular interaction between puerarin and PC liposomes. *Chinese Science Bulletin, 52*, 2612–2617. https://doi.org/10.1007/s11434-007-0285-8.

Saraf, A. S. (2010). Applications of novel drug delivery system for herbal formulations. *Fitoterapia, 81*, 680–689. https://doi.org/10.1016/j.fitote.2010.04.002.

Shen, Y. J., Zhang, Z. W., Luo, X. G., Wang, X. F., & Wang, H. L. (2008). Nanoparticles of traditional Chinese herbs inhibit thrombosis *in vivo*. *Hematology Journal, 93*, 1457. https://doi.org/10.1038/s41598-018-30741-0.

Tanwar, Y. S., Gupta, G. D., & Ramawa, K. G. (2006). Development and evaluation of microparticles of guggulipid. *The Pharma Review, 5*, 124–132. https://doi.org/10.1007/s11605-009-1042-1.

Terreno, E., Castelli, D. D., Cabella, C., Dastru, W., Sanino, A., Stancanello, J., Tei, L., & Aime, S. (2008). Paramagnetic liposomes as innovative contrast agents for magnetic resonance (MR) molecular imaging applications. *ChemBioChem, 5*(10), 1901–1912. https://doi.org/10.1002/cbic.200800083.

Trickler, W. J., Nagvekar, A. A., & Dash, A. K. (2008). A novel nanoparticle formulation for sustained paclitaxel delivery. *AAPS PharmSciTech, 9*, 486–493. https://doi.org/10.1208/s1530-9932(08)90032-1.

Tzu, H. W., Feng, L. Y., Liang, T. L., Tong, R. T., Chun, C. L., & Thau, M. C. (2008). Preparation, physicochemical characterization, and antioxidant effects of quercetin nanoparticles. *International Journal of Pharmaceutics, 346*, 160–168. https://doi.org/10.1016/j.ijpharm.2007.06.016.

Wang, F., Zhou, L., & Gu, F. (2010). Characterization of anticancer hypocrellin A encapsulated with silica nanoparticles. *Journal of Thermal Analysis and Calorimetry, 80*, 213–218. https://doi.org/10.1007/s10973-005-5982-0.

Wen, Z., Liu, B., Zheng, Z., You, X., Pu, Y., & Li, Q. (2010). Preparation of liposomes entrapping essential oil from *Atractylodes macrocephala* Koidz by modified RESS technique. *Chemical Engineering Research & Design, 88*, 1102–1107. https://doi.org/10.1016/j.cherd.2010.04.002.

Yen, F. L., Wu, T. H., Lin, L. T., Cham, T. M., & Lin, C. C. (2008). Nanoparticle formulation of *Cuscuta chinensis* prevents acetaminophen-induced hepatotoxicity in rats. *Food and Chemical Toxicology, 46*, 1771–1777. https://doi.org/10.1016/j.fct.2007.12.016.

Youfang, C., Xianfu, L., Hyunjin, P., & Richard, G. (2009). Study of artemisin in nanocapsules as anticancer drug delivery systems. *Nanomedicine: Nanotechnology, Biology, and Medicine, 5*, 316–322. https://doi.org/10.1016/j.nano.2008.10.005.

Zhao, H. R., Wang, K., Zhao, Y., & Pan, L. Q. (2002). Novel sustained-release implant of herb extract using chitosan. *Biomaterials, 23*, 4459–4462. https://doi.org/10.1016/S0142-9612(02)00284-5.

CHAPTER 12

Environmental Pollution and Technological Applications

KHOLA RAFIQUE,[1] ALVINA GUL,[2,*] VOLKAN ALTAY,[3,*] MUNIR OZTURK,[4] and PEDRO GARCIA CAPARROS[5]

[1]Pest Warning and Quality Control of Pesticides, Agriculture Department, Lahore, Punjab, Pakistan

[2]Atta-Ur-Rahman School of Applied Biosciences, National University of Sciences and Technology, Islamabad, Pakistan

[3]Hatay Mustafa Kemal University, Faculty of Arts and Sciences, Department of Biology, Hatay, Türkiye

[4]Ege University, Department of Botany and Center for Environmental Studies, Izmir, Türkiye

[5]Department of Superior School Engineering, University of Almería, Almería, Spain

*Corresponding author.

ABSTRACT

Technology has become an integral part of human life having beneficial as well as detrimental impacts on sustainability. It has brought industrial revolutionary improvements and assisted with economic progress whereas some have been shown to cause damage to the environment. Several technological applications have been developed to mitigate and monitor pollution mainly the GIS, remote sensing, and nanotechnology. Both technology and the environment have a complex connection, an understanding of which is critical for sustainability achievement. The impact of advanced technologies on environmental sustainability needs to be explored more for sustained future economic development. An investment must be made in environmental efficiency and

Agricultural Biotechnology: Issues, Challenges, and Recent Developments. Mohd. Sayeed Akhtar and Khalid Rehman Hakeem (Eds.)
© 2025 Apple Academic Press, Inc. Co-published with CRC Press (Taylor & Francis)

technologies simultaneously as this could lead to a significant reduction in pollution. This chapter focuses on a few of such technological applications, both aspects of technological applications on environmental pollution, and its causes and effects on environmental sustainability. The chapter will help to gain insight into current techniques, identification of advanced approaches, and further advancement in various aspects of research and development.

12.1 INTRODUCTION

The term "Environmental pollution" is stated as the contamination of both physical as well as biological components of the earth or atmosphere system to such an extent that normal environmental processes are severely affected (Iyyanki and Manickam, 2017). Environmental pollution is among the extremely life-threatening global challenges. It is rising gradually and becoming a serious threat to the environment, human beings, animals, plants, and all living organisms. The contributions, both natural and man-made, have resulted in environmental pollution. The natural contributors originate as a result of various repeated biochemical processes occurring in the biosphere like matter transformations by bacterial activity, geothermal and volcanic activity, photosynthetic process, the animal's life, and so on. The anthropogenic contributors are mostly human activities including energy production and use, raw materials exploration and production, the industrial conversion of raw material inputs into useful finished products, and so forth (Ghori et al., 2019; Nath and Cholakov, 2009; Ozturk et al., 2015, 2017a). Recently, it has been shown that activities by man via urbanization, industrialization, exploration, and mining are at the forefront of world environmental pollution (Altay and Ozturk, 2012; Ashraf et al., 2010; Haq et al., 2021; Ozturk et al., 2017b, 2019; Ukaogo et al., 2020).

The substances that are involved in causing pollution are termed as "pollutants." A pollutant can be of different categories such as chemicals (gases, toxic metal, organophosphorus compounds, radionuclides) or geochemical substances (sediment, dust), biological organisms or their products, or physical substances (heat, sound wave, radiation) that are either released intentionally or unintentionally by a human into the natural environment having actual or potentially harmful, adverse, unpleasant, or problematic effects. Such unwanted effects may be directly affecting humans or indirectly, and they are being targeted through a resource organism or climate change (Imanberdieva et al., 2018; Ozturk et al., 2015; Rai, 2016). The environmental pollutants or contributors, in case of an outcome of human activity, encompass various aspects and can be comprehended in distinctive ways based on the region of the world involved doing the observation.

The climate change scenario and the outcomes of global warming extremely affect numerous ecosystems, resulting in issues like food safety, melting of ice and icebergs, extinction of animals, and harm to plant communities (Imanberdieva et al., 2018; Marlon et al., 2019; Ozturk et al., 2016, 2021, 2022a). There is a great variation among the countries, existing in the patterns of environmental pollution as well as pollution-induced diseases. The national income of a country and its stage of development are crucial factors that are responsible for such strong differences (De Maio, 2011). In the case of advanced and well-developed countries, pollution generally signifies the emission of CO_2 gas or additional greenhouse gases (GHGs), whereas in the poorest and least-developed countries, the term "pollution" refers to the waste of basic activities of mankind (Borowski, 2017; Darwash et al., 2023). The quality decline of the ecosystem as an outcome of environmental pollution is evident by biological diversity, vegetation loss, huge amounts of harmful chemicals present in food grains and the surrounding atmosphere, and the development of higher risks of environment-related accidents and threats to life-supporting systems (Rai, 2016).

Exposure to pollution has resulted in direct harm to human health, flora, and fauna and disturbance in the ecosystem (Hameed et al., 2017; Ozturk et al., 2022b, 2023a; Orru et al., 2017; Sardar et al., 2018; Yalcin and Altay, 2023). It has caused increased mortality and morbidity rates. The pollutants enter the immune defense system and damage the vital body organs including the brain, lungs, and heart (Brook et al., 2017). It causes cardiovascular disorders, cerebrovascular disease, ischemic heart disease (Brook et al., 2017), asthma, blood pressure, lung cancer, chronic obstructive pulmonary disease, neurological and autism spectrum disorders (Costa et al., 2017), diabetes, hypertension, and various other ailments (Thiering and Heinrich, 2015). In early stage of life in infants and childhood causes death and various diseases as well as chronic and non-communicable diseases that may occur at any stage of life (Suk et al., 2016). Likewise, exposure during the critical gestational period causes harmful effects on an infant's life such as low body weight at birth, premature, and small gestational age births (Silva et al., 2016). Climate change has also affected soil nutrient quality, land-use patterns, water ecosystems, and atmospheric gaseous composition (Agache et al., 2022; Bai et al., 2019). From an economic point of view, the analysis showed that climate change affects the economic performance of enterprises globally (both at macro- and micro-levels; Aldieri and Vinci, 2020). As per the Environmental Performance Index (EPI, 2018), poor quality of air is the biggest threat to human health (Wolf et al., 2022). It has been reported that human diseases caused by air-borne pollutants led to the loss of 65% of life-years due to environmental pollution-associated human deaths and disabilities in the year 2016 (Friedrich, 2018). According to EPI (2022), it has been indicated a complete world environmental analysis and it leverages 40 performance indicators which are categorized into 11 issue categories that are further grouped into three policy objectives as shown in Figure 12.1 (Wolf et al., 2022).

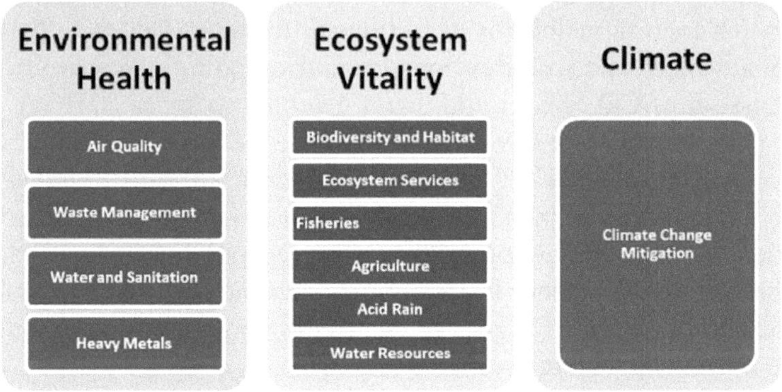

FIGURE 12.1 Policy objectives and issue categories under the EPI (2022) framework.

Undoubtedly, the entire aforesaid is closely related to climate change, and considering the imminent risk, the consequences can severely affect mankind (Moores, 2009). Consequently, such alarming evidence has resulted in attention to environmental pollution studies over the last few years. There is a positive interest in the protection of ecosystems via environmental analysis and contaminants determination (Hou et al., 2018; Wang et al., 2020). Monitoring pollutants has now become a serious concern globally due to the adverse effects of pollutants and limited resources (Artiola and Brusseau, 2019; Hojjati-Najafabadi et al., 2022) and there is a positive interest in the protection of ecosystems via environmental analysis and contaminants determination (Hou et al., 2018; Wang et al., 2020). Presently, the commonly used techniques including spectroscopy, chromatography, and hyphenated techniques proved to be reliable, precise, and sensitive. However, these techniques have presented shortcomings such as prolonged steps for sample preparations, less economical, use of potentially toxic solvents, and most importantly the demand for well-trained certified operators. All these approaches have certainly become limited in terms of instant, on-site, and *in situ* analysis (Huang et al., 2019; Wen et al., 2018). These global issues have also been tried to be tackled through clean and green technologies, which aim at protecting the environment and conserving resources, thus facilitating sustainable developments. Likewise, nanoparticles are considered useful for the mitigation of GHG, the use of optimized solvent in paints, purification, and filtration of water, detection of pollution and remediation, and generation of substitute green manufacturing technological applications. However, the production and utilization of nanoparticles raise numerous toxicity-related issues (Bensebaa, 2013).

12.2 NEGATIVE IMPACT OF TECHNOLOGICAL INNOVATION ON ENVIRONMENT

New technologies have emerged because of the industrial revolution. This was achieved by nonstop industrialization and additional technological advancements that occurred in developed countries worldwide. The impact of such technology on the global environment has resulted in mismanagement, threat, and harm to our natural planet. It has caused major impacts on the environment and ultimately led to climate change and global warming resulting in several consequences (Table 12.1). These technological innovations have also caused damage to our earth by two principal means *viz.* (i) pollution of the environment; and (ii) the depletion of natural resources (Bisong and Apologun, 2020).

TABLE 12.1 Impact of Technology on the Environment and Resulting Consequences of Climate Change and Global Warming

Impact of Technology on the Environment	Global Warming and Climate Change Consequences
• Increased global warming	• Toxic air pollution
• Increased pollution	• Increased heat
• Increased waste	• Rise in sea levels
• Increased deforestation	• Shrinkage of glaciers and sea ice
• Increased radiation	• Diseases epidemics and pandemics
• Increased power consumption	• Drought stress and wildfires
• Water quality affects	• Heavy rainfall
–	• Change in the amount of precipitation and snow patterns.
–	• Change in plants and animals' life cycles
–	• Change in migration of animals
–	• Loss of biodiversity
–	• Desertification
–	• Conflict and war
–	• Less food supply
–	• Increased volcanic activity

Over the last few years, the global community has under a continuous effort to attain the desired goals of sustained development and environmental protection. However, the key barrier to achieving the said goals is the unexpected rise in the release of GHGs as a result of the huge increase in consumption as well as production processes since the Industrial Revolution. Consequently, activities by man are widely considered a major culprit in the deterioration of the environment and disturbance of ecosystem balance (Lan et al., 2022; Li et al., 2022a, b; Usman et al., 2020). It has now been realized by the world, the presence of serious threats *viz.* climate change and global warming. Therefore, the emphasis has been shifted to predicting the outcomes of such environmental changes and the factors involved in mitigating these effects that are having impacts on the global environment (Yasara et al., 2019; Yen et al., 2021; Zhao et al., 2022a, b; Zhenyu and Sohail, 2022).

Global warming causes a rise in the average temperature of the atmosphere of the earth as well as oceans (Bisong and Sunday, 2014; Imanberdieva et al., 2018; Ozturk et al., 2015, 2021, 2022a). It is believed that global warming mainly results from an increase in GHG concentrations, which are produced by various human activities like deforestation and the burning of fossil fuels. Such fossil fuel burning mostly occurred via industrial and technological activities (Lu et al., 2018; Yalcin and Altay, 2023). Global warming causes various negative effects as shown in Table 12.1. It also causes other effects such as food security threats from reducing yield of crops and increasing humidity (Battista, 2009). All such changes are believed to be induced by humans (Kennedy et al., 2010; Ozturk et al., 2023a; Yalcin and Altay, 2023). Technology-induced activities by man since the Industrial Revolution have resulted in a continuous increase in atmospheric GHGs and led to an increased radioactive force from CO_2, tropospheric ozone, methane, nitrous oxide, and chlorofluorocarbons (Bisong and Apologun, 2020; Darwash et al., 2023; Ozturk et al., 2023a).

Emissions of gases notably nitrogen oxides from industries and vehicles have a very detrimental impact on the quality of water (Yalcin and Altay, 2023). The deposited nitrogen in water serves as a fertilizer that encourages algal growth in rivers, lakes, streams, estuaries, and oceans. Such growth of algae creates eutrophic conditions that result in the destruction of submerged aquatic vegetation and the hindrance of commercial fishing. The agricultural runoff including pesticides, fertilizers especially nitrates, and animal wastes appear to be more harmful to water quality. This runoff to rivers and streams causes their contamination thus, making their water unfit and polluted for aquatic life (Bisong and Apologun, 2020; Haq et al., 2021; Yalcin and Altay, 2023). It has also been considered that technology has drastically increased the rate and level of environmental pollution. Humans now travel more and unnecessarily than before as a result of technological advancement. Such an increase in traveling mediated by technology directly results in air pollution. The pollution produced from such travel is enormous and it appears in the form of air pollution, water pollution, and noise pollution. The main air pollutants are carbon monoxide, lead, ozone, sulfur dioxide, and nitrogen oxides. Most of these pollutants are generated from fossil fuel burning, such as in the case of coal-powered electric plants, in vehicles especially trucks and cars, and also during the processing of natural oil and gas processing (Bisong and Apologun, 2020; Yalcin and Altay, 2023).

Power consumption is also high because of technological innovations. Technology (like phones, television, radios, and so forth), is employed in schools, work environments, homes, and other places. The use of phones, television, and radio technologies, which are used at different places and powered by electricity, utilizes fossil fuels or nuclear fuel materials (which are non-renewable) on a huge scale (Ozturk et al., 2023a). It has been reported that the rate of the potential energy of the biosphere being spent by humans is 10 times higher than its accumulation by living organisms, which can absorb sunlight (Bisong, 2016). Similarly, the sources of waste are different in all regions around the world but are directly or indirectly related to technology. Technological up-gradation is one of the major sources or causes of waste in industry-based countries. Toxic wastes are produced in huge quantities on a regular basis via replacement in technologies. For example, in computer technology, shifting from large desktop computers to laptops and tablets has resulted in zero use and wastage of previous computer systems, and are thus disposed into the environment. The laptops and tablets are compact and portable, which makes desktop computers obsolete and thus generates waste products. The same is the case with tungsten bulbs which have been switched to fluorescent bulbs and are also now replaced by incandescent bulbs. All these used items enter the environment in the form of waste products, which are mostly non-biodegradable. Such waste products ultimately contaminate the soil, making it harmful to both plant's and animals' lives (Bisong and Apologun, 2020; Ozturk et al., 2023a). An increase in deforestation and depletion of forests globally is also now possible through the advancement of high-capacity machinery (Altay, 2019). Deforestation not only affects living organisms, but it also affects the global climate. This results in species extinction and climatic change which is now present throughout the world.

Humans can now easily deplete greenery in a short period of time and can dig through huge mountains and hills with minimal stress (Duke, 2020). Concludingly, it has been shown that unauthorized and uncontrollable use of technology negatively impacts the environment but if used in the right direction, it has a great potential and capability to restore the natural environment.

12.3 POTENTIAL OF TECHNOLOGY TO PROTECT THE ENVIRONMENT

Technological advancement has a close relationship with environmental sustainability. Earlier, it was often perceived that such advancement in technology has a negative impact on the environment. But, at present, it has been switched and leading towards a greener planet. In the mid-18th century, advanced technology-based innovations gained human attention, and since then people have used natural resources. Heavier utilization of raw materials has resulted in increased production and the world's population has resulted in CO_2 emission and depletion of resources. The technological advancement has enabled various sectors including government, various companies as well as individuals to implement energy-efficient and environmentally friendly practices, establish processes and utilize resources in a responsible manner aiming to attain long-lasting environmental sustainability. However, technology inflicts a negative impact on the environment, which must be considered in mind. A sustainable advancement in technology is essential for the growth and development of humankind. Several historical examples have also shown that technology can have profound impacts on our future (Okafor, 2020). Scientists in various studies have proposed that technology if it is used positively, can serve as a medium for helping the environment (Figure 12.2). For example, geo-engineering or climate engineering techniques have removed atmospheric CO_2 and neutralized the adverse effects of toxic gases through less radiation absorption (Ahuja, 2022). This technology includes two categories *viz.* management of solar radiation and removal of carbon dioxide (CO_2). The geo-engineering involving the removal of CO_2 removes atmospheric CO_2 and thus reduces the GHGs present in the atmosphere. The geo-engineering encompassing management of solar radiation allows neutralizing the damaging effects of GHGs by assisting in less absorption of solar radiation by the earth. This can be effectively achieved either through the deflection of sunlight away from the side of the earth or through the improvement of the surface reflectivity of the earth (Bisong and Apologun, 2020).

FIGURE 12.2 The potential of the technology to save the environment.

Bioremediation technologies that involve the removal of pollutants using microorganisms in order to reduce metal toxicity in soil or water have also been implemented (Ahuja, 2022). In the earlier days, various naturally prevailing microorganisms have been employed to break down organic wastes of different sectors including agricultural, industrial, and municipal. Later, genetically engineered organisms were being used to clear effluent from industries, contaminated soils, and petroleum spills (Lovley, 2003). In this respect, *Deinococcus radiodurans*, which is a known highly radio-resistant bacterium, was modified to devour and digest ionic mercury and toluene from most radioactive nuclear waste (Brim et al., 2000). Presently, it has been shown that bioremediation technology can treat around 5–10% of entire toxic chemical groups and wastes. It can efficiently treat oil, toluene, gasoline, pentachlorophenol, naphthalene, and diverse agricultural waste (Bisong and Apologun, 2020).

Technologies have also been used to improve and restore the streams and rivers' health. The restoration of rivers and streams involves a group of activities, which are organized to improve the health of a stream or river. Health improvement can be affected as a result of an increase in the habitat of distinct species and a decrease in erosion of stream banks. Improved stream or river health might also involve an improved quality of water, and the attainment of a self-sustaining, and efficient stream flow system in which there is no periodic intervention by man (like the construction of structures for controlling floods; Gilman and Jarrod, 2009; Haq et al., 2021; Yalcin and Altay, 2023).

Restoration activities might encompass river or stream bank stabilization, clearance of a disturbance inhibiting natural water flow, and installation of facilities for flood or storm-water management (Cronin, 2003). It has been shown that most streams and river contaminants enter from the runoff of fertilizers and pesticide chemicals used on farms (Altay and Ozturk, 2012). In the last few decades, geographic information systems (GIS) based technological developments through remotely sensed data have aided in the observation, identification, and control of pollution sources. GIS applications have resulted in an enhancement and encouragement of precise farm practices through computer-based, real-time, and comprehensive information about plant health. GIS tools also encompass pesticide chemicals distribution analysis, endurance in the surrounding environment, and effects on human health (McKelvey et al., 2004; Thapinta and Hudak, 2003). Remote sensing is also a novel technology that can effectively detect and monitor marine pollution. The advancement and development in software applications and new sensors have enabled remote sensing to monitor the water quality of marine, thus making it a powerful tool (Loughland and Saji, 2008). It can also be used to monitor the quality of air at a micro-scale level (Narashid and Mohd, 2010). Remote sensing also assists farmers to identify crop rows that require irrigation, and also for pesticide and fertilizer applications to reduce chemical use and achieve a higher crop yield. Remote sensors help to detect pest infestation and disease attacks on crops (Prabhakar et al., 2022).

12.4 RECENT TECHNOLOGICAL APPLICATIONS FOR MONITORING AND MITIGATING POLLUTION AND MAINTAINING ENVIRONMENT SUSTAINABILITY

12.4.1 SENSING TECHNOLOGY

The use of wireless sensor networks is a technological approach to monitor air pollution (Khedo et al., 2010). These sensors are small and less expensive used for the collection and spreading of environmental data. Such sensors can be efficiently used for monitoring of environment, surveillance, indoor climate control, structural monitoring, air monitoring, medical diagnosis, emergency response and disaster management, and information collection from inhospitable sites (Ukhade and Parachuru, 2022).

Remote sensing serves as a vital approach for monitoring atmospheric conditions on a large scale. It characterizes the distribution of air pollution (both temporal and spatial) and its multiple effects on the environment. The sensor's spatial resolution provides a vital source to satisfy the picture of the pollution and ecology interactions either at a regional level or a global scale. In turn, the occurrence of variation in the ecosystem and environment affects the pollutant's chemistry and remote sensing monitoring. Various studies have documented that air pollution has a great influence on the aerosol optical properties, atmospheric vertical structure, the acidity and alkalinity properties of soil and water, precipitation, and solar radiation, which ultimately affects the precision level of observations generated from remote sensing and the environment (Wang et al., 2022).

Increasing pollution levels in the marine environment have also demanded the application of numerous preventative measures in maintaining the world's oceans' water quality. Remote sensing is a useful technology to monitor and detect the pollution existing in the marine environment, and advancements in new sensor development and software applications have presented remote sensing as an efficient tool for monitoring the quality of marine water (Loughland and Saji, 2008). These sensors also provide digital data regarding the infestation of pests through air, ground, or space-borne platforms. This technique calculates the variation in electromagnetic radiation due to pests' infestation and generates an accurate mean of objective quantification of biotic stresses as compared to visual assessment procedures besides continuously employed to assemble sample measurements without any destruction or invasion. This helps in the reduction of pesticides and fertilizers applications, better crop yield, and ecosystem safety (Prabhakar et al., 2022).

12.4.2 SPECTROSCOPIC TECHNIQUES

Spectroscopic techniques have been developed and are being used for monitoring and detecting air pollution (Hodgeson et al., 1973), among which chemiluminescence was the first to be applied. It acts as an analytical technique that takes place in exothermic chemical reactions that involve the excitement of electrons along with their vibrational states in the material and it helps to detect O_3 and NO in any required area. Another technique is Molecular Gas Phase Fluorescence, which helps to measure pollutants including NO, NO_2, SO_2, and Cl_2 present in the atmosphere. The visible absorption bands of such pollutants are utilized to stimulate the fluorescence of a molecule. Moreover, the ultraviolet absorption method uses electronic circuitry, dual monitors containing distinctive referencing systems, a stabilized source of mercury, and signal integration (Ukhade and Parachuru, 2022).

Spectroscopy has been considered an important technology for the investigation of the Earth's atmosphere and has been identified as the only tool to collect remote atmospheric observations. Such remote measurements are essential to check the spatial heterogeneity of the chemistry and physical features of the atmosphere, and also the technical and financial problems to directly probe the earth's atmosphere with varied types of *in situ* sensors. *In situ,* spectroscopy techniques have also been broadly applied in atmospheric studies, such as the measurement of short-lived reactive species involved in controlling the atmospheric chemical composition, isotopologue concentrations, and the optical properties of particles. The data collected from these spectroscopic tools is important for the determination of the quality of air, detection of air pollutants emissions, and quantification of the magnitude and fluctuations of climate forcing agents, ensuring regulations compliance, and notifying environmental policies at both national and international platforms (Cuisset et al., 2016).

12.4.3 NANOTECHNOLOGY AND NANOMATERIALS

Nanomaterials are small-sized materials that have a minimum of one-dimension that measures below 100 nm (Hadef, 2018). They serve as suitable adsorbents and catalysts because of their smaller size and bigger reactive surfaces (Barik et al., 2021; Lu et al., 2016). There are three categories of nanomaterials including nanoparticles, nano-emulsions, and nano-clay (Figure 12.3). These are further divided into two groups *viz.* organic and inorganic (Li et al., 2019). Compared to conventional technological applications, the advancement of nanotechnology has led to the application of nanomaterials in environmental pollution remediation. These unique features of nanotechnology in addition to cost-effectiveness have enabled to use of it efficiently for the removal and degradation of pollutants from the environment (Ozturk et al., 2023b; Sánchez et al., 2011).

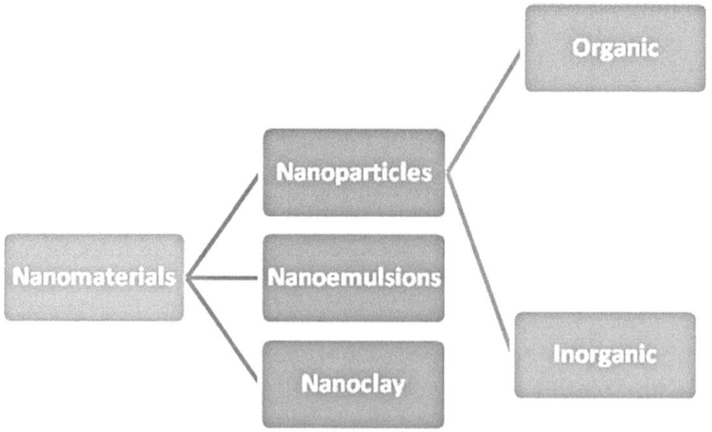

FIGURE 12.3 Categories of nanomaterials include nanoparticles, nano-emulsions, and nano-clay.

Nanotechnology serves as a novel remediation technology that has been used to maintain a natural clean environment (Bhawana and Fulekar, 2012). Nanoparticles possess a unique anti-bacterial property, higher surface area to mass ratio, hardness, and magnetic as well as electric properties (Corsi et al., 2018). Diverse types of nanotechnologies have been used to mitigate the effects of different environmental pollutants. In a study (Yunus et al., 2012), nanotechnology has been applied to alleviate the negative effects of atmospheric air pollution. This study suggested the potential of nanotechnology in improving existing technologies as well as in creating novel technologies. This study has reported three main features of this technology *viz.* (i) remediation and purification; (ii) contaminants detection; and (iii) pollution prevention. It has been examined that various nanotechnologies can be effectively applied to target certain substances present in the environment to prevent further degradation of the environment (Ukhade and Parachuru, 2022). A very useful application example of nanotechnology is carbon nanotubes. Carbon nanotubes were reported to target NO_x by passing NO and O_2 through them and containing NO which is being oxidized to NO_2, which is then adsorbed on the surface of nitrate species. Likewise, fullerene is another nanotechnology that targets CO_2. It has been reported that fullerene has a high adsorption capacity for CO_2 through strong chemisorption. Moreover, graphene is a nanotechnology that targets CO_2, SO_2, NH_3, N_2, and H_2S (Ukhade and Parachuru, 2022).

Nanotechnology proved to be an efficient treatment technology to manage and remediate air pollution by different means such as by benefiting from nano-material properties and using them as catalysts, adsorbents, sensors, and membranes (Zhao, 2009). Recently two nanomaterials *viz.* nano-fibers and nanocomposites have been identified as useful in the remediation of pollution from wastewater. However, their production on a large-scale and more widespread application is still undetermined. There is still a scope for improving the development of nanomaterials, which can efficiently and selectively remediate pollutants, offer better resistance as well as stability, and less toxicity (Thangavelu et al., 2022). Although there are several remarkable properties of nanomaterials including higher efficiency and reactivity, but still, possess certain limitations that discourage their use on a large scale for environmental remediation. It has been considered that they express harmful effects on the natural environment. Their toxic properties and effects on the ecosystem as well as on public health are the main risks. Additionally, challenges involving the nano-materials regeneration and re-use will significantly disturb the treatment cost. Also, the prevalent toxic and harmful conditions in environmental applications would cause the degradation of the structure of the applied nano-sized particles. Furthermore, a precise classification of nanomaterials used in pollution remediation is necessary for improved knowledge encompassing their distinct properties and functions (Thangavelu et al., 2022).

12.4.4 POLLUTION MONITORING USING GEOGRAPHIC INFORMATION SYSTEM (GIS)

Geographic information system (GIS) represents an efficient and practical tool for the analysis and planning of environmental data. GIS collects spatial data in the form of a digital mapping environment (Abbas and Ali, 2012; Adamu et al., 2014). The generated digital-based map can be overlaid onto a map along with data and other information to view and observe spatial information and relationships. It allows enhanced observations and deep knowledge of physical features as well as the relationships that have a great impact on a given environmental state (Hacisalihoglu et al., 2001). Various factors like slope steepness, vegetation, and other aspects can be observed and overlaid to assess numerous environmental parameters and impact analysis (Abechi et al., 2010).

The Web-based technology GIS is applied in the software development of monitoring and detecting environmental pollution because of its easy access and operation simplicity for users belonging to diverse professional backgrounds. The use of such web-based programs anticipates the dearth of GIS-based desktops through easy accessibility and use (Anna and Fikriyah, 2019). GIS tool has also been utilized for monitoring, analyzing, and modeling purposes regarding chemical pesticide migration in the environment. It has been proved that GIS analysis serves as a valuable tool for studying the environment and public health and generating significant results that may ultimately lead to prevention against over and uncontrollable exposure to pesticides. Despite all such advantages, GIS is still not commonly practiced for such studies mainly in the developing countries of the world that have limited technology knowledge and access (Kaminska et al., 2004).

12.5 CONCLUSION

Global issues including environmental pollution, climate change, and global warming have drastically increased as a result of various natural causes and some activities by humans. This scenario has gained the interest of different sectors working on environment conservation, developing clean and green technologies that might remediate environmental pollution much more effectively. Some of these technologies have been highlighted in this chapter along with positive and negative impacts on the environment. Environmental pollution and resulting climate changes and global warming are not influenced by technology but by its mismanagement. With advancements in technologies, the concerned scientists or agencies must assess the significance of both existing as well as new technological innovations. The techniques must guarantee cost-effectiveness, eco-friendliness, consumer safety, prolonged durability, and no hazardous waste release. Insight knowledge and continuous efforts are still needed to make certain right designing, cost control, and implementation of innovative ideas, plans, and novelties. This could require collaboration between researchers and environmentalists so that we can limit the troubles at the earliest.

KEYWORDS

- **climate change**
- **cost-effectiveness**
- **environment sustainability**
- **environmental pollution**
- **geographic information system**
- **global warming**

REFERENCES

Abbas, M., & Ali, A. (2012). Effect of the extract of *Schanginia aegyptiaca* on seed germination and seedling growth of roselle (*Hibiscus sabdariffa* L.). *Asian Journal of Agricultural Research, 6*(2), 83–90. https://doi.org/10.3923/ajar.2012.83.90.

Abechi, E. S., Okunola, O. J., Zubairu, S. M. J., Usman, A. A., & Apene, E. F. (2010). Evaluation of heavy metals in roadside soils of major streets in Jos metropolis, Nigeria. *Journal of Environmental Chemistry and Ecotoxicology, 2*(6), 98–102. https://doi.org/10.5897/JECE10.013.

Adamu, S. A., Ayuba, M. K., Murtala, A. M., & Uriah, L. A. (2014). Assessment of potentially toxic metals in soil and sediments of the Keana brine field in the Middle Benue Trough,

North Central Nigeria. *American Journal of Environmental Protection, 3*(6), 77–90. https://doi.org/10.11648/j.ajep.20140306.11.
Agache, I., Sampath, V., Aguilera, J., Akdis, C. A., Akdis, M., Barry, M., & Nadeau, K. C. (2022). Climate change and global health: A call to more research and more action. *Allergy, 77*(5), 1389–1407. https://doi.org/10.1111/all.15217.
Ahuja, T. (2022). Impact of technology on the environment. *Times of India*. Retrieved from: https://timesofindia.indiatimes.com/readersblog/taneesha-ahuja/impact-of-technology-on-environment-43973 (accessed on 12 September 2024).
Aldieri, L., & Vinci, C. P. (2020). Climate change and knowledge spillovers for cleaner production: New insights. *Journal of Cleaner Production, 271*, 122729. https://doi.org/10.1016/j.jclepro.2020.122729.
Altay, V. (2019). Ecology of *Pinus sylvestris* L. forests–A case study from Istanbul (Turkey). *Pakistan Journal of Botany, 51*(5), 1711–1718. https://doi.org/10.30848/PJB2019-5(21).
Altay, V., & Ozturk, M. (2012). Land degradation and halophytic plant diversity of Milleyha wetland ecosystem (Samandağ-Hatay), Turkey. *Pakistan Journal of Botany, 44*, 37–50. https://doi.org/10.30848/PJB2012-1(5).
Anna, A. N., & Fikriyah, V. N. (2019). Environmental pollution monitoring using a web-based GIS in Surakarta. In *IOP Conference Series: Earth and Environmental Science* (Vol. 332, pp. 012066). IOP Publishing. https://doi.org/10.1088/1755-1315/332/1/012066.
Artiola, J. F., & Brusseau, M. L. (2019). The role of environmental monitoring in pollution science. In M. L. Brusseau, L. M. Gan, T. J. Smith, & T. E. McKibben (Eds.), *Environmental and Pollution Science* (pp. 149–162). Academic Press. https://doi.org/10.1016/B978-0-12-814803-1.00010-2.
Ashraf, M., Ozturk, M., & Ahmad, M. S. A. (2010). *Plant Adaptation and Phytoremediation*. Springer Science. https://doi.org/10.1007/978-1-4419-5704-8.
Bai, Y., Ochuodho, T. O., & Yang, J. (2019). Impact of land use and climate change on water-related ecosystem services in Kentucky, USA. *Ecological Indicators, 102*, 51–64. https://doi.org/10.1016/j.ecolind.2019.02.033.
Barik, T. K., Maity, G. C., Gupta, P., Mohan, L., & Santra, T. S. (2021). Nanomaterials: An introduction. In T. S. Santra & L. Mohan (Eds.), *Nanomaterials and Their Biomedical Applications* (pp. 1–33). Springer. https://doi.org/10.1007/978-3-030-52978-7_1.
Battista, D. (2009). Historical warnings of future food insecurity with unprecedented seasonal heat. *Science, 323*(5911), 240–424. https://doi.org/10.1126/science.1164378.
Bensebaa, F. (2013). Environmental and societal applications and implications. In S. J. Park & M. K. Seo (Eds.), *Interface Science and Technology* (pp. 481–516). Elsevier. https://doi.org/10.1016/B978-0-12-404627-6.00014-5.
Bhawana, P., & Fulekar, M. H. (2012). Nanotechnology: Remediation technologies to clean up environmental pollutants. *Research Journal of Chemical Sciences, 2*(2), 90–96. https://doi.org/10.7314/ajls.2012.5.1.87.
Bisong, P. B. (2016). *An Ethic of Sustainable Environment from Ibuanyidanda's Perspective*. Lambert Academic Publishing. https://doi.org/10.13140/RG.2.1.5112.3767.
Bisong, P. B., & Apologun, S. (2020). Technology can save the environment. *International Journal of Humanities, Management and Social Sciences, 3*(1), 11–19. https://doi.org/10.11648/j.ijhms.20200301.12.
Bisong, P. B., & Sunday, A. S. (2014). The noetic propaedeutic pedagogy as a panacea to environmental degradation. *Canadian Social Science, 10*(4), 37–46. https://doi.org/10.3968/4592.

Borowski, P. F. (2017). Environmental pollution is a threat to the ecology and development of Guinea Conakry. *Environmental Protection and Natural Resources, 28*(4), 27–32. https://doi.org/10.1515/epn-2017-0004.

Brim, H., McFarlan, S. C., Fredrickson, J. K., Minton, K. W., Zhai, M., Wackett, L. P., & Daly, M. J. (2000). Engineering *Deinococcus radiodurans* for metal remediation in radioactive mixed waste environments. *Nature Biotechnology, 18*(1), 85–90. https://doi.org/10.1038/71063.

Brook, R. D., Newby, D. E., & Rajagopalan, S. (2017). Air pollution and cardiometabolic disease: An update and call for clinical trials. *American Journal of Hypertension, 31*, 1–10. https://doi.org/10.1093/ajh/hpw002.

Corsi, I., Winther-Nielsen, M., Sethi, R., Punta, C., Della Torre, C., Libralato, G., & Buttino, I. (2018). Ecofriendly nanotechnologies and nanomaterials for environmental applications: Key issues and consensus recommendations for sustainable and eco-safe nanoremediation. *Ecotoxicology and Environmental Safety, 154*, 237–244. https://doi.org/10.1016/j.ecoenv.2018.01.060.

Costa, L. G., Chang, Y. C., & Cole, T. B. (2017). Developmental neurotoxicity of traffic-related air pollution: Focus on autism. *Current Environmental Health Reports, 4*, 156–165. https://doi.org/10.1007/s40572-017-0137-4.

Cronin, A. (2003). Restoring paradise in Moscow, Idaho. *Land & Water, 47*(2), 18. https://www.jstor.org/stable/27675435 (accessed on 12 September 2024).

Cuisset, A., Venables, D. S., Gao, X., & Aroui, H. (2016). Applications of spectroscopy in environmental monitoring of gases and aerosols. *Journal of Spectroscopy, 2016*, 2575782. https://doi.org/10.1155/2016/2575782

Darwash, M., Yousuf, A., Ozturk, M., & Bhat, R. A. (2023). Phytosequestration–Strategies for mitigation of aerial carbon dioxide and aquatic nutrient pollution. *Springer Briefs*. https://doi.org/10.1007/978-3-030-31612-7.

De Maio, F. G. (2011). Understanding chronic non-communicable diseases in Latin America: Towards an equity-based research agenda. *Global Health, 7*(1), 36. https://doi.org/10.1186/1744-8603-7-36.

Duke, E. O. (2020). From Christian spirituality to eco-friendliness. *International Journal of Humanities Innovations, 3*(1), 34–38. https://doi.org/10.11648/j.ijhi.20200301.16.

Friedrich, M. J. (2018). Air pollution is the greatest environmental threat to health. *JAMA, 319*(11), 1085. https://doi.org/10.1001/jama.2018.1181.

Ghori, N. H., Ghori, T., Hayat, M. Q., Imadi, S. R., Gul, A., Altay, V., & Ozturk, M. (2019). Heavy metal stress and responses in plants. *International Journal of Environmental Science and Technology, 16*, 1807–1828. https://doi.org/10.1007/s13762-019-02250-0.

Gilman, J., & Jarrod, K. (2009). Challenges of stream restoration as a stormwater management tool; Part 1: A designer's perspective. *Stormwater, 10*(3), 36.

Hacisalihoglu, G. R., Hart, J. J., & Kochian, L. V. (2001). High- and low-affinity zinc transport systems and their possible role in zinc efficiency in bread wheat. *Plant Physiology, 125*(1), 456–463. https://doi.org/10.1104/pp.125.1.456.

Hadef, F. (2018). An introduction to nanomaterials. In N. Dasgupta (Ed.), *Environmental Nanotechnology* (pp. 1–58). Springer. https://doi.org/10.1007/978-3-319-76055-6_1.

Hameed, A., Qadri, T. N., Mahmooduzzafar, T. O. S., Ozturk, M., Altay, V., & Ahmad, P. (2017). Physio-biochemical and nutritional responses of *Abelmoschus esculentus* (L.) Moench (Okra) under mercury contamination. *Fresenius Environmental Bulletin, 26*, 5814–5823.

Haq, S., Bharose, R., Ozturk, M., Altay, V., Bhatti, A. A., Dervash, M. A., & Hakeem, K. R. (2021). Impact of treated sewage water on nutrient status of alfisols and vegetable crops. *Notulae Botanicae Horti Agrobotanici Cluj-Napoca, 49*(2), 12255. https://doi.org/10.15835/nbha49212255.

Hodgson, J. A., McClenny, W. A., & Hanst, P. L. (1973). Air pollution monitoring by advanced spectroscopic techniques: A variety of spectroscopic methods are being used to detect air pollutants in the gas phase. *Science, 182*(4109), 248–258. https://doi.org/10.1126/science.182.4109.248.

Hojjati-Najafabadi, A., Mansoorianfar, M., Liang, T., Shahin, K., & Karimi-Maleh, H. (2022). A review on magnetic sensors for monitoring of hazardous pollutants in water resources. *Science of the Total Environment, 824*, 153844. https://doi.org/10.1016/j.scitotenv.2022.153844.

Hou, H., Zeinu, K. M., Gao, S., Liu, B., Yang, J., & Hu, J. (2018). Recent advances and perspective on the design and synthesis of electrode materials for electrochemical sensing of heavy metals. *Energy & Environmental Materials, 1*(3), 113–131. https://doi.org/10.1002/eem2.12036.

Huang, D. L., Wang, J., Fan, C., Ali, A., Guo, H. S., Xiao, Y., & Liu, H. Y. (2019). Correction to: Synergistic effect of a cobalt fluoroporphyrin and graphene oxide on the simultaneous voltammetric determination of catechol and hydroquinone. *Microchimica Acta, 186*(6), 186. https://doi.org/10.1007/s00604-019-3674-2.

Imanberdieva, N., Imankul, B., & Severoğlu, Z. (2018). Potential impacts of climate change on plant diversity of Sary-Chelek Biosphere Reserve in Kyrgyzstan. In D. Egamberdieva & M. Ozturk (Eds.), *Vegetation of Central Asia and Environs* (pp. 349–364). Springer Nature Switzerland AG. https://doi.org/10.1007/978-3-319-78914-3_15.

Iyyanki, V. M., & Manickam, V. (2017). Introduction. In V. M. Iyyanki & M. Valli (Eds.), *Environmental Management* (pp. 1–4). Butterworth-Heinemann. https://doi.org/10.1016/B978-0-12-804631-8.00001-2.

Kaminska, I. A., Oldak, A., & Turski, W. A. (2004). Geographical information system [GIS] as a tool for monitoring and analyzing pesticide pollution and its impact on public health. *Annals of Agricultural and Environmental Medicine, 11*(2), 181–184. https://www.aaem.pl/Geographical-information-system-GIS-as-a-tool-for-monitoring-and-analyzing-pesticide-pollution-and-its-impact-on-public-health,78027,0,2.html (accessed on 12 September 2024).

Kennedy, J., Thorne, P. W., Peterson, T. C., Ruedy, R. A., Stott, P. A., Parker, D. E., Good, S. A., Titchner, H. A., & Willett, K. M. (2010). How do we know the world has warmed? State of the climate in 2009. *Bulletin of the American Meteorological Society, 91*(7), 27. https://doi.org/10.1175/2010BAMSSTATEOFTHECLIMATE.1.

Khedo, K. K., Perseedoss, R., & Mungur, A. (2010). A wireless sensor network air pollution monitoring system. *International Journal of Wireless & Mobile Networks, 2*(2), 31–45. https://doi.org/10.5121/ijwmn.2010.2204.

Lan, H., Cheng, C., & Sohail, M. T. (2022). Asymmetric determinants of CO_2 emissions in China: Do government size and economic size matter? *Environmental Science and Pollution Research, 29*, 47225–47232. https://doi.org/10.1007/s11356-022-23598-3.

Li, W., Cao, Z., Liu, R., Liu, L., Li, H., Li, X., & Liu, Y. (2019). AuNPs as an important inorganic nanoparticle applied in drug carrier systems. *Artificial Cells, Nanomedicine, and Biotechnology, 47*(1), 4222–4233. https://doi.org/10.1080/21691401.2019.1592684.

Li, Y., Chen, J., & Sohail, M. T. (2022a). Does education matter in China? Myths about financial inclusion and energy consumption. *Environmental Science and Pollution Research, 29*(48), 73542–73551. https://doi.org/10.1007/s11356-022-19837-8.

Li, Y., Chen, J., & Sohail, M. T. (2022b). Financial inclusion and their role in renewable energy and non-renewable energy consumption in China: Exploring the transmission channels. *Research Square*. https://doi.org/10.21203/rs.3.rs-1355688/v1.

Loughland, R. A., & Saji, B. (2008). Remote sensing: A tool for managing marine pollution in the Gulf. In A. H. Abuzinada et al. (Eds.), *Protecting the Gulf's Marine Ecosystems from Pollution* (pp. 131–145). Springer Birkhäuser Basel. https://doi.org/10.1007/978-3-7643-8476-6_11.

Lovley, D. R. (2003). Cleaning up with genomics: Applying molecular biology to bioremediation. *Nature Reviews Microbiology, 1*, 35–44. https://doi.org/10.1038/nrmicro731.

Lu, H., Wang, J., Stoller, M., Wang, T., Bao, Y., & Hao, H. (2016). An overview of nanomaterials for water and wastewater treatment. *Advances in Materials Science and Engineering, 2016*, 4964828. https://doi.org/10.1155/2016/4964828.

Lu, J., Gabriel, V., & Thomas, R. (2018). Expansion of the Hadley cell under global warming. *Geophysical Research Letters, 34*(6), 1. https://doi.org/10.1029/2007GL030200.

Marlon, J. R., Bloodhart, B., Ballew, M. T., Rolfe-Redding, J., Roser-Renouf, C., Leiserowitz, A., & Maibach, E. (2019). How hope and doubt affect climate change mobilization. *Frontiers in Communication, 4*, 20. https://doi.org/10.3389/fcomm.2019.00020.

McKelvey, W., Brody, J. G., Aschengrau, A., & Swartz, C. H. (2004). Association between residence on Cape Cod, Massachusetts, and breast cancer. *Annals of Epidemiology, 14*, 89–94. https://doi.org/10.1016/j.annepidem.2003.09.010.

Moores, F. C. (2009). Climate change and air pollution: Exploring the synergies and potential for mitigation in industrializing countries. *Sustainability, 1*, 43–54. https://doi.org/10.3390/su1010043.

Narashid, R. H., & Mohd, W. M. N. W. (2010). Air quality monitoring using remote sensing and GIS technologies. In *International Conference on Science and Social Research (CSSR 2010)* (pp. 1186–1191). https://www.researchgate.net/publication/256364726_Air_Quality_Monitoring_Using_Remote_Sensing_and_GIS_Technologies (accessed on 12 September 2024).

Nath, B., & Cholakov, G. S. (2009). Pollution Control Technologies – Vol. II. EOLSS Publications.

Okafor, J. (2020). Negative impact of technology on the environment. *Trvst*. https://www.trvst.world/environment/negative-impact-of-technology-on-the-environment/ (accessed on 12 September 2024).

Orru, H., Ebi, K. L., & Forsberg, B. (2017). The interplay of climate change and air pollution on health. *Current Environmental Health Reports, 4*(4), 504–513. https://doi.org/10.1007/s40572-017-0147-0.

Ozturk, M., & Altay, V. (2021). Role of *Quercus coccifera* (=Q. calliprinos) in the light of climate change scenarios in the Mediterranean Basin. *Plant and Fungal Research, 4*(2), 8–20.

Ozturk, M., Altay, V., & Aksoy, A. (2016). Ecology of some endangered endemic plant taxa of Turkiye in relation to climate change. In *Conservation and Sustainable Use of Genefund of Plant World in Eurasia at the Present Stage–International Scientific Conference within "Day of Kazakhstan," at EXPO 2016, Antalya-Turkey* (pp. 12–15).

Ozturk, M., Altay, V., & Dimeyeva, L. A. (2022a). Climate change, biodiversity and ecology of Caspian Sea: An overview. In M. A. Dervash & A. A. Wani (Eds.), *Climate Change Alleviation for Sustainable Progression* (pp. 78–94). CRC Press. https://doi.org/10.1201/9780429283084.

Ozturk, M., Altay, V., & Karahan, F. (2017b). Studies on trace elements distributed in *Glycyrrhiza* taxa in Hatay, Turkey. *International Journal of Plant and Environment, 3*(2), 1–7.

Ozturk, M., Altay, V., Kucuk, M., & Yalçın, I. E. (2019). Trace elements in the soil-plant systems of copper mine areas: A case study from Murgul Copper Mine in the Black Sea region of Turkey. *Phyton, 88*(3), 223–238.

Ozturk, M., Ashraf, M., Aksoy, A., & Ahmad, M. S. A. (2015). *Phytoremediation for Green Energy*. Springer.

Ozturk, M., Hakeem, K. R., Faridah-Hanum, I., & Efe, R. (2015). *Climate Change Impacts on High-Altitude Ecosystems*. Springer.

Ozturk, M., Metin, M., Altay, V., Bhat, R., Toderich, K., Ejaz, M., Gul, A., Unal, B. T., Hasanuzzaman, M., Nibir, L., Nahar, K., & Kawano, T. (2022b). Arsenic and human health: Genotoxicity, epigenomic effects, and cancer signaling. *Biological Trace Element Research, 200*, 988–1001. https://doi.org/10.1007/s12011-021-02812-6.

Ozturk, M., Metin, M., Altay, V., Prasad, M. N. V., Gul, A., Bhat, R. A., Darvash, M. A., Hasanuzzaman, M., Nahar, K., Unal, D., Unal, B. T., Caparrós, P. G., Kawano, T., Toderich, K., & Bukhari, A. (2023a). Role of rare earth elements in plants. *Plant Molecular Biology Reporter*. https://doi.org/10.1007/s11105-023-01369-7.

Ozturk, M., Roy, A., Bhat, R. A., Vardar-Sukan, F., & Tonelli, F. M. P. (2023b). *Synthesis of Bionanomaterials for Biomedical Applications*. Elsevier.

Ozturk, M., Saba, N., Altay, V., Iqbal, R., Hakeem, K. R., Jawaid, M., & Ibrahim, F. H. (2017a). Biomass and bioenergy: An overview of the development potential in Turkey and Malaysia. *Renewable and Sustainable Energy Reviews, 79*, 1285–1302. https://doi.org/10.1016/j.rser.2017.05.158.

Prabhakar, M., Thirupathi, M., & Mani, M. (2022). Principles and application of remote sensing in crop pest management. *Trends in Horticultural Entomology, 2022*, 157–183. https://doi.org/10.1016/j.treent.2022.05.002.

Rai, P. K. (2016). Particulate matter and its size fractionation. *Biomagnetism and Monitoring of Particulate Matter, 1*, 1–13. https://doi.org/10.1007/s10000-016-0376-2.

Sánchez, A., Recillas, S., Font, X., Casals, E., González, E., & Puntes, V. (2011). Ecotoxicity of, and remediation with, engineered inorganic nanoparticles in the environment. *Trends in Analytical Chemistry, 30*(3), 507–516. https://doi.org/10.1016/j.trac.2010.11.003.

Sardar, M. F., Ahmad, H. R., Zia-Ur-Rehman, M., Ozturk, M., & Altay, V. (2018). Absorption of foliar-applied lead (Pb) in rice (*Oryza sativa* L.): A hydroponic experiment. *Fresenius Environmental Bulletin, 27*(10), 6634–6639.

Silva, R. A., Adelman, Z., Fry, M. M., & West, J. J. (2016). The impact of individual anthropogenic emissions sectors on the global burden of human mortality due to ambient air pollution. *Environmental Health Perspectives, 124*, 1776–1784. https://doi.org/10.1289/ehp.1509757.

Suk, W. A., Ahanchian, H., Asante, K. A., Carpenter, D. O., Diaz-Barriga, F., Ha, E. H., Huo, X., King, M., Ruchirawat, M., da Silva, E. R., & Sly, L. (2016). Environmental pollution: An under-recognized threat to children's health, especially in low-and middle-income countries. *Environmental Health Perspectives, 124*(3), A41–A45. https://doi.org/10.1289/ehp.124-A41.

Thangavelu, L., Veeraragavan, G. R., Mallineni, S. K., Devaraj, E., Parameswari, R. P., Syed, N. H., Dua, K., Chellappan, D. K., Balusamy, S. R., & Bhawal, U. K. (2022). Role of nanoparticles in environmental remediation: An insight into heavy metal pollution from dentistry. *Bioinorganic Chemistry and Applications, 2022*, 1946724. https://doi.org/10.1155/2022/1946724.

Thapinta, A., & Hudak, P. F. (2003). Use of geographic information systems for assessing groundwater pollution potential by pesticides in Central Thailand. *Environmental International, 29*, 87–93. https://doi.org/10.1016/S0160-4120(02)00144-5.

Thiering, E., & Heinrich, J. (2015). Epidemiology of air pollution and diabetes. *Trends in Endocrinology & Metabolism, 26*, 384–394. https://doi.org/10.1016/j.tem.2015.04.006.

Ukaogo, P. O., Ewuzie, U., & Onwuka, C. V. (2020). Environmental pollution: Causes, effects, and remedies. In P. Chawdhary et al. (Eds.), *Microorganisms for Sustainable Environment and Health* (pp. 419–429). Elsevier. https://doi.org/10.1016/B978-0-12-821539-1.00021-1.

Ukhade, R., & Parachuru, R. (2022). A comprehensive look at the detection, monitoring, and control of environmental pollution with a particular focus on new and emerging solutions to address the problem. *Journal of Materials Science and Engineering, 11*(01), 1. https://doi.org/10.4172/2169-0022.1000e135.

Usman, A., Ullah, S., Ozturk, I., Chishti, M. Z., & Zafar, S. M. (2020). Analysis of asymmetries in the nexus among clean energy and environmental quality in Pakistan. *Environmental Science and Pollution Research, 27*, 20736–20747. https://doi.org/10.1007/s11356-020-08357-8.

Wang, H., Shen, L., Zhang, J., Xie, X., & Guan, X. (2022). Air pollution remote sensing and the subsequent interactions with ecology on regional scales. *Frontiers in Environmental Science, 10*, 1014003. https://doi.org/10.3389/fenvs.2022.1014003.

Wang, L., Peng, X., Fu, H., Huang, C., Li, Y., & Liu, Z. (2020). Recent advances in the development of electrochemical aptasensors for the detection of heavy metals in food. *Biosensors and Bioelectronics, 147*, 111777. https://doi.org/10.1016/j.bios.2019.111777.

Wen, G. L., Zhao, W., Chen, X., Jia, Q., Wang, Y., Zhang, Y., & Wu, Y. C. (2018). N-doped reduced graphene oxide/MnO_2 nanocomposite for electrochemical detection of Hg^{2+} by square wave stripping voltammetry. *Electrochimica Acta, 291*, 95–102. https://doi.org/10.1016/j.electacta.2018.09.082.

Wolf, M. J., Emerson, J. W., Esty, D. C., Sherbinin, A. D., & Wendling, Z. A. (2022). *Environmental Performance Index (EPI) Results*. Yale Center for Environmental Law & Policy.

Yalcin, I. E., & Altay, V. (2023). Investigation of water-soil-plant relationships based on hazardous and macro-micro element concentrations on Orontes River, Türkiye. *International Journal of Phytoremediation*. https://doi.org/10.1080/15226514.2023.2202241.

Yasara, A., Farooqa, T., Tabindaa, A. B., Sohailb, M. T., Mahfooza, Y., & Malika, A. (2019). Macrophytes as a potential indicator of heavy metals in river water. *Desalination and Water Treatment, 142*, 272–278. https://doi.org/10.5004/dwt.2019.24463.

Yen, Y., Zhao, P., & Sohail, M. T. (2021). The morphology and circuit of walkable, bikeable, and drivable street networks in Phnom Penh, Cambodia. *Environment and Planning B: Urban Analytics and City Science, 48*, 169–185. https://doi.org/10.1177/23998083211013430.

Yunus, I. S., Harwin, K. A., Adityawarman, D., & Indarto, A. (2012). Nanotechnologies in water and air pollution treatment. *Environmental Technology Reviews, 1*(1), 136–148. https://doi.org/10.1080/21622515.2012.736738.

Zhao, J. (2009). Turning to nanotechnology for pollution control: Applications of nanoparticles. *Dartmouth Undergraduate Journal of Science, 2009*, 1.

Zhao, W., Chang, M., Yu, L., & Sohail, M. T. (2022a). Health and human wellbeing in China: Do environmental issues and social change matter? *Frontiers in Psychology, 13*, 860321. https://doi.org/10.3389/fpsyg.2022.860321.

Zhao, W., Huangfu, J., Yu, L., Li, G., Chang, Z., & Sohail, M. T. (2022b). Analysis of price game and supervision of natural gas pipeline tariff under the background of pipeline network separation in China. *Polish Journal of Environmental Studies, 31*, 2961–2972. https://doi.org/10.15244/pjoes/150438.

Zhenyu, W., & Sohail, M. T. (2022). Short- and long-run influence of education on subjective well-being: The role of ICT in China. *Frontiers in Psychology, 13*, 927562. https://doi.org/10.3389/fpsyg.2022.927562.

Index

B

C

D

H

J

K

L

N

T

X

Y

Z